Monographs in Computational Science and Engineering

1

Editors

Timothy J. Barth
Michael Griebel
David E. Keyes
Risto M. Nieminen
Dirk Roose
Tamar Schlick

Joakim Sundnes   Glenn Terje Lines   Xing Cai
Bjørn Fredrik Nielsen   Kent-Andre Mardal
Aslak Tveito

Computing the Electrical Activity in the Heart

With 99 Figures and 23 Tables

 Springer

Joakim Sundnes
Glenn Terje Lines
Xing Cai
Bjørn Fredrik Nielsen
Kent-Andre Mardal
Aslak Tveito

Simula Research Laboratory
P.O. Box 134
1325 Lysaker, Norway
email: sundnes@simula.no
     glennli@simula.no
     xingca@simula.no
     bjornn@simula.no
     kent-and@simula.no
     aslak@simula.no

Aslak Tveito has received financial support from the NFF – Norsk faglitterær forfatter- og oversetterforening

Library of Congress Control Number: 2006927369

Mathematics Subject Classification: 35Q80, 65M55, 65M60, 92C50, 92C0510

ISBN-10 3-540-33432-7 Springer Berlin Heidelberg New York
ISBN-13 978-3-540-33432-3 Springer Berlin Heidelberg New York

Springer is a part of Springer Science+Business Media
springer.com
© Springer-Verlag Berlin Heidelberg 2006
Printed in The Netherlands

Typesetting: by the authors and techbooks using a Springer LaTeX macro package
Cover design: *design & production* GmbH, Heidelberg

Printed on acid-free paper   SPIN: 11737407   46/techbooks   5 4 3 2 1 0

Preface

The heart is a fantastic machine; during a normal lifetime it beats about 2.5 billion times and pumps 200.000 tons of blood through an enormous system of vessels extending 160.000 kilometres throughout the body.

For centuries, man has tried to understand how the heart works, but there remain many unsolved problems, problems that have captured the attention of thousands of researchers worldwide. There is, for example, a huge amount of research being devoted to the analysis of single heart cells. Other areas of research include trying to understand how it works as a complete muscle, and how blood flows through the heart. The entire process is extremely complex.

The history of bioelectricity can be traced back to the late eighteenth century and the experiments of Luigi Galvani. A century later, in 1887, Augustus Wallers managed to measure the electrical signal generated by the heart at the surface of the body [142]. His dog Jimmy earned a place in history by being the first to have his heart measured in this way; see Figure 1.1. In 1903 Willem Einthoven [34] developed the first commercial device for recording electrocardiograms (ECGs); see Figure 1.2.

This book is about computing the electrical activity in the heart. In order to do so, we will need mathematical models of how the electrical signals are generated in the heart. Furthermore, we will need mathematical models of how the signals are transported through the heart and distributed in the body. All these models are formulated in terms of differential equations. Based on these models, we will derive discrete models suitable for numerical simulations. We will base our methods on a finite element approach and will solve the equations that arise on parallel computers.

In order to understand this book, you will need to have a basic course in partial differential equations, and it is definitely an advantage to know a bit about finite element methods; but that is about it. We will explain all the biology you will need. We will provide a detailed introduction to parallel computing and how to solve linear systems of equations.

This book is about computational science; in fact, its main objective is to present a large project in computational science. We will begin by giving you the necessary physiological background. What is going on in the heart and how can we use a recording on the body surface to say anything about the condition of the heart? These questions are discussed briefly in Chapter 1.

In Chapter 2 we jump straight to the mathematical models. A wonderful feature of mathematics is its ability to phrase extremely complex phenomena in rather simple equations. Yet despite this ability, the equations we describe in Chapter 2 are rather complicated. These models are absolutely necessary in order to be able to compute the electrical activity on a computer. We guide you through the basic physics of the process, ranging from models of what happens within a single cell to a complete mathematical model of the electrical activity in the entire heart.

In Chapter 3 we discuss how to discretize the equations derived in Chapter 2. We will do this using the finite element method. The approach is straightforward, but matters become complicated due to the complexity of the mathematical models. Chapter 3 will motivate three specific problems that have to be considered carefully. First, the finite element method leads to large systems of linear equations. It is well known from other fields that these equations may be very challenging to solve, and in Chapter 4 we introduce and analyze preconditioning techniques to solve the linear system that arises from the finite element discretizations. Second, systems of ordinary differential equations that model the processes going on in single cells have to be solved. We discuss appropriate methods for this in Chapter 5. Third, the problem that we are going to solve is huge, and therefore has to be solved on parallel computers. We introduce such computers in Chapter 6 and demonstrate their usefulness for the problems at hand.

Our primary goal is to contribute to the development of machinery that can compute the location and geometry of a myocardial infarction. This is a huge task occupying the minds of many researchers around the world, but a technologically feasible solution still lies in the future. In Chapter 7 we initiate a discussion of how to utilize our ability to simulate the electrical activity in the heart and the body in order to detect an infarction. This is, in fact, an inverse problem; we measure the result of a process and we then compute its cause. In the present setting, the measurements are electrocardiograms and the cause is the infarcted area of the heart. The process is the transport of electrical signals in the heart and the distribution of these in the body.

We hope this book can serve as an introduction to this field for applied mathematicians and computational scientists and also for researchers in bioengineering who are interested in using the tools of computational science in their research. The book presents the authors' view of the field, with a strong emphasis on the mathematical models and how to solve them numerically. There are few new scientific results in the book. Most of the material presented here is based on already published material.

Fornebu, Norway
March 2006

Joakim Sundnes
Glenn Terje Lines
Xing Cai
Bjørn Fredrik Nielsen
Kent-Andre Mardal and Aslak Tveito

Table of Contents

Chapter 1

Physiological Background

Our knowledge about the heart dates back more than two millenia. Already in the days of Aristotle (350 b.c.) the importance of the heart was recognized, and it was, in fact, considered to be the most important organ in the body. Other vital organs, such as the brain and lungs, were thought to exist merely to cool the blood. Over two thousand years later, the heart maintains its position as one of the most important, and also most studied, organs in the human body.

During the last few decades, the quest for knowledge of the heart has been motivated, not only by a desire to uncover the secrets of this vital organ, but also by its growing clinical importance. In the western world, heart failure is by far the most frequent cause of death, and the related financial and personal costs are huge. An improved understanding of how the heart works may lead to new techniques for the diagnosis and treatment of heart problems, and this serves as motivation for the enormous resources that are invested in heart-related research.

The pumping function of the heart is the result of a rhythmic cycle of contraction and relaxation of about 10^{10} muscle cells, a process that is controlled by a complex pattern of electrical activation. Electrical activity is essential for the function of the heart, and many heart problems are closely linked to disturbances of the electrical activity. In fact, most serious heart problems either result from, or cause, abnormalities in the electrical activity.

This close link between the electrical activity and heart problems is the basis of the diagnostic power of the electrocardiogram, or ECG, which is the oldest non-invasive tool for diagnosing heart conditions. The ECG is a recording of electrical potential differences on the body surface that result from the electrical activity in the heart. These potential differences are caused by sources of electrical current in the active heart muscle, and they change in characteristic patterns when the electrical activity is disturbed. Worldwide, it is estimated that about one million ECG recordings are performed every day, which makes it the most widely-used tool for heart diagnosis. It may be used to detect a variety of different pathological conditions, such as rhythm disturbances caused by conduction system abnormalities, infarctions, and ischemic heart disease.

The electrical activity of the heart is a well-studied process, and there exists a vast store of knowledge, both of the small scale processes occurring in the cells and of the organ-level pattern of electrical activation. Even so, there are a large number of mechanisms that are not fully understood. The organ-level electrical activity of the heart is the result of billions of small-scale processes occurring in the cells, and the existing knowledge of how all these processes interact is very limited. A very promising technique to extend our knowledge in this field is the use of mathematical modelling and computer simulations. By formulating precise, quantitative descriptions of the small-scale processes, these models can be combined to form mathematical models of larger systems. This field is often referred to as integrative physiology, and has the potential to improve significantly our understanding of how small-scale processes interact to form the functioning organ.

In addition to their efficacy, experiments based on mathematical models and computations often represent a simpler, and less expensive, alternative to experiments with real hearts. As noted in the Financial Times on February 6, 2004 [90]:

> Researchers are experimenting with virtual hearts partly because it is easier than tinkering with or peering into a living, beating one. As Alan Garfinkel, a cardiologist at the University of California, Los Angeles, says, "You can't get the light into the meat."

For the case of the electrical activity of the heart, improved understanding may lead to improved treatment techniques and diagnosis, for instance in the form of more precise ECG analysis.

The focus of this book is on computer simulations of the electrical activity of the heart. Particular emphasis is placed on simulating ECGs, but the simulation results have a number of other applications as well. We do not attempt to give a detailed description of the general physiology of the heart, but focus on the mathematical models and related computational issues. However, to motivate the mathematical models, we will here provide a very brief description of the ECG and the electrophysiology of the heart. Readers interested in a more thorough discussion of these fields may consult, for instance, [71] or a general textbook on physiology, such as [11].

1.1   The Electrocardiogram

The first human ECG was published in 1887 by Augustus D. Waller [142]. Waller gave several demonstrations of his techniques, many of them involving his bulldog Jimmy standing with his paws in buckets of saline, see Figure 1.1. The conducting solutions in the buckets acted as electrodes, and were connected to a device that recorded the electrical potential difference between the two electrodes. The potential difference was seen to pulsate in rhythm with Jimmy's heart beats, and Waller presented evidence to support the idea that the potential differences resulted from electrical activity in the heart. Waller was also the first to use the term electrocardiogram.

One of Waller's demonstrations was attended by Dutch physiologist Willem Einthoven. He refined Waller's technique and invented a more robust and sensitive device for recording the potential differences (the string galvanometer). To record the human electrocardiogram, Einthoven used three electrodes, attached to the left and right arm and to the left leg. Early methods used saline-filled buckets as electrodes, as seen in Figure 1.2, but these were later replaced by more convenient

Fig. 1.1. The famous demonstration by Augustus D. Waller, recording the electrocardiogram of his bulldog Jimmy.

electrodes. Strictly speaking, it is not possible to measure the electrical potential in a point; only potential differences may be measured. As illustrated in Figure 1.1, Waller recorded the potential difference between two of Jimmy's legs, and similarly, Einthoven used the three limb electrodes to record three potential differences. The potential differences were called leads I, II, and III, and are illustrated in Figure 1.2.

Fig. 1.2. An illustration of the original leads defined by Einthoven.

If we introduce the notation ϕ_{LA}, ϕ_{RA}, and ϕ_{LL} for the potential on the left arm, right arm, and left leg, respectively, the leads are defined by

$$I = \phi_{LA} - \phi_{RA}, \tag{1.1}$$
$$II = \phi_{LL} - \phi_{RA}, \tag{1.2}$$
$$III = \phi_{LL} - \phi_{LA}. \tag{1.3}$$

The leads form a triangle on the torso, as illustrated in Figure 1.2.

Since the Einthoven leads record changes in the potential difference between two point electrodes, they are called bipolar leads. The potential changes recorded by one lead will be the combined result of potential variations in the two electrodes that define that lead. For instance, we may assume that a given electrical source in the body causes an increase in the potential ϕ_{LA} and ϕ_{RA}, and a decrease of ϕ_{LL}. It is easy to see from (1.1)–(1.3) that the result will be a fairly large, negative signal in lead II and lead III, while for lead I the potential will increase in both electrodes, resulting in a small or zero signal being recorded by that lead.

A schematic view of a typical ECG, i.e. a recording from one of the leads, is shown in Figure 1.3. The figure shows how the potential difference recorded by one lead changes during a heart cycle. Time is shown along the horizontal axis and the potential difference on the vertical axis. The straight line segments in the ECG, where the recorded potential difference is zero, corresponds to intervals in the heart cycle when there are no sources of electrical current in the heart. The five deflections in the ECG are results of current sources arising in the heart during the electrical activation of the tissue. The five deflections were first identified by Einthoven, who introduced the notation P, Q, R, S, and T. The first signal is the P-wave, while the next three signals form a combined signal called the QRS-complex. The last signal of the normal ECG is called the T-wave. Understanding the physiological origin of each signal requires some basic knowledge of the signal propagation in the heart, and will be described below.

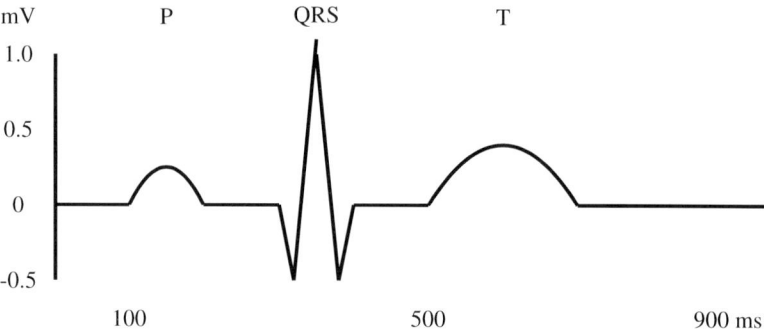

Fig. 1.3. Sketch of a typical ECG, showing the potential variations recorded by one lead during a heart cycle.

V1

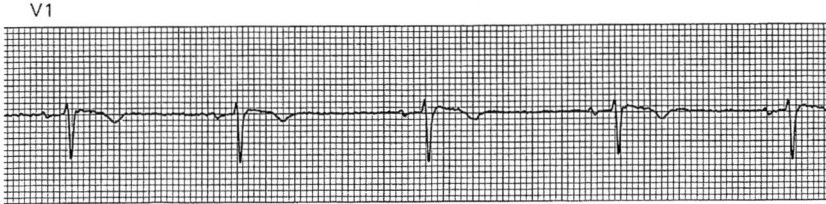

Fig. 1.4. A real ECG, from a healthy 30-year old male.

Figure 1.3 shows a very schematic and idealized ECG. In practice, the shape and amplitude of each signal will be different in the three leads, and it might be that not all five signals are visible in all leads. Figure 1.4 shows a real ECG, from lead V1, of a healthy 30-year old male. In this figure, showing several heart beats, the QRS-complexes are easily identified. The P-wave and the T-wave are visible, although they both have fairly low amplitude in this lead.

1.1.1 Physics and Physiology

The current sources that give rise to the ECG are caused by the electrical activation of the heart muscle cells. Because of the conductive properties of the body, these sources result in currents and potential variations that can be recorded on the body surface. The ECG is hence closely connected to cellular electrophysiology, but very useful interpretations of the ECG can be made with little attention to the underlying physiology. For instance, Waller and Einthoven introduced the very powerful simplification of viewing the heart as a dipole embedded in a volume conductor. This approach enables very simple and powerful interpretations of the ECG, and is still an essential part of modern electrocardiology.

The idea is based on the assumption that the body has the properties of a volume conductor, and that the sources of electrical current in the heart can be viewed as dipoles. A volume conductor is simply a three-dimensional conducting medium, while a dipole is a pair of opposite electrical charges with equal magnitude $(-q, q)$, separated by a small distance d. Strictly speaking, a dipole has only two point charges, but various arrangements of multiple charges or charge distributions may also have properties similar to a dipole. A dipole generates an electrical field, and in a volume conductor this electrical field leads to currents throughout the medium. The relation between the electrical field and the current will be described in more detail in Chapter 2. The strength of a dipole, and the strength of the electrical field it generates, is characterized by the dipole moment, which is the product of the charge in each pole and the distance between the poles. The dipole moment has an associated direction, which is the direction from the negative to the positive pole. Denoting the dipole moment by **p**, we have

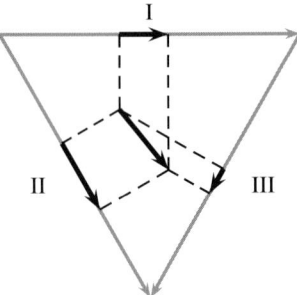

Fig. 1.5. The Einthoven triangle, formed by the three lead vectors I, II, and III. The signal recorded in each lead is the projection of the heart vector onto the lead vector.

$$\mathbf{p} = q\mathbf{d},$$

where $\mathbf{d}$ is the vector pointing from the negative to the positive pole.

The current sources present during the activation of the heart muscle can be approximated by a number of dipoles, with associated dipole moments. The sum of all the dipole moments gives a vector describing a single dipole, which characterizes the sources of electrical current in the heart. This vector is called the *heart vector*, and has been a central part of ECG analysis since it was introduced by Einthoven in the early 20th century. The position of the heart vector is assumed to be static, but its strength and orientation varies during the heart cycle.

The potentials recorded by the three Einthoven leads can be interpreted as projections of the heart vector onto the three lead vectors of the Einthoven triangle, as illustrated in Figure 1.5. Hence, the signal amplitude recorded in each lead can be used to construct the heart vector, and in this way it is possible to identify changes in the activation pattern. For instance, the current sources present in the heart during the QRS complex (see Figure 1.3) can be approximated by a dipole that is oriented downwards and to the left. The resulting heart vector is illustrated in Figure 1.5. We see that the vector is almost parallel to lead II, and nearly perpendicular to lead III. As illustrated by the vector projections, we therefore expect to see a large amplitude of the QRS complex in lead II, and a small amplitude in lead III. The amplitude in lead I will have an intermediate value. The directions of the three Einthoven leads have been chosen so that a normal activation pattern gives positive R-waves (the largest signal of the QRS-complex) in all three leads.

A surprisingly large amount of information can be extracted from the three limb leads used by Einthoven. In fact, since the three leads are sufficient to find the direction and strength of the heart vector, one may think that no additional information can be obtained by adding more leads. This would be the case if the heart were truly a dipole oriented in the frontal plane of the body, i.e. the plane defined by the position of the three limb electrodes. However, this simplified view of the heart is not always sufficient. Specifically, the heart vector is not always oriented in the frontal plane of the body, and the dipole approximation does not fully reproduce the complicated electrical activity in the heart. Therefore, more leads have been added, to produce a more accurate picture of the condition of the heart.

As noted above, the electrical potential at a point must always be measured relative to some reference potential, and the three limb leads defined by Einthoven are all bipolar leads. To obtain a good picture of the potential changes in a single point, it will be useful to have an independent reference, or a zero electrode, which changes very little during the course of a heart cycle. This concept was introduced by Wilson [144] and his group, who constructed an independent reference by connecting the three limb electrodes of Einthoven. The idea is that since no electrical charge enters or leaves the body during the heart cycle, the sum of all changes of potential in the body must be zero. Ideally, therefore, one would construct a zero electrode by connecting a large number of electrodes, distributed over the entire body. This would obviously be an inconvenient solution, and the connection of the three limb electrodes has been shown to give a reference electrode with sufficiently small variations. The obtained reference potential is referred to as the Wilson central terminal, and is used to construct unipolar leads, i.e. leads that characterize the potential changes at a single point only.

The six unipolar leads V1-V6 were constructed by placing six electrodes on the front of the chest, as illustrated in Figure 1.6, and recording the potential difference between these electrodes and the Wilson central terminal. Together with the original limb leads of Einthoven, these leads formed a standard nine-lead ECG in 1938. In 1942, three additional leads were introduced by Goldberger [48]; the augmented limb leads aVR, aVL, and aVF. These leads are defined by comparing the potential in each of the three limb electrodes to a reference defined by connecting the two other limb electrodes. They are normally referred to as unipolar leads, although the zero electrode is constructed using only two leads. These 12 leads still form the standard ECG, although there is an ongoing debate regarding the possible advantages of using additional leads; see e.g. [20]. Table 1.1 shows the definition of the 12 leads. The left column shows the name of each lead, the middle column shows the so-called exploring electrode, while the rightmost column lists the leads used to construct the zero electrode. The concept of a lead that consists of an exploring electrode and a zero electrode is common for unipolar leads, and for simplicity we use the same notation for the Einthoven leads.

Figure 1.7 shows a standard 12-lead ECG. One heart cycle is shown for all leads, and the position of each lead is standardized. The bottom line shows the recording of lead II for several heart cycles, included to give a better picture of the heart rhythm. The box-shaped signal to the right on every curve is a calibration artifact resulting from a 1 mV pulse. The position of each lead on the paper is standardized, as is the paper itself. Although barely visible in the figure, the ECG paper is divided into squares. The smallest squares are 1 mm x 1 mm and the larger squares are 5 mm × 5 mm. (The ECG shown in the figure has been scaled, so the dimensions do not match here.) In the horizontal direction each large square represents 0.2 seconds, while in the vertical direction one large square represents 0.5 mV. This standard, with 1 mV being represented by 1 cm, is so well-established that ECG signals are very often described in terms of millimeters rather than millivolts.

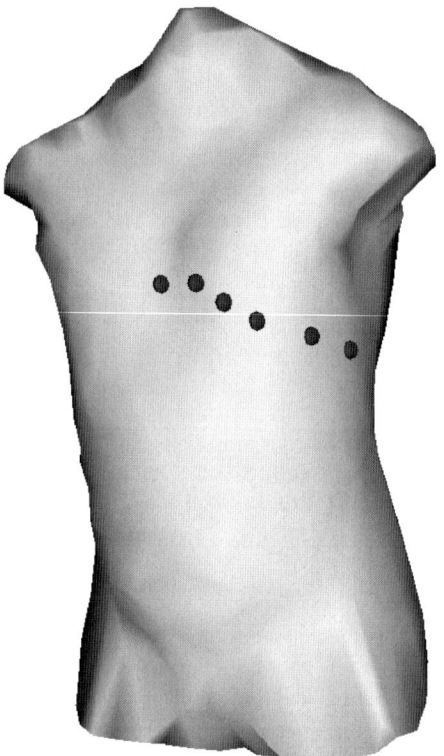

Fig. 1.6. The position of the chest electrodes. The leads V1-V6 are constructed by measuring the potential difference between each of these electrodes and the Wilson central terminal.

Table 1.1. The leads of the 12-lead ECG. The notation LL, LA, and RA is used for the left leg, left arm, and right arm electrodes, respectively. The six chest electrodes are denoted 1-6.

Lead	Exploring	Zero
I	LA	RA
II	LL	RA
III	LL	LA
aVR	RA	LA and LL
aVL	LA	RA and LL
aVF	LL	RA and LL
V1-V6	1-6	RA, LA, and LL

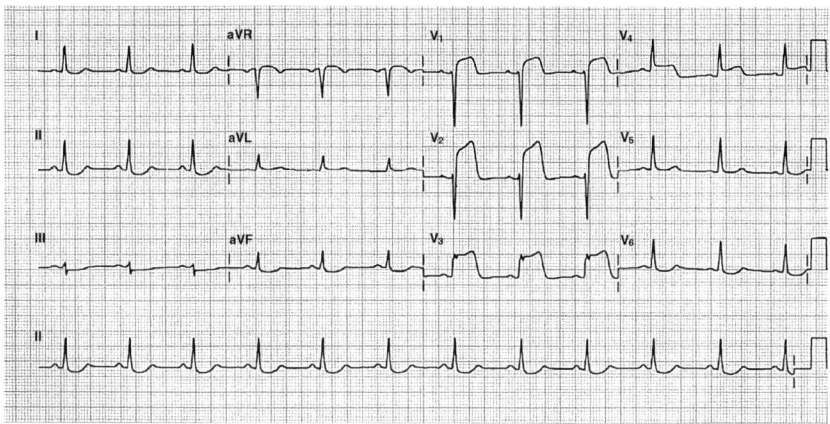

Fig. 1.7. A 12-lead ECG, showing indications of ischemic heart disease. (T.B. Garcia and N.E. Holtz, 12-lead ECG – The art of interpretation, 2001: Jones and Bartlett Publishers, Sudbury, MA, www.jbpub.com. Reprinted with permission.)

The ECG shown in Figure 1.7 is not normal; it shows signs of an acute heart infarction. We will provide a brief discussion of these signs below.

1.1.2   Cellular Electrical Activity

Approximating the heart as a rotating dipole with varying strength is a very useful simplification, but a true understanding of the ECG must be based on the underlying physiology. Specifically, since the electrical activity is the result of electrochemical reactions in the heart cells, an understanding of the ECG must be based on knowledge of the fundamentals of cellular electrophysiology. This topic will be discussed in more detail in Chapter 2, where we present mathematical models for the cellular processes. However, to ease the understanding of the ECG, we will here give a very brief introduction to the essential behaviour of heart cells. Readers interested in a more detailed description are referred to, e.g., [72,64]

From the perspective of electrical activity, the most important property of cardiac cells is that they are excitable, i.e. they have the ability to respond actively to an electrical stimulus. Under resting conditions, the cells maintain internal ionic concentrations different from those of their surroundings. Because of the electrical charge of the ions, this results in a potential difference across the cell membrane, which is called the *transmembrane potential*, or simply the membrane potential. More specifically, the potential in the internals of the cells will be negative compared to their surroundings, with typical transmembrane potential values in the range of -70 to -100 mV. The actual value of the resting transmembrane potential varies between different species and different cell types.

If an electrical stimulus is applied to a cell, the result is a change in the transmembrane potential. An excitable cell may respond to such a potential change in one of two ways. If the change is small, the conductive properties of the membrane remain unchanged, and the potential quickly returns to its resting value when the applied current is removed. However, if the applied stimulus is strong enough to raise the transmembrane potential above a certain threshold value, the response will be different. In this case the conductive properties of the cell membrane change, resulting in a rapid flux of positive ions into the cell. This causes a *depolarization* of the membrane, where the transmembrane potential increases from its negative resting value to a value around, or significantly above, zero.

After the depolarization, the potential returns to its negative resting state, a process referred to as repolarization. The complete process of depolarization and repolarization is called an action potential. In many excitable cells, such as nerve cells and skeletal muscle cells, the repolarization occurs almost immediately after the depolarization, producing very short action potentials. Heart muscle cells however, stay at their depolarized value for a significant period of time, called the plateau phase.

Figure 1.8 shows a typical action potential for a ventricular muscle cell. The figure shows the transmembrane potential (in mV) plotted along the y-axis, while time (in ms) is shown along the x-axis. We see that fast depolarization, often called the upstroke phase, is followed by a fairly long plateau phase, before repolarization returns the cell to the resting state. The detailed mechanisms of the action potential will be discussed in more detail in Chapter 2. From a computational point of view, the rapid upstroke of the action potential generates a number of challenges. When modelling a single cell only, very small time steps are needed in the upstroke region to capture the rapid change. Additional challenges arise when we attempt to model the electrical activation of the complete heart. This will be described in greater detail below.

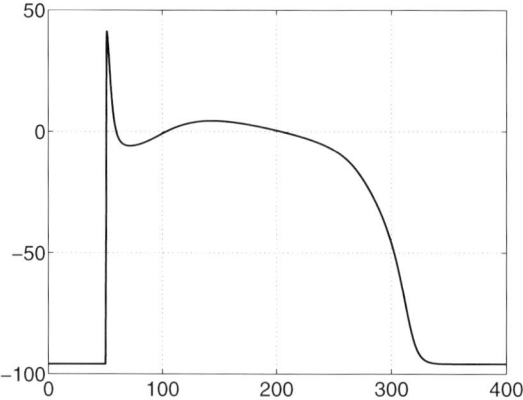

Fig. 1.8. An action potential of a ventricular muscle cell from a dog. The potential is computed with a model by Winslow et al. [145].

1.1.3 Signal Conduction

The pumping function of the heart relies on a collective, systematic contraction of billions of muscle cells. This operation can not be achieved without significant communication and synchronization between the cells. To ensure that each part of the heart contracts at the correct time, the activation is controlled by a complex system for signal conduction. The mathematical models we derive in this book will not pay particular attention to the conduction system, but a basic understanding of the signal propagation in the heart is useful for understanding the characteristics of the ECG.

The electrical signal in the heart starts in the sinoatrial node, which is located above the right atrium; see Figure 1.9. The cells in the sinoatrial node are so-called self-oscillatory cells, often referred to as pacemaker cells. This means that they spontaneously produce action potentials, not relying on any external stimulus. The frequency of the action potentials in the sinoatrial node are controlled by external signals, for instance to adjust the heart rate to different activity levels.

The electrical activation of the sinoatrial node stimulates neighbouring atrial muscle cells. The muscle cells are connected by so-called gap junctions, which are large proteins that form channels between adjacent muscle cells. These channels allow a flux of electrical current in the form of ions, and hence provide a direct electrical connection between the internals of neighbouring cells. When one cell is depolarized this coupling will affect the potential in neighbouring cells, and may raise the transmembrane potential above the threshold value. Stimulating a small region of the atria hence results in a propagating wavefront of depolarization, which activates the complete atria and causes them to contract.

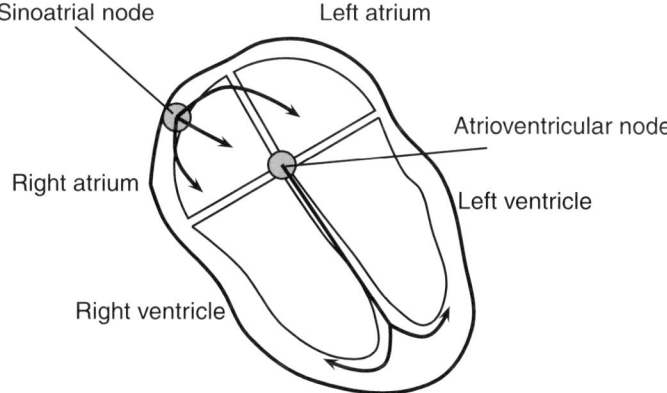

Fig. 1.9. A schematic view of the pathways of electrical conduction in the heart. The activation starts in the sinoatrial node, then activates the atria, and the signal is then conducted through the atrioventricular node to activate the ventricles.

The atria are separated from the ventricles by a non-conductive layer, so the depolarization wavefront does not propagate directly into the ventricles. Instead, the only place where the electrical signal can be transmitted to the ventricles is through the atrioventricular node; see Figure 1.9. Conduction through the atrioventricular node is quite slow, which results in a small time delay between the activation of the atria and the ventricles. The effect of this delay is that the atria contract while the ventricles are still relaxed, which improves the filling of the ventricles and the pumping function of the heart.

From the atrioventricular node, the signal enters the atrioventricular bundle, also referred to as the common bundle or the bundle of His. This bundle branches out into a tree-like structure of Purkinje fibres, which provide rapid conduction of the electrical signal. The muscle cells are stimulated at the ends of the Purkinje network, which are located near the endocardiac (inner) surface of the ventricles. Like the atrial cells, the ventricular muscle cells are also connected by gap junctions. The stimulus hence causes wavefronts of depolarization, which eventually activates the complete mass of the ventricles, and triggers the contraction of the cells.

As shown in Figure 1.8, the depolarization of the cell is a very fast process. At the tissue level, this results in the propagating wavefront of depolarization being very sharp, with a variation in the transmembrane potential of approximately 100 mV over a spatial range of less than 1 mm. This steep, moving gradient represents a significant source of electrical current, and because of the volume conductor properties of the body it results in currents and potential variations that can be recorded on the body surface. The current source that Waller and Einthoven modelled as a single dipole is hence a moving, curved wavefront of depolarization.

We noted above that the rapidity of the upstroke generates computational challenges, in that a high temporal resolution is needed to resolve the fast depolarization. When modelling signal propagation in the tissue, the steepness of the wavefront places strict demands on the spatial resolution as well. This issue will be described in more detail below.

Having provided a brief overview of the electrical activation pattern in the heart, it is now possible to relate the signals in the ECG (see, e.g., Figure 1.3), to the specific events. The P-wave occurs first, and is the result of the depolarization of the atria. The next signal, the QRS-complex, corresponds to the depolarization of the ventricles. The larger amplitude of this signal compared to the P-wave is a result of the larger mass of the ventricular muscles. The final event that produces a current source strong enough to be visible on ECG is the repolarization of the ventricles, which shows up as the T-wave. The smoothness and smaller amplitude of this signal compared to the QRS-complex is a result of the repolarization phase being less steep than the depolarization, as shown in Figure 1.8.

1.1.4 Diagnosis and the Inverse Problem

The diagnostic power of the ECG lies in the fact that the signal changes characteristically in response to a number of different heart conditions. The field of heart diseases and their relation to the ECG is huge, and the description we give here is nowhere near being complete. However, a couple of examples are useful, to illustrate the diagnosis of heart problems based on the ECG, and to serve as a background for computer simulations that will be presented later.

The ECG is a recording of the electrical activity of the heart, and it therefore seems obvious that abnormalities in the sinoatrial node or the conduction system will change the ECG. For instance, abnormal activity in the sinoatrial node may cause an increased or decreased heart rate, which is easily detected on the ECG. Other disturbances of the electrical activity are also fairly easy to detect. For instance, the presence of P-waves with missing QRS-complexes is a sign that the atria are activated successfully, but that the signal is blocked before it can activate the ventricles. This indicates a conduction block in the atrioventricular node or bundle (AV block).

Many pathological conditions in the heart are identified by using the lead potentials to reconstruct the imaginary heart vector that was described above. We mentioned that during the activation of the ventricles, i.e. during the QRS-complex of the ECG, the current sources in the heart may be approximated by a dipole, the orientation of which is down and to the left, as shown in Figure 1.5. Abnormalities in the activation pattern will lead to a shift in the heart vector, which can be identified by changes in the signal amplitudes recorded in each lead.

Conditions such as myocardial infarction and ischemic heart Disease are also commonly diagnosed using the ECG. Although they are not always easy to detect, several changes of the ECG are characteristic for these conditions. One characteristic change is a shift in the ST segment, which is the interval between the QRS-complex and the T-wave. In the normal ECG, this is a straight line that is on a level with the TP segment, i.e. the interval between the T-wave and the P-wave. When a portion of the heart lacks a sufficient blood supply, electrophysiological changes in this area will cause a shift in the ST-segment. Depending on the size and location of the ischemic area, the ST-segment will shift either up or down compared to the TP-segment. In the 12-lead ECG shown in Figure 1.7, there are large ST elevations in lead V1-V4, and also small signs of elevation in aVL. In leads II, III, and aVF we see a small depression of the ST-segment. As described in [46], this ECG is typical for an anteroseptal infarction, i.e. an infarction that affects the frontal wall of the heart and the wall between the left and right ventricle. The topic of ST-segment shifts will be discussed further in Chapter 3, where we will present simulations of the ECG changes that result from ischemic heart disease.

The list of heart conditions mentioned here is far from complete. A thorough description of the relation between heart conditions and the ECG can be found in a general textbook on heart physiology, e.g. [71], or a book devoted to ECG interpretation, such as [46].

In the examples listed here we have mentioned various changes in the electrical activity of the heart, and then described the results that show up on the ECG. Such a situation, in which we want to find the effect of a known cause, is commonly referred to as a forward problem. The situation faced everyday by physicians analyzing ECGs is the exact opposite. *From* an observed effect (ECG results), they have to make inferences to determine the cause (the underlying condition of the heart).

In mathematical terms, this kind of problem is referred to as an inverse problem. As will be discussed below, both the forward problem and the inverse problem may be formulated mathematically. Most of this book will be devoted to mathematical models and solution techniques for the forward problem, but the inverse problem will be discussed in Chapter 7.

1.2   Computer Simulations

Over the last few decades, there has been a huge increase in the amount of available information about the mechanisms of the heart, and of biological systems in general. Techniques to observe cellular, and even molecular level, processes are refined continuously, leading to remarkable progress in discovering the innermost secrets of living organisms. In parallel with this development, there has been a rapid development of techniques for non-invasive examination of the function of the heart. Although the ECG is still the most widely-used tool for heart diagnosis, the use of other techniques that improve our ability to produce functional images of the heart, such as ultrasound and magnetic resonance imaging (MRI), is increasing rapidly.

Despite the above mentioned advances, there remain several unresolved questions. A good example is defibrillation; the application of a large electrical shock to end ventricular fibrillation. Ventricular fibrillation is a state in which leaves the heart muscle cells contract in a seemingly random manner, which the heart unable to pump blood. Defibrillation shocks are applied regularly, both by implanted devices and by the well-known external defibrillators used by rescue squads and in the emergency room. The success rate is remarkably high, but our understanding of the phenomenon is very limited. The Financial Times [90], quoting Richard Gray, a biomedical engineer at the University of Alabama, Birmingham, describes the situation as follows:
It is a medical miracle, and no one can explain how it works. "We don't even know how the electric current gets into the heart," says Mr Gray. Nor does anyone really know how ventricular fibrillation gets started, or why a big shock ends it.

1.2.1   Why Simulate?

Initially, our limited understanding of a widely-used process such as defibrillation would seem to be somewhat unexpected. However, if one considers the complexity of the process, the situation becomes much less surprising. Although there exists a remarkable amount of knowledge about the processes on a cellular and sub-cellular level, it is very difficult to understand the details of how these processes interact to form the functioning organ. Even if we only consider a single cell, the electrical activity is the result of a complex interaction between a variety of electrochemical phenomena. Furthermore, as described above, the electrical activity of the complete heart is the result of the collective activity of billions of cells that form a complex heterogeneous system for conduction and contraction.

The result of this complexity is that it becomes very difficult to integrate knowledge of small-scale processes into an understanding of complete organ function.

In fact, although complete understanding can be obtained of a number of separate, small-scale processes, it may be almost impossible to understand how these processes will interact to form a complex system. The behaviour of the heart under pathological conditions such as fibrillation, or the extreme conditions induced by defibrillation, is even more difficult. As noted in the introduction to this chapter, a promising method by which this problem may be overcome is to describe the processes in terms of mathematical models. If the small-scale processes are described in terms of precise, quantitative models, the task of combining these models into a model for the complete organ is at least feasible, although still a serious challenge.

The field of mathematical modelling in physiology is rapidly gaining popularity, and the potential both for increasing general knowledge and for clinical applications is huge. Although mathematical modelling has been applied to a large number of different biological phenomena, the function of the heart is a field that has received particular attention. As mentioned above, the quest for knowledge in this field is driven both by the curiosity that has inspired physiologists for millennia, and by the growing health problems related to the heart. In fact, the promising aspects of mathematical models applied to analyze the behaviour of the heart is the main motivation for writing this text.

1.2.2 State of the Art

Computer simulations of physical phenomena, as discussed in this text, are based on mathematical models. For the case of the electrical activity of the heart, a number of different mathematical models have been developed that describe the electrophysiology on many different scales. The concept of a heart vector, which was introduced more than 100 years ago, is one example of a mathematical model of the heart. However, this model was based on a top-down approach and did not take into account the underlying physiology. A significant breakthrough in this respect was the work of Hodgkin and Huxley [66,65] in the 1950s. Their model for signal propagation along an axon (part of a nerve cell) was based on detailed models for several ionic currents in the cell, which were combined to give a description of the complete electrical activity of the cell. Models for the electrical activity of heart cells, which are all based on the framework introduced by Hodgkin and Huxley, will be described in Chapter 2.

Since the publication of the Hodgkin-Huxley model more than 50 years ago, the modelling and simulation of biological phenomena has grown much more complex and realistic. The development of experimental techniques provides more and more information about biological systems, and this information drives the development of successively more sophisticated models. This development has been accompanied by a rapid development in hardware and numerical techniques, which allow us to perform simulations based on the increasingly sophisticated models.

The Hodgkin-Huxley model includes only four ordinary differential equations: one that describes the transmembrane potential and three others that characterize the permeability of the membrane. Models with a similar structure, for other types of excitable cells, were developed subsequently [104,9]. More and more properties of the single cell have been incorporated into the mathematical models, and the number of variables has increased more than tenfold compared to the Hodgkin-Huxley model. The increasing physiological accuracy increases the possible impact of the models, but also gives rise to computational challenges. In general, it is not very difficult to solve a system of, say, 40 ordinary differential equations. However, when a cellular model is used for full-scale simulations of the electrical activity in the heart, the equations have to be represented in millions of grid points. The resulting computation is a challenge for any existing computer.

One may wonder what the purpose is of all the variables included in recent models for cellular activity. Many are related to the permeability of the membrane, just as in the Hodgkin-Huxley model. In the cell membrane there are embedded proteins that control the traffic of substances across the membrane. These proteins are typically very specific, allowing only one kind of substance to pass. In models of electrophysiology it is necessary to track the flow of ions through such channels, and many variables are used to characterize the dynamics of channels as they open and close.

The intracellular calcium concentration is important in muscle cells, because it forms the link between electrical activation and the contraction of the cell. The concentration is controlled by intracellular buffers and compartments. The more recent models include the dynamic control of intracellular calcium [87,68].

Most or all variables in the electrophysiology models are thus used to describe membrane properties and ionic concentrations, often with special emphasis on the intracellular calcium.

As more information becomes available, the models will continue to be refined. New ionic channels will probably be discovered and modelled. A further consideration is that the heart muscle is not homogeneous, but consists of different cell types. It is well known that the cells in the atria are different from ventricular cells, and there are also regional differences in the ventricular wall [2]. Cells close to the inner surface of the ventricles have different action potentials from the mid-wall cells, which in turn are different from the cells near the outer surface of the muscle. These differences are important for the signal conduction pattern in the heart, and must be included in realistic, organ level simulations. From a clinical point of view, it is important to make models specific to humans, and also models for how the cells behave under pathological conditions. Most recent models are based on measurements from non-human cells, but several groups have started to develop models based on human cells [115,125]. Models of cellular behaviour under conditions such as insufficient blood supply (ischemia) are also emerging as adjustments of the normal models.

A further improvement of the cell models would be to track energy consumption. Some models for the metabolism in heart muscle cells have been proposed [123,26]. Combining models of electrical and metabolic activity could be important for the study of the behaviour of the heart during ischemic conditions.

As the level of detail in the models increases, it is possible to create simulations that not only reproduce data from experiments, but that also have predictive power.

As described in Section 1.2.1, one of the key features of mathematical models is their ability to predict complex interactions between different processes. Because of the complexity of biological systems, it is often difficult to predict every effect that results from, for instance, a drug-induced or pathological change in one cellular process. The increasing realism and complexity of the mathematical models enable simulations of secondary effects that are otherwise difficult to predict.

In order to simulate the electrical activity of the complete heart and its relation to the ECG, the models for electrical activity of the cells need to be combined to form tissue models and, eventually, models of complete organs. Because of the large number of cells, it is difficult to base organ-level simulations directly on the models for the behaviour of single cells. In fact, present-day computers are not powerful enough to facilitate such simulations. Instead, the mathematical models described in this book use a continuum approach to describe the tissue. The basic assumption is that the heart consists of two separate, continuous domains: the intracellular and the extracellular spaces. The domains are separated by the cell membrane, and the model is derived by requiring conservation of current and a quasi-static assumption on the electrical fields in the two domains. The model, referred to as the bidomain model, was first proposed by Tung [137] in the seventies, and has proven to be a lasting framework for studying the flow of current in the heart. We will return to the derivation of the bidomain model in Chapter 2.

A complete description of the heart will include both physiological and anatomical data, and the mathematical models must, of course, be completed with such data in order to perform realistic simulations. For electrophysiological simulations, such data include the shape of the heart and the orientation of its muscle fibres. To simulate electrical signals outside the heart, such as ECG signals, the conductive properties of organs such as lungs and skeletal muscle must be included in the model. The development of imaging techniques, such as MRI, has facilitated the acquisition of accurate anatomical data. The techniques enable accurate descriptions of the geometry of internal organs, and even internal structures, such as the orientation of muscle fibres in the heart, can be obtained using diffusion tensor MRI [136].

Performing realistic simulations depends not just on setting up accurate models with suitable data, but also on the ability to solve the resulting equations. The complex models that describe the heart are impossible to solve analytically, so in order to be of any use the equations must be solved on a computer using numerical techniques. This involves a discretization of the equations in space and time, using, for instance, finite difference or finite element methods. We noted above that the sharp depolarization wavefront in the heart leads to strict requirements on resolution when the equations are discretized. The required resolution depends on the actual cell model used and the desired accuracy. For realistic heart cell models the upstroke is very fast, and resolving the fast dynamics properly requires a spatial discretization of about 0.2 mm and a temporal resolution of about 0.1 ms. For the entire volume of the heart this translate to 40 million points, and a complete heart cycle corresponds to approximately 5000 time steps. It is clear that the resulting problem will be very large and that the most efficient solution techniques must be employed. Techniques for solving the ODE systems describing the cellular activity will be presented in 5, while techniques for discretizing PDEs are discussed in Chapter 3 and Chapter 4.

At present, it is not possible to perform accurate full-scale simulations of the heart's electrical activity using existing scalar computers. A number of simplifications can be made to make the problem less demanding, such as studying a small slab of tissue, or restricting the problem to two space dimensions. Another option is to use simpler models for the cellular activity, which involve fewer variables, and where the requirements on discretization are less strict. However, for organ-scale simulations based on physiologically accurate models, the only option is to use fast, parallel computers in combination with efficient numerical methods. Parallel solution techniques for the bidomain model will be described in Chapter 6

Numerical techniques give an approximate solution to the equations of the mathematical model. Typically, the approximate solution is simply a huge amount of numbers that describe the value of the involved quantities in a number of discrete points. Hence, the solution must normally be visualized in order to be of any use. Scientific visualization is a field of research in itself, and for complicated problems the presentation of the obtained information is, indeed, a non-trivial task. For the case of the electrical activity of the heart, we are mostly interested in visualizing electrical potentials and currents. This is easily achieved in two space dimensions, but for full three-dimensional datasets, the task is much more difficult. Maps of the electrical potential on the surface of the heart and the body are fairly easy to produce, and give results that can be compared to real measurements. However, one of the advantages of using computations is the ability to see what is going on inside the tissue, and this requires much more advanced visualization techniques. The topic of scientific visualization is of great relevance for biological modelling, but will not be discussed in detail in this book.

Following the visualization of the numerical solution, the results have to be validated by comparing them to physical measurements. For a complex phenomenon such as the electrical activity of the heart, it is likely that significant differences will be observed between measured and simulated values. If the equations of the mathematical model are solved correctly, the differences must be caused by limitations in the mathematical model. In this case, a refinement of the mathematical model is required, and in most cases we will enter an iterative process of mathematical modelling, numerical simulations, and comparison with physical measurements. This process is illustrated in Figure 1.10.

The bidomain model is widely accepted as a good model for describing the electrical activity in muscle tissue. However, because of the high resolution requirements, which result in huge computational demands for accurate simulations, the model has not yet been validated for a complete organ. Accurate full-scale simulations based on the bidomain model will therefore be an important achievement, valuable both for validating the model and for improving our understanding of heart physiology.

Although forward simulations based on the bidomain model have the potential to have a strong impact, the inverse problem may be even more interesting from a clinical point of view. A precise formulation of the inverse problem, followed by a

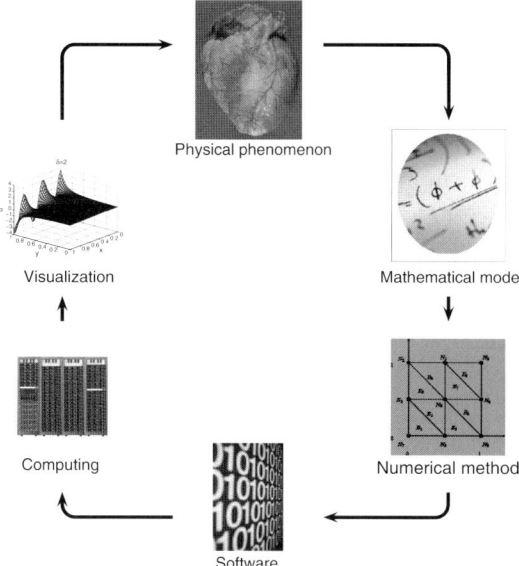

Fig. 1.10. The mathematical models are developed through a continuous interaction between modelling, computer simulations, and physical experiments.

sufficiently accurate solution, has the potential to revolutionize the field of electro-cardiology. Ideally, the condition of the heart can then be determined directly from the electrical potentials on the surface, by means of mathematics and large-scale computations. However, it is not clear that this scenario will be realized in the near future, because of the difficulty of solving the inverse problem. The forward problem is itself difficult to solve, and the inverse problem poses a number of additional difficulties. Even to give a precise mathematical formulation of the problem may be difficult, and the mathematical problem will be ill-posed, which makes it very challenging to solve. The inverse problem and techniques for its solution are discussed in Chapter 7.

Chapter 2

Mathematical Models

2.1 Modelling the Body as a Volume Conductor

As described in the previous chapter, the human body consists of billions of cells, which may be connected by various coupling mechanisms depending on the type of tissue under consideration. When constructing mathematical models for electrical activity in the tissue, one possible approach would be to model each cell as a separate unit, and couple them together using mathematical models for the known coupling mechanisms. However, the large number of cells will prohibit using this type of model for anything but very small samples of tissue. When studying electrical phenomena on the level of complete organs or even organisms, the level of detail provided by such an approach also goes far beyond what it is necessary, or even possible, to utilize.

In continuum mechanics, a standard technique is to study averaged volume quantities to avoid the difficulties of modelling the molecular structure of fluids and solids; see e.g. [45,67]. With this kind of volume-averaging approach, a quantity at a point P is viewed as an average over a small volume around P. The chosen volume over which the quantity is averaged must be small compared to the scales of the problem under study, but large compared to the molecular size of the material.

This averaging of volumes can also be used at the level of cells, to obtain a continuous description of biological tissue. The chosen volume over which the averaging is performed should now be small compared to the dimension of the problem under study, but large compared to the volume of a single cell. At each point P within the tissue, a quantity is defined as the average over a small but multicellular volume around P. In this way we avoid the difficulties of modelling the discrete nature of the tissue.

2.1.1 A Volume Conductor Model

For the case of electrophysiology, the local volume averaging introduced above allows us to model the body as a volume conductor. From Maxwell's equations we have that for a volume conductor, the relation between the electric and magnetic fields is given by

$$\nabla \times E + \frac{\partial B}{\partial t} = 0, \tag{2.1}$$

where E and B are the strengths of the electric and magnetic fields, respectively. For a description of Maxwell's equations and electromagnetism in general, see e.g. [32].

Throughout this book we describe the electrical activation of the heart as a fast process, but in the context of volume conductor theory, the variations in the resulting electric and magnetic fields are fairly slow. In fact, for the frequencies and conductance values that are relevant for the heart, the effects of the temporal variations may be disregarded, and for each moment in time the fields may be treated as static [113]. The fields are termed *quasi-static*, and the assumption that the time variations are insignificant is called the *quasi-static condition*.

Assuming quasi-static fields, (2.1) becomes

$$\nabla \times E = 0,$$

which entails that the electric field E may be written as the gradient of a scalar-valued potential. Denoting the potential by u, we have

$$E = -\nabla u, \tag{2.2}$$

where the negative sign is a convention. Note that following the ideas introduced above, the potential u, the electrical field E, and all other quantities introduced in this section must be viewed as quantities averaged over a small multicellular volume. The current J in a conductor is given by the general relation

$$J = ME,$$

where M is the conductivity of the medium. Inserting (2.2) gives

$$J = -M\nabla u. \tag{2.3}$$

We now assume that there are no current sources or sinks in the medium, and that there is no build-up of charge in any point. For a small volume V, we then have that the net current leaving the volume must be zero. Denoting the surface of V by S, we have

$$\int_S n \cdot J \, dS = 0,$$

where n is the outward surface normal of S. Application of the divergence theorem gives

$$-\int_V \nabla \cdot J \, dV = 0.$$

This relation must hold for all volumes V, which implies that the integrand itself must be zero throughout the domain. We get

$$\nabla \cdot J = 0,$$

and if we insert (2.3) in this equation, we get

$$\nabla \cdot (M\nabla u) = 0, \tag{2.4}$$

which is the relation describing the electric potential in a volume conductor with no current sources.

Equation (2.4) is only valid for a volume conductor with no current sources. The active heart tissue does generate such sources, and to simulate this activity the equation has to be adjusted slightly. A commonly used approach is to model the current sources in the active heart as one or more dipoles. The dipoles give rise to a source term on the right hand side of (2.4), and the resulting equation is valid throughout the body, including the heart muscle.

It is natural to assume that the body is surrounded by air or some other electrically insulating medium. This implies that on the surface of the body, the normal component of the current must be zero. We have

$$n \cdot J = 0,$$

where n is the outward unit normal on the surface of the body. Inserting (2.3) in this expression gives

$$n \cdot M \nabla u = 0$$

on the surface of the body. The electrical activity in the body is now described by

$$\nabla \cdot (M \nabla u) = f, \quad x \in \Omega, \tag{2.5}$$
$$n \cdot M \nabla u = 0, \quad x \in \partial \Omega. \tag{2.6}$$

Here Ω is the body including the heart, $\partial \Omega$ is the surface of the body, and f is a given source term. For the boundary value problem to have a solution the source term has to satisfy

$$\int_\Omega f dx = 0.$$

The value of the conductivity M will vary throughout the body, because different types of tissue have different conductive properties. Conductivity values for a number of different tissue may be found in, e.g., [76] and [51].

A different approach to computing the potential distribution in the body is to employ (2.4) only to those areas of the body immediately surrounding the heart. Since this domain is viewed as a passive conductor, there is no need to include a source term on the right hand side. We denote this domain by T, and the respective conductivity and potential by M_T and u_T. The outer boundary of the domain, i.e. the surface of the body, is denoted by ∂T. Since the domain does not include the heart, we also have an inner boundary, which is the interface between the heart and the surrounding torso. This part of the boundary is denoted by ∂H; see Figure 2.1. As above, we assume that the torso is surrounded by air, so that on the outer boundary the normal component of the current is zero. However, to solve the equations we also need boundary conditions on the inner boundary ∂H. A natural condition here is to assume that the potential distribution is known. We then have

$$\nabla \cdot (M_T \nabla u_T) = 0 \quad x \in T, \tag{2.7}$$
$$n \cdot M_T \nabla u_T = 0 \quad x \in \partial T, \tag{2.8}$$
$$u_T = u_{\partial H} \quad x \in \partial H, \tag{2.9}$$

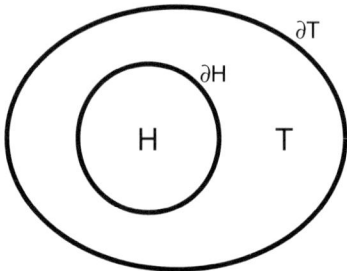

Fig. 2.1. Schematic view of the heart domain H and the surrounding torso T. The outer boundary of the torso is denoted by ∂T, while ∂H denotes the interface between the heart and the torso.

where $u_{\partial H}$ is the known potential distribution on the surface of the heart. For practical purposes, this potential distribution may be derived either from computations based on mathematical models for the electrical activity in the heart, or from direct measurements of the surface potential on the heart. The system (2.7)–(2.9) may then be solved to compute the potential distribution in the surrounding body. Of particular interest is the potential distribution on the body surface, which is directly related to ECG recordings. Assuming that we know the potential distribution $u_{\partial H}$ for a series of time points through a heart cycle, we may also compute approximate ECG curves by solving a series of problems (2.7)–(2.9).

Another interesting problem, especially from a clinical point of view, is the inverse problem. This involves finding the potential distribution on ∂H that results in a given potential distribution on ∂T. This is the classical inverse problem of electrocardiology, and has great potential for clinical applications. We will discuss this problem in Chapter 7.

2.2   A Model for the Heart Tissue

2.2.1   Excitable Tissue

The heart muscle cells belong to a class of cells known as excitable cells, which have the ability to respond actively to an electrical stimulus. As described in Chapter 1 the heart cells are also connected, so that a stimulated cell may pass the electrical signal on to neighbouring cells. This ability enables an electric stimulation of one part of the heart to propagate through the muscle and activate the complete heart.

The signal propagation in excitable tissue takes the form of a so-called *depolarization* of the cells. When the cells are at rest, there is a potential difference across the cell membrane. The potential inside the cells, called the *intracellular* potential, is negative compared to the *extracellular potential*, which is the potential in the space between the cells. When excitable cells are stimulated electrically they depolarize, i.e. the difference between intracellular and extracellular potential changes from its normal negative value to being positive or approximately zero. The depolarization is a very fast process, and it is followed by a slower *repolarization* that restores the potential difference to its resting value. The complete cycle of depolarization and repolarization is called an *action potential*. It is this action potential that initiates the contraction of muscle cells, and which allows both nerve cells and heart muscle cells to respond to an incoming electrical signal by passing the signal on to neighbouring cells.

Because the potential difference across the cell membrane is essential for the behaviour of excitable tissue, we need to construct a mathematical model that is able to describe this difference. The simple volume conductor model introduced in Section 2.1.1 is, therefore, not suitable. A natural approach would be to model each cell as a separate unit, thus enabling complete control of all variations in the intracellular and extracellular potentials. However, as discussed above, the number of cells in the heart is too large to model each cell separately. Hence, we are constrained to consider continuous approximations of the tissue, which must be able to distinguish between the intracellular and extracellular domains.

2.2.2 The Bidomain Model

The mathematical model for the heart tissue is based on a volume-averaged approach similar to the one introduced in the previous section. However, to include the effects of the potential difference across the membrane, the tissue is now divided into two separate domains: the intracellular and the extracellular. Both domains are assumed to be continuous, and they both fill the complete volume of the heart muscle. The justification for viewing the intracellular space as continuous is that the muscle cells are connected via so-called *gap junctions*. These are small channels embedded in the cell membrane, which form direct contact between the internals of two neighboring cells. Because of the gap junctions, substances such as ions or small molecules may pass directly from one cell to another, without entering the space between the two cells (the extracellular domain).

In each of the two domains we define an electrical potential, which at each point must be viewed as a quantity averaged over a small volume. An important consequence of these definitions is that every point in the heart muscle is assumed to be in both the intracellular and the extracellular domains, and consequently is assigned both an intracellular and an extracellular potential.

The intracellular and extracellular domains are separated by the cell membrane. We have assumed that both domains are continuous and fill the complete volume of the heart, and this assumption must also apply to the cell membrane, which is also viewed as a distributed continuum that fills the complete tissue volume. The membrane acts as an electrical insulator between the two domains, since otherwise we could not have a potential difference between the intracellular and extracellular domains. However, although the resistance of the cell membrane itself is very high, it allows electrically charged molecules (ions), to pass through specific channels embedded in the membrane. An electrical current will therefore cross the membrane, the magnitude of which depends on the potential difference across the membrane and on its permeability to the ions. The potential difference across the membrane is called the *transmembrane potential*. It is defined for every point in the heart, as the difference between the extracellular and intracellular potential.

The quasi-static condition introduced in the previous section also applies to the heart tissue. The currents in the two domains are hence given by

$$J_i = -M_i \nabla u_i, \tag{2.10}$$
$$J_e = -M_e \nabla u_e, \tag{2.11}$$

where J_i is the intracellular and J_e the extracellular current, M_i and M_e are the conductivities in the two domains and u_i and u_e are the respective potentials. In Section 2.1.1 we assumed that there was no build-up of charge at any point. This is a reasonable assumption, since any build-up of charge will generate large electrostatic forces, which tend to restore the uniform distribution. In the bidomain model, the cell membrane acts as an insulator between the two domains, and therefore has the ability to separate charge. It is therefore natural to assume that there may be some accumulation of charge in each domain. However, because of the small thickness of the membrane, any accumulation of charge on one side of the membrane immediately attracts an opposite charge on the other side of the membrane. This balance in charge accumulation implies that the total charge accumulation in any point is zero, described mathematically by

$$\frac{\partial}{\partial t}(q_i + q_e) = 0, \tag{2.12}$$

where q_i is intracellular charge and q_e is extracellular charge.

In each domain, the net current into a point must be equal to the sum of the rate of charge accumulation at that point and the ionic current exiting the domain at that point. This is expressed as

$$-\nabla \cdot J_i = \frac{\partial q_i}{\partial t} + \chi I_{ion}, \tag{2.13}$$
$$-\nabla \cdot J_e = \frac{\partial q_e}{\partial t} - \chi I_{ion}, \tag{2.14}$$

where I_{ion} is the ionic current across the membrane. The ionic current is most conveniently measured per unit area of the cell membrane, while densities of charge and current are measured per unit volume. The constant χ represents the area of cell membrane per unit volume. Therefore, while I_{ion} is ionic current per unit cell membrane area, χI_{ion} is ionic current per unit tissue volume. The positive direction of the ionic current is defined to be from the intracellular to the extracellular domain.

Combining (2.12) with (2.13) and (2.14) gives

$$\nabla \cdot J_i + \nabla \cdot J_e = 0,$$

which states that the total current is conserved. Inserting (2.10) and (2.11) into this equation gives

$$\nabla \cdot (M_i \nabla u_i) + \nabla \cdot (M_e \nabla u_e) = 0. \tag{2.15}$$

The amount of charge that may be separated by the cell membrane depends on the transmembrane potential and the capacitive properties of the membrane. The transmembrane potential v, defined as $v = u_i - u_e$, is related to the amount of separated charge by the relation

$$v = \frac{q}{\chi C_m},\tag{2.16}$$

where C_m is the capacitance of the cell membrane and

$$q = \frac{1}{2}(q_i - q_e).\tag{2.17}$$

Again, we must introduce the membrane area to volume ratio χ, since C_m is measured per unit membrane area while the quantities v and q are defined with respect to volume. Combining (2.16) and (2.17) and taking the time derivative yields

$$\chi C_m \frac{\partial v}{\partial t} = \frac{1}{2} \frac{\partial (q_i - q_e)}{\partial t},$$

and from (2.12) we get

$$\frac{\partial q_i}{\partial t} = -\frac{\partial q_e}{\partial t} = \chi C_m \frac{\partial v}{\partial t}.$$

Inserted into (2.13), this gives

$$-\nabla \cdot J_i = \chi C_m \frac{\partial v}{\partial t} + \chi I_{ion},$$

and using (2.10) gives

$$\nabla \cdot (M_i \nabla u_i) = \chi C_m \frac{\partial v}{\partial t} + \chi I_{ion}.\tag{2.18}$$

Equations (2.15) and (2.18) describe the variations in the three potentials u_i, u_e and v. Using the definition of v, we are able to eliminate the intracellular potential from the equations. We have $u_i = u_e + v$, which gives

$$\nabla \cdot (M_i \nabla (u_e + v)) = \chi C_m \frac{\partial v}{\partial t} + \chi I_{ion},$$

$$\nabla \cdot (M_i \nabla (u_e + v)) + \nabla \cdot (M_e \nabla u_e) = 0,$$

and a rearrangement of the terms gives

$$\nabla \cdot (M_i \nabla v) + \nabla \cdot (M_i \nabla u_e) = \chi C_m \frac{\partial v}{\partial t} + \chi I_{ion},\tag{2.19}$$

$$\nabla \cdot (M_i \nabla v) + \nabla \cdot ((M_i + M_e) \nabla u_e) = 0.\tag{2.20}$$

This is the standard formulation of the bidomain model, which was introduced by Tung [137] in the late 70s.

The conductive properties of heart muscle tissue are strongly Anisotropic, which implies that the parameters M_i and M_e are tensor quantities. The anisotropy results from the fact that the heart muscle consists of fibres, and the conductivity is higher in the direction of the fibres than in the cross-fibre direction. Furthermore, the muscle fibres are organized in sheets, which gives three characteristic directions for the conductive values of the tissue: parallel to the fibres, perpendicular to the fibres but parallel to the sheet, and perpendicular to the sheet. The fibre directions, and hence the conductivity tensors, will vary throughout the heart muscle. At a given point, we may define a set of perpendicular unit vectors a_l, a_t, and a_n, where a_l is directed along the fibres, a_t is perpendicular to the fibres in the sheet plane, and a_n is normal to the sheet plane. Expressed in the basis formed by these three unit vectors, the local conductivity tensor M^* is diagonal:

$$M^* = \begin{bmatrix} \sigma_l & 0 & 0 \\ 0 & \sigma_t & 0 \\ 0 & 0 & \sigma_n \end{bmatrix}. \tag{2.21}$$

For an electrical field $E^* = (e_1, e_2, e_3)^T$, defined in terms of the local basis vectors a_l, a_t, and a_n, Ohm's law gives the corresponding current as

$$J^* = M^* E^* = (\sigma_l e_1, \sigma_t e_2, \sigma_n e_3)^T = (j_1, j_2, j_3)^T. \tag{2.22}$$

The current vector is mapped to the global coordinates by

$$J = j_1 a_l + j_2 a_t + j_3 a_n = A J^*, \tag{2.23}$$

where A is a matrix having a_l, a_t, and a_n as columns. An electrical field E expressed in the global coordinate system is mapped to the local coordinates through the inverse mapping $E^* = A^{-1} E$. Since the column vectors of A are perpendicular unit vectors, we have $A^{-1} = A^T$. Combining this relation with (2.22) and (2.23), we obtain the relation between the electrical field E and the current J, both expressed in the global coordinate system:

$$J = A M^* A^T E. \tag{2.24}$$

The global conductivity tensors M_i and M_e are hence defined as

$$M_i = A M_i^* A^T, \tag{2.25}$$
$$M_e = A M_e^* A^T, \tag{2.26}$$

where M_i^* and M_e^* are the intracellular and extracellular conductivity tensors expressed in the local coordinate system. These are diagonal tensors defined as in (2.21), and this may be used to simplify expressions (2.25)–(2.26). An entry M_{ij} in the global conductivity tensor in (2.24) is given by

$$M_{ij} = a_l^i a_l^j \sigma_l + a_t^i a_t^j \sigma_t + a_n^i a_n^j \sigma_n, \tag{2.27}$$

for $i, j = 1, 2, 3$. Values of the local conductivities σ_l, σ_t, and σ_n, are given in Table 2.1. The table also specifies the surface to volume ratio χ and the membrane capacitance C_m. The conductivity values are taken from Klepfer et al. [76], while the surface to volume ratio is taken from Henriquez et al. [62], and the membrane capacitance from Pollard et al. [114].

Table 2.1. Values of the parameters used in the bidomain equations.

C_m	$1.0 \ \mu\mathrm{F}/cm^2$
χ	2000 cm^{-1}
σ_l^i	3.0 mS/cm
σ_t^i	1.0 mS/cm
σ_n^i	0.31525 mS/cm
σ_l^e	2.0 mS/cm
σ_t^e	1.65 mS/cm
σ_n^e	1.3514 mS/cm

For a given ionic current I_{ion}, and a distribution of fibre directions, i.e. a specification of the local basis vectors a_l, a_t, and a_n at each point, we now have a complete specification of the parameters in (2.19)–(2.20). However, to be able to solve these equations we also need boundary conditions for u_e and v. Assuming that the heart is surrounded by a non-conductive medium, we require the normal component of both the intracellular and extracellular current to be zero on the boundary. We have

$$n \cdot J_i = 0,$$
$$n \cdot J_e = 0, \tag{2.28}$$

where n is the outward unit normal vector of the boundary of the heart. Using the expressions for the two currents, and eliminating u_i, we get

$$n \cdot (M_i \nabla v + M_i \nabla u_e) = 0, \tag{2.29}$$
$$n \cdot (M_e \nabla u_e) = 0. \tag{2.30}$$

For the ionic current term I_{ion} in (2.19), the simplest choice is to assume that it is a function of the transmembrane potential only:

$$I_{ion} = f(v). \tag{2.31}$$

A popular choice is to define f to be a cubic polynomial in v:

$$f(v) = A^2(v - v_{rest})(v - v_{th})(v - v_{peak}), \tag{2.32}$$

where A is a parameter that determines the *upstroke velocity*, i.e. the rate of change of the transmembrane potential in the depolarization phase. Furthermore, v_{rest} is the

Table 2.2. Values for the parameters in the cubic ionic current model.

A	0.04
v_{rest}	$-85\,\text{mV}$
v_{th}	$-65\,\text{mV}$
v_{max}	$40\,\text{mV}$

resting potential, v_{th} is the so-called threshold potential, and v_{max} is the maximum potential. We require $v_{max} > v_{th} > v_{rest}$. Possible values of the parameters are listed in Table 2.2.

With the ionic current term given by (2.31)–(2.32), the equations (2.19)–(2.20) and boundary conditions (2.29)–(2.30) form a complete system, which may be solved for the potentials u_e and v. The given choice of ionic current function yields two stable stationary points for v; v_{rest} and v_{max}, and one unstable stationary point v_{th}. Let us, for a moment, disregard the diffusive terms in (2.19)–(2.20), i.e. let us assume that the conductivities M_i and M_e are both zero. Then, at a given point, v will approach either the resting potential or the maximum potential, depending on whether the initial value of v is above or below v_{th}.

With non-zero conductivities, diffusive effects may bring the transmembrane potential at a point that lay, initially, below the threshold to a value above the threshold v_{th}, making v approach the stable stationary point v_{max}. If we have initially $v = v_{rest}$ in most of the tissue, but $v > v_{th}$ in a small region Ω_i, the value of v in Ω_i will rapidly approach the maximum potential v_{max}. For sufficiently high conductivity values, diffusive effects will bring v above v_{th} in a small region surrounding Ω_i, and this process continues until $v = v_{max}$ throughout the tissue. Stimulating the tissue in a small region hence creates a wavefront of depolarization that propagates through the tissue. The sharpness of the wavefront, and also the propagation speed , depends on the value of the parameter A. A more detailed discussion of this model, with a specification of the conductivity values necessary to ensure propagation, is given in [72].

Since v_{max} is a stable stationary point for v, the ionic current specified by (2.31)–(2.32) is not able to reproduce the repolarization phase, where v is supposed to return to the resting value v_{rest}. Hence, this simplified ionic current model is not suitable for simulating the complete cardiac cycle. However, it is still a popular choice for some applications of modelling, because it captures essential parts of the tissue behaviour with a very simple model.

2.2.3 The Monodomain Model

The bidomain model for the electrical activity in the heart is a system of partial differential equations, which is difficult to solve and analyze. By making an assumption on the conductivity tensors M_i and M_e, it is possible to simplify the system to a scalar equation, describing only the dynamics of the transmembrane potential v. If we assume *equal anisotropy rates*, i.e., $M_e = \lambda M_i$, where λ is a constant scalar, then M_e can be eliminated from (2.19)–(2.20), resulting in

$$\nabla \cdot (M_i \nabla v) + \nabla \cdot (M_i \nabla u_e) = \chi C_m \frac{\partial v}{\partial t} + \chi I_{ion} \qquad (2.33)$$

$$\nabla \cdot (M_i \nabla v) + (1 + \lambda)\nabla \cdot (M_i \nabla u_e) = 0. \qquad (2.34)$$

From (2.34) we get

$$\nabla \cdot (M_i \nabla u_e) = -\frac{1}{1 + \lambda}\nabla \cdot (M_i \nabla v),$$

and if we insert this into (2.33) we get

$$\nabla \cdot (M_i \nabla v) - \frac{1}{1 + \lambda}\nabla \cdot (M_i \nabla v) = \chi C_m \frac{\partial v}{\partial t} + \chi I_{ion}.$$

A simple rearrangement of the terms gives the standard formulation of the monodomain model,

$$\frac{\lambda}{1 + \lambda}\nabla \cdot (M_i \nabla v) = \chi C_m \frac{\partial v}{\partial t} + \chi I_{ion}. \qquad (2.35)$$

With the assumption of equal anisotropy rates, the boundary conditions (2.29)–(2.30) become

$$n \cdot (M_i \nabla v + M_i \nabla u_e) = 0,$$
$$n \cdot (\lambda M_i \nabla u_e) = 0.$$

The second equation gives
$$n \cdot (M_i \nabla u_e) = 0,$$

and inserting this into the first equation gives

$$n \cdot (M_i \nabla v) = 0. \qquad (2.36)$$

The monodomain model, given by (2.35) and (2.36), is a significant simplification of the original bidomain equations, with advantages both for mathematical analysis and computation. However, the model also has some important limitations. One weakness is directly related to the assumption of equal anisotropy rates. Measurements of intracellular and extracellular conductivities contradict this assumption, and it is difficult to specify the parameter λ so as to obtain the closest approximation of the physiological conductivities. Furthermore, some important electrophysiological phenomena vanish when we have equal anisotropy rates in the intracellular and extracellular domains. Therefore, while the monodomain model is useful for analysis and simplified computational studies, realistic simulations of many important phenomena must still be based on the complete bidomain model.

2.3 Coupling the Heart and the Body

In Section 2.2.2 we presented the bidomain model for the electrical activity in the heart, under the assumption that the heart was electrically insulated from its surroundings. While this may be a good approximation for some experimental settings, the normal physiological situation is that the heart is surrounded by a body, which in Section 2.1.1 was modelled as a passive volume conductor. This is also representative for many experimental situations, where, for instance, a small tissue sample is surrounded by a conductive bath. In the following we will refer to the domain surrounding the heart as the extracardiac domain. The electrical potential in the extracardiac domain is described by (2.7)–(2.8), but the condition on ∂H, given by (2.9), will be modified slightly because the equation must be coupled with the bidomain equations (2.19)–(2.20).

With the heart surrounded by a conductor, the normal component of the total current must be continuous across the boundary of the heart. More precisely, on the boundary we have that the normal component of the current in the heart must be equal to the normal component of the current in the surrounding tissue. This follows from the arguments of conservation of charge and current, which were used to derive mathematical models for both passive and excitable tissue. The total current in the heart tissue is the sum of the intracellular and extracellular current. Using the notation from Section 2.1.1 for the extracardiac current, we have

$$n \cdot (J_i + J_e) = n \cdot J_T,$$

where n is the outward unit normal of the surface of the heart. Inserting the expressions for J_i, J_e, and J_T, and writing the currents in terms of the main variables u_e, v, and u_T, we get

$$n \cdot (M_i \nabla v + (M_i + M_e)\nabla u_e) = n \cdot (M_T \nabla u_T) \qquad (2.37)$$

This condition is not sufficient to close the equation system (2.19)–(2.20), so we need to make additional assumptions on the coupling between the heart and the surrounding tissue. Several different choices have been made for this coupling (see e.g., [62,119]), but we restrict our discussion to the original boundary conditions proposed by Tung [137].

This set of boundary conditions is based on the assumption that the extracellular domain is in direct contact with the extracardiac domain, while the intracellular domain is completely insulated from its surroundings. If two volume conductors are directly connected there cannot be a discontinuity in their potentials on the boundary, so the first condition implies that on the boundary of the heart the extracellular potential must be equal to the extracardiac potential:

$$u_e = u_T. \qquad (2.38)$$

The assumption that the intracellular domain is completely insulated implies that the normal component of the intracellular potential must be zero on the heart surface. We have

$$n \cdot J_i = 0,$$

and written in terms of u_e and v we get

$$n \cdot (M_i \nabla v + M_i \nabla u_e) = 0 \qquad (2.39)$$

on the heart surface. Inserting this expression into (2.37) gives

$$n \cdot (M_e \nabla u_e) = n \cdot (M_T \nabla u_T). \qquad (2.40)$$

The three boundary conditions (2.38), (2.39), and (2.40) couple the equation system (2.19)–(2.20) to (2.7)–(2.8). Combined with an expression of the form (2.31) and (2.32) for the ionic current term, the equations form a complete system that may be solved for the unknown potentials u_e, v, and u_T.

2.4 Models for the Ionic Current

In Section 2.2.2 we introduced a simplified model for the ionic current term I_{ion}, in the form of a cubic polynomial in v. Although this model conveniently reproduces essential parts of the behaviour of the heart tissue, other important phenomena, such as the repolarization of the tissue, are neglected. Furthermore, the model was chosen based on observations of the macroscopic behaviour of excitable cells, rather than on a correct physiological description of the cell membrane. For more realistic simulations of the process of cardiac activation and deactivation, we need to use more realistic models of ionic current. A number of such models exist, which are commonly grouped into three categories:

- First, there are phenomenological models, which are simple models constructed to reproduce the macroscopically observed behaviour of the cells. The cubic polynomial introduced above falls into this model category.
- Second, there is a group of models often referred to as first generation models. These attempt to describe not only the observed cellular behaviour, but also the underlying physiology. They describe the ionic currents that are most important for the action potential, but use a simplified formulation of the underlying physiological details.
- The third group of models, referred to as second generation models, offer a very detailed description of the physiology of the cells. The models are based on advanced experimental techniques, enabling observations of very fine-scale details of cell physiology.

The purely phenomenological models, like the cubic polynomial used above, may be useful for a number of applications. However, the full potential of the simulations can only be reached by using more sophisticated models, because an important purpose of mathematical modelling and simulation is to investigate how changes in physiology on a cellular, or even subcellular, level affect the function of the muscle tissue and the complete heart. In this section, we describe some of the most well-known models for cardiac cells, and also provide a short description of the basic physiology of the cells, and of the cell membrane in particular.

The models will, in this section, be discussed in the context of modelling a single cell. In this situation, all the charge transported by the ionic current accumulates at the membrane to change the transmembrane potential v. We have

$$C_m \frac{dv}{dt} = -I_{ion} + I_{app},$$

where I_{ion} is the ionic current that also occurs in the bidomain equations. The second term I_{app} is an applied stimulus current,which is used to trigger the action potential of the cell.

2.4.1 The FitzHugh-Nagumo Model

As described above, the cubic polynomial introduced for the ionic current term is not capable of reproducing the repolarization phase of the cardiac cycle. To obtain a qualitatively correct description of this phase, the cubic model (2.31)–(2.32) must be extended by introducing a second variable w, called a recovery variable. The resulting two-variable model is the FitzHugh-Nagumo model [40], which in its original formulation is given by

$$\frac{dv}{dt} = c_1 v(v-a)(1-v) - c_2 w + i_{app}, \tag{2.41}$$

$$\frac{dw}{dt} = b(v - c_3 w). \tag{2.42}$$

Here a, b, c_1, c_2, and c_3 are given parameters, which may be adjusted to simulate different cell types. The parameter values used in the original formulation of the model are specified in Table 2.3. These parameter choices give a normalized action potential, with the resting potential being zero and the peak potential approximately 0.9. The applied current i_{app} is also scaled to match this normalized model. Plots of v and w are shown in the left panel of Figure 2.2. The action potential shown in the figure results from applying a stimulus current i_{app} of strength 0.05, lasting from $t = 50$ to $t = 60$.

Several variations have been derived to address various limitations of the original model. One detail of the original formulation that does not match well with physiological data is that the cell hyperpolarizes in the repolarization phase. This

Table 2.3. The values of the parameters used in the original formulation of the FitzHugh-Nagumo model.

a	0.13
b	0.013
c_1	0.26
c_2	0.1
c_3	1.0

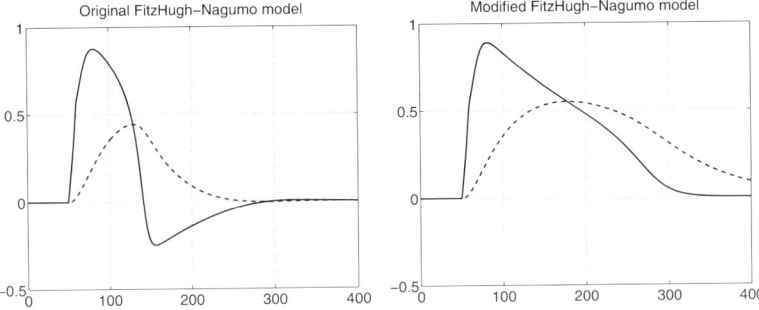

Fig. 2.2. Plots of v (solid) and w (dashed) for the original and modified FitzHugh-Nagumo model.

is seen clearly in the left panel of Figure 2.2, where v reaches values significantly below the resting potential before returning to the resting state. A modification of the equations to overcome this problem was suggested by Rogers and McCulloch [118]. By a slight modification of the last term in (2.41), the model becomes

$$\frac{dv}{dt} = c_1 v(v-a)(1-v) - c_2 vw + i_{app}, \qquad (2.43)$$

$$\frac{dw}{dt} = b(v - c_3 w). \qquad (2.44)$$

The right panel of Figure 2.2 shows the action potential computed with the modified model. We see that the undershoot in the transmembrane potential is eliminated, giving a more physiologically realistic solution. However, the actual values of the transmembrane potential are incorrect. To give realistic values for the transmembrane potential, it is necessary to change the parameters for the model. Following the notation introduced for the polynomial ionic current model, we introduce a resting potential v_{rest} and a peak potential v_{peak}. The total amplitude of the action potential is then $v_{amp} = v_{peak} - v_{rest}$. We define new variables V and W, given by

$$V = v_{amp} v + v_{rest}, \qquad (2.45)$$

$$W = v_{amp} w. \qquad (2.46)$$

Since the original variable v has a resting value of zero, and peaks at approximately 1.0, the new scaled potential will have a resting value equal to v_{rest} and a peak value close to v_{peak}. The scaled recovery variable W is introduced for convenience of notation only, since w has no physiological interpretation. Solving (2.45)–(2.46) for v and w, and inserting into (2.43)–(2.44) gives

$$\frac{dV}{dt} = \frac{c_1}{v_{amp}^2}(V - v_{rest})(V - v_{th})(v_{peak} - V)$$
$$- \frac{c_2}{v_{amp}}(V - v_{rest})W + I_{app}, \tag{2.47}$$

$$\frac{dW}{dt} = b(V - v_{rest} - c_3 w). \tag{2.48}$$

Here, we have defined the threshold potential $v_{th} = v_{rest} + av_{amp}$, and the rest of the parameters are the same as in the original model. The new applied current I_{app} is given by $I_{app} = v_{amp}i_{app}$. Figure 2.3 shows a plot of the action potential computed with the scaled model, where we have chosen $v_{rest} = -85$ and $v_{peak} = 40$. With the parameter choices listed in Table 2.3 this gives a threshold potential $v_{th} = -68.75$. The shape and amplitude of the action potential is fairly realistic, but the upstroke velocity is lower than the realistic value for heart muscle cells. However, by tuning the parameters of the model it is possible to obtain an action potential with approximately correct amplitude, duration, and upstroke velocity.

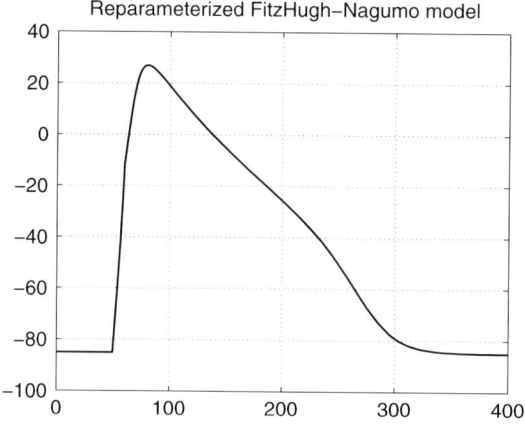

Fig. 2.3. Plots of v and w for the reparameterized FitzHugh-Nagumo model.

2.4.2   The Cell Membrane

We have seen that it is possible to reproduce the most important characteristics of the action potential with a simple two-variable model. However, an important goal for the modelling of physiological phenomena is to investigate how changes in the physiology on a cellular, or sub-cellular, level affects the function of tissues and complete organs. To attain this goal, it is necessary to employ models that give a more detailed description of the cell physiology than the simplified FitzHugh-Nagumo model. In order to understand the foundation of this type of model, it is necessary to provide a brief introduction to the physiology of cardiac cells, and in particular the cell membrane.

The cell membrane separates the extracellular space from the *myoplasma*, the intracellular space that contains the contractile units. The cell membrane consists mainly of lipids, which have a polar head that is attracted to water, and a nonpolar tail that repels water. When brought into contact with water, a lipid will tend to spread out until the layer has a thickness of one molecule, with the polar heads facing the water and the nonpolar tails directed away from the water. The cell membrane consists of a double layer of such lipids, which are organized such that the polar heads are directed towards the internals of the cell and the space surrounding the cell, while the tails form the centre of the membrane. The hydrophobic lipid tails act as an insulator, preventing the passage of ions through the membrane.

Although the cell membrane itself is impermeable to ions, it has embedded in it a number of large proteins that form channels through the membrane where the ions can pass. Figure 2.4 gives a schematic view of the structure of the cell membrane, with an embedded transport protein. Some transport proteins form pumps and exchangers, which are important for maintaining the correct ionic concentrations in the cells. Both pumps and exchangers have the ability to transport ions in the opposite direction of the flow generated by concentration gradients and electrical fields. This process, which increases the potential energy of the ions, is accomplished either by using the concentration gradient of a different ion (exchangers) or by consuming chemically stored energy in the form of ATP. For a general discussion of cellular reactions and physiology in general, see, e.g., [11].

In addition to the pumps and exchangers, certain transport proteins form channels in the membrane, through which ions may flow. The flow of ions through these channels is passive, being driven by concentration gradients and electric fields. Still, the channels are extremely important for the behaviour of excitable cells. One reason for this is that most of the channels are highly selective regarding which ions

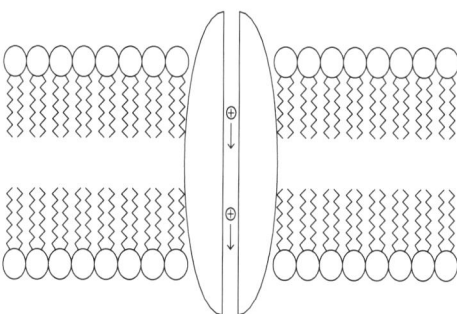

Fig. 2.4. Schematic view of the cell membrane, with an embedded protein forming a channel through which ions can pass.

are allowed to pass, and as we will see later, this specificity of the channels is essential for generating and maintaining the potential difference across the membrane. The channels also have the ability to open and close in response to changes in the electrical field and ionic concentrations, and this ability is essential for the signal propagation in excitable tissue.

To illustrate how the specificity of the ionic channels generates a potential difference across the membrane, consider two compartments i and e, separated by a thin membrane. Ions are added to both compartments, for example, in the form of NaCl. In compartment i we add NaCl to a concentration of 100 mM, while in compartment e the concentration is only 10 mM. Now, if the membrane had been completely permeable to all ions, diffusive effects would drive ions from compartment i to compartment e, until the concentrations in the two compartments were equal. Assume now that the membrane is only permeable to Na^+ ions. In this case, diffusive forces will drive Na^+ ions over to compartment e, where the concentration is low, but the Cl^- ions remain in compartment i. The result is a surplus of positive Na^+ ions in compartment e, and of negative Cl^- ions in compartment i. This charge difference sets up an electric field across the membrane, which tends to drive the Na^+ ions back to compartment i. An equilibrium will be reached where the flux of Na^+ caused by diffusion is equal and in a direction opposite to the Na^+ flux caused by the electric field. In this situation of equilibrium, there will be a potential difference between the two compartments.

Because the ion channels in the cell membrane are specific with respect to which ions are allowed to pass, the same situation will occur across the cell membrane. The result is that in the equilibrium situation there will be a potential difference across the membrane, which we refer to as the transmembrane potential.

2.4.3   The Nernst Equilibrium Potential

In this section, we consider the flux of only one type of ion, and derive a mathematical expression for the equilibrium situation where the diffusive flux is equal to the electrically driven flux. Assuming first that there are no concentration gradients, the ion flux J_E due to an electrical field is given by Planck's equation; see, e.g., [72]:

$$J_E = m \frac{z}{|z|} cE,$$

where m is the mobility, z the charge and c the concentration of the ion. Furthermore, E is the electrical field, which may be written as the negative gradient of a scalar potential ϕ, which gives

$$J_E = -m \frac{z}{|z|} c \nabla \phi. \tag{2.49}$$

In the presence of concentration gradients, the electrically driven flux will be supplemented by ionic flux caused by diffusion. Ions will tend to move from areas with high concentration towards areas with low concentration. The diffusive flux J_D is given by Fick's law:

$$J_D = -D \nabla c,$$

where D is the diffusion coefficient for the ion.

For a functioning cell, we normally have differences in both electrical potential and ionic concentrations across the membrane, and the total flux will be given by

$$J = J_D + J_E. \tag{2.50}$$

The mobility m is related to the diffusion coefficient D through the following relation:

$$m = D\frac{|z|F}{RT},$$

where R is the ideal gas constant, T is the temperature, and F is Faraday's constant. Inserting this expression into (2.49), (2.50) gives

$$J = -D\left(\nabla c + \frac{zF}{RT}c\nabla\phi\right). \tag{2.51}$$

This equation for the total flux is called the Nernst-Planck equation.

We are interested in the equilibrium situation in which the total current J is zero. Note that we may still have fluxes caused both by concentration gradients and electrical fields, as long as we have $J_E = -J_D$. For the cell membrane, it is natural to assume that the variations in potential and concentrations occur only in the direction across the membrane. It is, therefore, natural to study the flux in only one space dimension. Setting $J = 0$, we then get from (2.51):

$$\frac{dc}{dx} + \frac{zF}{RT}c\frac{d\phi}{dx} = 0.$$

Dividing this equation by c and integrating from $x = 0$ to $x = L$ gives

$$\int_0^L \frac{1}{c}\frac{dc}{dx}dx + \int_0^L \frac{zF}{RT}\frac{d\phi}{dx}dx = 0.$$

Here, L is the thickness of the membrane, and the coordinate system is oriented so that $x = 0$ is the inside of the membrane and $x = L$ is the outside. Performing the integration gives

$$\ln(c)|_{c(0)}^{c(L)} = -\frac{zF}{RT}(\phi(L) - \phi(0)) = \frac{zF}{RT}v,$$

where we have used the definition of the transmembrane potential, $v = \phi_i - \phi_e$, i.e. the difference between the intracellular and extracellular potential. The value of the transmembrane potential that gives zero flux, for given intra- and extracellular concentrations, is now given by

$$v_{eq} = \frac{RT}{zF}\ln\left(\frac{c_e}{c_i}\right), \tag{2.52}$$

where c_e and c_i are the extra- and intracellular concentrations, respectively. The potential v_{eq} is referred to as the Nernst equilibrium potential.

2.4.4 Models for Ionic Flux

Although we were able to find an expression for the equilibrium potential from the Nernst-Planck equation (2.51), it is more difficult to use the equation to compute the actual flux for non-equilibrium situations. The difficulty is that we do not know the details of how the concentration and the electrical potential vary across the membrane. For the equilibrium situation, this problem was avoided because the flux is known to be zero, which enabled us to integrate the rest of the terms in the equation over the membrane thickness. However, any expression for passive ionic flux should at least satisfy (2.52), giving zero flux for $v = v_{eq}$. The simplest formulation which satisfies this condition is a linear model, giving the flux as

$$J = G(v - v_{eq}), \tag{2.53}$$

where G is the permeability of the membrane for the particular ion under study. Depending on the type of ionic flux we want to model, G may be constant or a function of time, membrane potential and, in some cases, ionic concentrations.

Equation (2.53) is motivated only by the need to satisfy the Nernst equilibrium potential, and has very little relation to the Nernst-Planck equation (2.51) itself. By making an assumption about the properties of the electric field across the membrane, it is possible to derive an alternative expression for the ionic flux that is more directly based on (2.51). We assume that the electric field across the membrane, expressed by the potential gradient $\nabla \phi$, is constant, with respect to both the spatial variable x and the flux J. In reality, any flux of ions will transport charge across the membrane and will, therefore, affect the transmembrane potential and hence the electric field. The assumption of a constant field is therefore justified only in the steady state situation, i.e. where we have more than one ionic current and the net current is zero, or if the flux J is too small to cause significant changes in the transmembrane potential. If the electrical field is constant, we can write $\nabla \phi = v/L$, and inserted into (2.51) this gives

$$\frac{dc}{dx} - \frac{zFv}{RTL}c + \frac{J}{D} = 0.$$

This is an ordinary differential equation in c, with known values at both endpoints and J as an additional unknown. Solving the equation, we get

$$J = \frac{D}{L}\frac{zFv}{RT}\frac{c_i - c_e \exp\left(\frac{-zFv}{RT}\right)}{1 - \exp\left(\frac{-zFv}{RT}\right)}. \tag{2.54}$$

This expression is commonly referred to as a Goldman-Hodgkin-Katz (GHK) formulation for the ionic flux.

To see that the fairly complex expression (2.54) also satisfies the Nernst equilibrium potential, we insert $v = v_{eq}$ into (2.54), with v_{eq} given by (2.52). We have

$$
\begin{aligned}
J &= \frac{D}{L} \frac{zFv_{eq}}{RT} \frac{c_i - c_e \exp(\frac{-zFv_{eq}}{RT})}{1 - \exp(\frac{-zFv_{eq}}{RT})} \\
&= \frac{D}{L} \ln(\frac{c_e}{c_i}) \left(\frac{c_i - c_e \exp(-\ln(\frac{c_i}{c_e}))}{1 - \exp(-\ln(\frac{c_i}{c_e}))} \right) \\
&= \frac{D}{L} \ln(\frac{c_e}{c_i}) \left(\frac{c_i - c_e(\frac{c_i}{c_e})}{1 - \frac{c_i}{c_e}} \right) \\
&= \frac{D}{L} \ln(\frac{c_e}{c_i}) \left(\frac{c_i - c_i}{1 - \frac{c_i}{c_e}} \right) = 0.
\end{aligned}
$$

If we only require that the flux should satisfy the Nernst equilibrium, it is of course possible to derive a number of alternative expressions for the ionic flux. However, the most commonly used expressions are (2.53) and (2.54), which are both used for important ionic currents in models for cardiac cells. The choice of which formulation to use for a specific current is made by considering which model best reproduces the experimental data. A more detailed discussion of expressions (2.53) and (2.54) is given in [72].

In the discussion so far, we have only considered the flux of one type of ion. In the physiological situation, there are a number of different ions, both in the intracellular and extracellular domains, all of which affect the transmembrane potential. To illustrate the effect of this, recall the situation described in Section 2.4.2 where we had two compartments, i and e, separated by a thin membrane. The compartments contained NaCl in different concentrations. Assume now that instead of being completely impermeable to Cl$^-$ ions, the membrane allows both ions to pass, but with different permeabilities. The difference in permeability for the two ions will cause a potential difference across the membrane, and the flux of each ion may reach an equilibrium situation where the net flux is zero. The equilibrium potential for both ions may be computed from (2.52). Introducing the notation v_{Na} and v_{Cl} for the two equilibrium potentials, we have

$$
v_{Na} = \frac{RT}{F} \ln \left(\frac{[Na]_e}{[Na]_i} \right),
$$

$$
v_{Cl} = -\frac{RT}{F} \ln \left(\frac{[Cl]_e}{[Cl]_i} \right).
$$

The two equilibrium potentials will, in general, not be equal. It is therefore impossible to reach a state of equilibrium for both ions simultaneously. Instead, we may have a different type of equilibrium situation, where the net current of both Cl$^-$ and Na$^+$ is non-zero, but the total current is zero. We may compute an expression for the transmembrane potential in this equilibrium state, based on applying the linear model (2.53) for each ionic flux. We have

$$J_{Na} = G_{Na}(v - v_{Na}),$$
$$J_{Cl} = G_{Cl}(v - v_{Cl}),$$

where G_{Na} and G_{Cl} describe the permeability of the membrane to the two ions. To have a non-trivial equilibrium situation, with $v \neq 0$, we must have $G_{Na} \neq G_{Cl}$. We want to compute the value of v which gives zero total current. We have

$$J_{tot} = J_{Na} + J_{Cl} = 0,$$

which gives

$$G_{Na}(v - v_{Na}) + G_{Cl}(v - v_{Cl}) = 0.$$

This equation is easily solved for v, giving

$$v = \frac{G_{Na}v_{Na} + G_{Cl}v_{Cl}}{G_{Na} + G_{Cl}}. \tag{2.55}$$

We see that if, for example, the membrane's permeability to Cl^- is zero, the total equilibrium potential is equal to v_{Na}. Furthermore, changes in the permeability of one ion, e.g. as a result of channels opening and closing, causes changes in the total equilibrium potential. The result is a flux of ions across the membrane, which brings the transmembrane potential to the new equilibrium value.

2.4.5   Channel Gating

As described in Section 2.4.2, ion channels that open and close in response to an electrical stimulus are essential for the behaviour of excitable cells. We have also stated that the ion channels are formed by large proteins embedded in the membrane. The mechanisms by which these channels open and close lie beyond the scope of this text, and the reader is referred to, e.g., Hille [64] or Plonsey and Barr [112] for a more detailed description. We will here focus on mathematical models for the ion channels, which are normally constructed by considering the channels to be composed of several sub-units. Each sub-unit may be either open or closed, and ions can only pass when all units are open. Consider first a channel that consists of only one sub-unit. We denote the concentration of channels in the open and closed state by $[O]$ and $[C]$, respectively. We assume that the total concentration of channels, $[O] + [C]$, is constant. The change between the open and closed state may be written as

$$C \underset{\beta}{\overset{\alpha}{\rightleftharpoons}} O, \tag{2.56}$$

where α is the rate of opening and β is the rate of closing. These rates typically depend on the transmembrane potential v The law of mass action (see, e.g., [72]), says that in a reaction like (2.56), the rate of change from the open state to the closed state is proportional to the concentration of channels in the open state. Similarly, the rate of change from closed to open is proportional to the concentration of closed channels. We get

$$\frac{d[O]}{dt} = \alpha(v)[C] - \beta(v)[O].$$

This equation may be divided with the total concentration $[O] + [C]$ to give

$$\frac{dg}{dt} = \alpha(v)(1 - g) - \beta(v)g, \tag{2.57}$$

where $g = [O]/([O] + [C])$ is the portion of open channels. If we look at a single channel, g may be viewed as the probability that the channel is open. The total number of channels is assumed to be large enough for these two interpretations of g to be completely equivalent.

Since α and β depend on v, it is not possible to find a general solution of (2.57). However, to illustrate the behaviour of the reaction, let us for a moment assume that α and β are constants. Equation (2.57) can be rewritten as

$$\frac{dg}{dt} = (g_\infty - g)/\tau_g, \tag{2.58}$$

with

$$g_\infty = \alpha/(\alpha + \beta) \text{ and } \tau_g = 1/(\alpha + \beta).$$

With α and β constant, the solution of (2.58) is

$$g(t) = g_\infty + (g_0 - g_\infty)e^{-t/\tau_g}.$$

Here, g_0 is the initial value for g, and we see that g approaches the steady state value g_∞ as t increases. The rate at which g approaches g_∞ is determined by the magnitude of the time constant τ_g. In our setting the rates are not constants, and (2.58) is therefore not a correct solution to (2.57). However, if the time constant is small, g may be assumed to reach its steady-state value almost immediately, so that g_∞ may be used as a reasonable approximation to $g(t)$. For many important ionic channels, the time constant is small enough for this approximation to be used, while for other channels τ_g is too large for this simplification to be valid, and the activity of the channel must be modelled with equations such as (2.57).

Most ionic channels consist of several subunits, which may all be either open or closed. In this context, (2.57) is viewed as the probability that one subunit is in the open state. Assuming that the subunits open and close completely independent of each other, the probability that a channel is open is equal to the product of the probability for each subunit being open. For a channel that consists of n equal subunits, the probability O that the channel is open is given by

$$O = g^n, \tag{2.59}$$

with g given from (2.57). From the definition of g and O it is clear that we have $O, g \in [0, 1]$.

A channel may also consist of several different subunits. The dynamics of each unit is then described by (2.57), but the rate functions α and β are different for each unit. For example, we may have a channel that consists of two different subunits. We denote the probability of one unit being open by g, and the probability of the other unit being open is denoted by h. The dynamics of both g and h are described by expressions like (2.57), with respective rates α_g, β_g and α_h, β_h. If the channel consists of m units of type g and n units of type h, the probability that the channel is open is given by

$$O = g^m h^n. \qquad (2.60)$$

The current through the membrane may be computed as the product of the maximum current, i.e. the current we would have if all channels were open, and the proportion of open channels. The maximum current is computed from expressions such as (2.53) and (2.54), derived in Section 2.4.4. If, for instance, we study an ionic current described by the linear model (2.53), the ionic current I is given by

$$I = G_{max}O(v - v_{eq}), \qquad (2.61)$$

where O is defined as products on the form (2.59) and (2.60), v_{eq} is the equilibrium potential for the specific ion, and G_{max} is the maximum conductance, i.e. the conductance if all channels are open.

The simple channel-gating expressions derived here are all based on the assumption that the opening and closing of each subunit is independent of the state of the other subunits. Other types of channel-gating models exist, which are not based on this assumption. Examples of such models may be found in, e.g., [68], [73].

2.4.6   The Hodgkin-Huxley Model

In a normally functioning cell, several ionic currents contribute to the changes in the transmembrane potential. Among the most important ions are sodium, potassium and calcium. A model for the total ionic current across the membrane may be constructed by describing each current with an expression of the form (2.61), and combining these expressions to form a model for the total current. The first model of this kind was that of Hodgkin and Huxley [65], which was published in 1952 and for which they won the Nobel Prize in Medicine in 1963. The model describes the action potential of a *squid giant axon*, a particularly large nerve cell found in squid. The model is, therefore, not directly relevant for studies of the heart, but it serves as a nice example of how the cell models are constructed. The formulations of the ionic currents introduced in the Hodgkin-Huxley model also serve as building blocks in subsequent models for both nerve and heart cells.

The Hodgkin-Huxley model describes three ionic currents; a sodium current, a potassium current, and one unspecified current called a leakage current. The currents are denoted by I_{Na}, I_K, and I_L, respectively, and are illustrated in Figure 2.5. The arrows indicate the main direction of the current during the action potential. The sodium current and the potassium current are both gated, while the leakage current

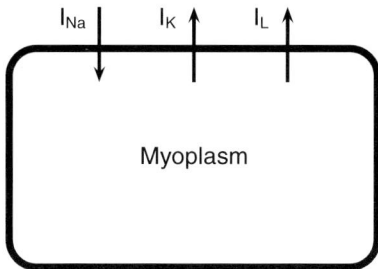

Fig. 2.5. The ionic currents in the Hodgkin-Huxley model.

is time independent and formulated as a simple linear function of the transmembrane potential. The currents are given by

$$I_{Na} = \bar{G}_{Na} m^3 h (\nu - \nu_{Na}),$$
$$I_K = \bar{G}_K n^4 (\nu - \nu_K),$$
$$I_L = G_L (\nu - \nu_L).$$

Here, ν is the deviation from the resting potential, defined as $\nu = v - v_{eq}$. Similarly $\nu_{Na}, \nu_K,$ and ν_L are shifted equilibrium potentials for each current, defined by

$$\nu_{Na} = v_{Na} - v_{eq}, \ \nu_K = v_K - v_{eq}, \ \nu_L = v_L - v_{eq},$$

where $v_{Na}, v_K,$ and v_L are the true equilibrium potentials for the three currents. Furthermore $m, h,$ and n are gate variables described by equations of the form (2.57), and $\bar{G}_{Na}, \bar{G}_K,$ and $\bar{G}_L$ are the maximum conductances for each current, i.e. the currents we would have if all channels were open. In view of the discussion of ion channels in Section 2.4.5, we see that the model describes the sodium channels as consisting of four subunits, three of them identical, while the potassium channel consists of four identical subunits. When the model was developed there existed no method for actually examining the detailed structure of the ion channels, and the expressions used for the ionic current were based on observations of the overall gating behaviour for the currents.

The total ionic current is given as the sum of the three currents above,

$$I_{ion} = I_{Na} + I_K + I_L,$$

and may be plugged directly into models like the bidomain model (2.19)–(2.20), derived in Section 2.2.2. However, in the context of single cell models, we have that the transmembrane potential is governed by

$$C_m \frac{dv}{dt} = -I_{ion} + I_{app}.$$

This may be seen as a special case of (2.19), where the diffusive effects are ignored. Inserting the expressions for the currents, the rate of change of the shifted membrane potential ν becomes

$$C_m \frac{d\nu}{dt} = -\bar{G}_{\text{Na}} m^3 h(\nu - \nu_{\text{Na}}) + \bar{G}_{\text{K}} n^4 (\nu - \nu_{\text{K}}) + G_L (\nu - \nu_L), \qquad (2.62)$$

with the gate variables given by equations of the form

$$\frac{dg}{dt} = \alpha_g(\nu)(1 - g) - \beta_g(\nu)g, \qquad (2.63)$$

for $g = m, h, n$. The rate functions α_g, β_g are specified in Appendix B. A detailed discussion of the Hodgkin-Huxley model is given in [72].

The action potential computed with the Hodgkin-Huxley model is shown in Figure 2.6. We see that the shape of the action potential is substantially different from the action potential shown in Figure 2.3, which we described as being fairly realistic for cardiac cells. The difference between the two models is not surprising, since they describe entirely different cell types. Therefore, while the membrane current of the Hodgkin-Huxley model may, in principle, be plugged directly into the bidomain equations, the results will not be very meaningful. Although we would be able to reproduce both the depolarization and the repolarization of the tissue, both the propagation velocity and the time interval between depolarization and repolarization would be incorrect. Hence, to obtain physiologically meaningful results we must use models that give a more accurate description of the behaviour of heart cells. One such model is the reparameterized FitzHugh-Nagumo model, but for many simulations it is attractive to use models with a more detailed representation of the underlying cellular physiology.

2.4.7   A Model for Cardiac Cells

The first model to describe the action potential in cardiac cells was proposed by Noble [104] in 1962, to describe Purkinje fibre cells. The model is based on the

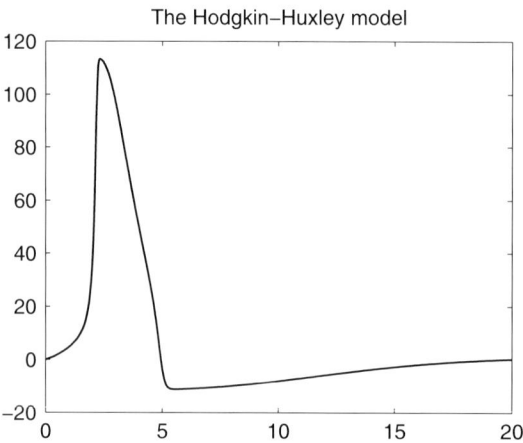

Fig. 2.6. The action potential produced by the Hodgkin-Huxley model.

Hodgkin-Huxley model, but the parameters have been refitted to reproduce the action potential of the Purkinje cells, which is markedly different from that of the squid giant axon. In particular, the action potential has a pronounced plateau phase, and the duration is 300-400 ms, compared to about 3 ms for the squid giant axon. The Noble model is still very similar to the Hodgkin-Huxley model, in that it describes a transmembrane current that carries two different ions: sodium and potassium. The passive leakage current present in the Hodgkin-Huxley model is assumed to be zero in the Noble model. The ionic currents included in the model are shown in Figure 2.7. We see that, in contrast to the Hodgkin-Huxley model, the Noble model assumes two separate potassium currents, I_{K1} and I_{K2}.

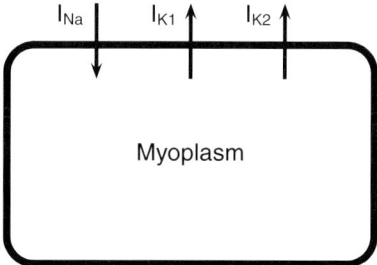

Fig. 2.7. The ionic currents in the 1962 Noble model.

Both the sodium and the potassium current are assumed to be linear in v, and the equation for v is given by

$$-C_m \frac{dv}{dt} = g_{Na}(v - v_{Na}) + (g_{K_1} + g_{K_2})(v - v_K) + I_{app}.$$

Here, g_{Na} is the sodium conductance, and g_{K_1} and g_{K_2} are the conductances for the two potassium currents. As usual, v_{Na} and v_K are the respective Nernst equilibrium potentials for the two ions. The conductance g_{K_2} is given by an expression similar to that used in the Hodgkin-Huxley model,

$$g_{K_2} = \bar{g}_{K_2} n^4,$$

where $\bar{g}_{K_2}$ is the maximum conductance and n is a gate variable governed by an equation of the form (2.63). However, the parameters are adjusted so that the response of n is much slower than in the Hodgkin-Huxley model, in order to prolong the action potential. The conductance g_{K_1} is assumed to be dependent only on voltage, and given by

$$g_{K_1} = 1.2 \exp\left(-\frac{v + 90}{50}\right) + 0.015 \exp\left(\frac{v + 90}{60}\right).$$

The sodium conductance is also similar to that of the Hodgkin-Huxley model:

$$g_{Na} = \bar{g}_{Na} m^3 h + g_i,$$

where $\bar{g}_{Na}$ is the maximum conductance and the constant term g_i represents a leakage current of sodium, which has the effect of prolonging the action potential. The dynamics of the gate variables n, m, and h are described by equations on the form (2.63). The rate functions, and the constants in the model, are specified in Appendix B.

A plot of the action potential computed with the model is shown in Figure 2.8. The prolonged action potential is clearly visible when compared to the Hodgkin-Huxley action potential in Figure 2.6. In addition to prolonging the action potential, the constant g_i added to the sodium conductance has the effect of making the model self-oscillatory, which is realistic for Purkinje cells. See, e.g., [72] for details. As seen in Figure 2.8, the potential gradually increases after the cell has repolarized, and this eventually brings the membrane potential above the threshold value, triggering a new action potential.

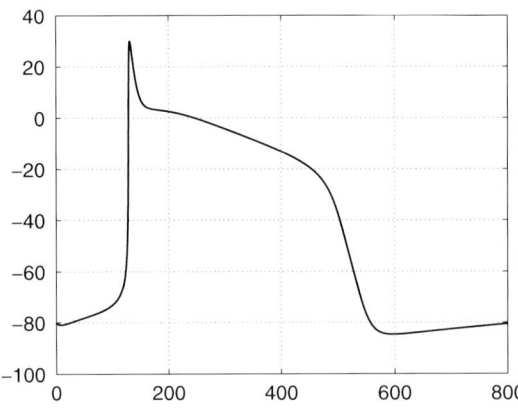

Fig. 2.8. Action potential for a Purkinje fibre cell, computed with the Noble model.

The Noble model was a successful attempt to simulate the action potential of Purkinje fibres with a simple model of Hodgkin-Huxley type. However, because the model was created before detailed data on ionic currents in cardiac cells became available, the underlying physiology is incorrect. An improvement of the model was presented by McAllister, Noble and Tsien [102] in 1975. The new model includes more ionic currents, and gives a much more accurate description of the underlying activity of different channels than the original model. However, the action potential computed with the improved model is essentially the same as that generated by the original Noble model.

2.4.8 Models for Ventricular Cells

For the purpose of simulating the electrical activity of the whole heart, and in particular its relation to ECG recordings, the most important cells are probably not the

Purkinje fibres, but rather the ventricular muscle cells. The first model to describe these cells was proposed by Beeler and Reuter [9] in 1975. An important difference between Purkinje cells and ventricular muscle cells is the role of calcium, which is essential for the contractile mechanism of the muscle cells.

The Beeler-Reuter model is based on experimental data from the guinea pig, and describes four ionic currents: the usual fast inward current carried by sodium, a slow inward current carried mostly by calcium, and two outward potassium currents. The currents are illustrated in Figure 2.9. The currents are controlled by six gating variables, and the model describes the intracellular calcium concentration in addition to the transmembrane potential. This gives a total of eight state variables, described by

$$-C_m \frac{dv}{dt} = I_{\text{Na}} + I_{\text{K}} + I_x + I_s + I_{app}$$
$$\frac{dc}{dt} = 0.07(1 - c) - I_s$$
$$\frac{dg}{dt} = \alpha_g(1 - g) - \beta_g g \tag{2.64}$$

with $g = m, h, j, d, f, x$, since all the gating variable equations have the same structure. The two outward potassium currents are denoted by I_{K} and I_x, and I_s denotes the inward current of calcium. The intracellular calcium concentration is described by the variable c, which has been scaled for convenience of notation, $c = 10^7[\text{Ca}_i]$. The ionic currents, rate functions, and constants are specified in Appendix B.

Although the Beeler-Reuter model was the first mathematical model for ventricular cells, it is still widely used. The main reason for this is its relative simplicity compared to more recent models. Increasingly complex models are not greatly problematic when simulating the behaviour of a single cell, but computational efficiency becomes very important when the cell models are used for simulating the electrical activity of the complete heart. In this setting, relatively simple models, such as the Beeler-Reuter model, have a significant advantage over more recent models, which tend to be far more computationally demanding.

Another classic ventricular cell model that is still widely used is the Luo-Rudy model [88] from 1991. The model is a development of the Beeler-Reuter model, and includes six ionic currents, controlled by a total of seven gate variables. The

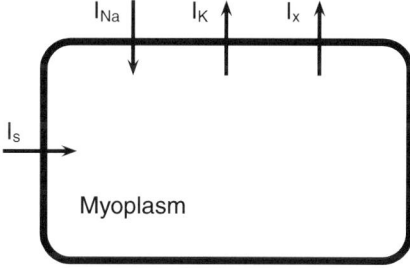

Fig. 2.9. The ionic currents in the Beeler-Reuter model.

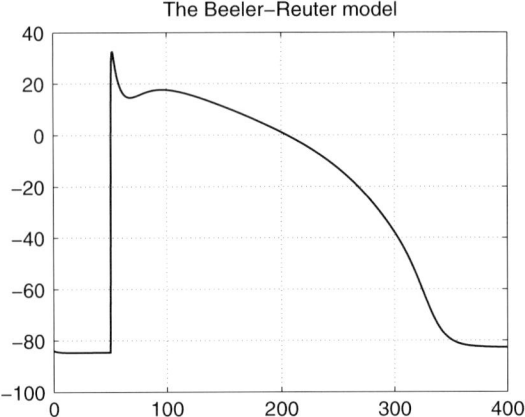

Fig. 2.10. Action potential for a ventricular cell, computed with the Beeler-Reuter model.

two additions to the Beeler-Reuter model are a time-independent potassium current called a plateau current, and a passive, linear background current. The ionic currents in the model are illustrated in Figure 2.11.

Similar to the Beeler-Reuter model, the Luo-Rudy model describes the dynamics of the intracellular calcium concentration in addition to the transmembrane potential. With the seven gate variables, the model consists of nine differential equations:

$$-C_m \frac{dv}{dt} = I_{\text{Na}} + I_{si} + I_{\text{K}} + I_{\text{K1}} + I_{\text{Kp}} + I_b + I_{app},$$

$$\frac{d[\text{Ca}]_i}{dt} = -0.0001 I_{si} + 0.07(0.0001 - [\text{Ca}]_i),$$

$$\frac{dg}{dt} = \alpha_g(1 - g) - \beta_g g,$$

with $g = m, h, j, d, f, X, X_i$ being the gate variables that control the ionic currents. Here, I_{Na} is the normal fast sodium current and I_{si} is a slow inward current of

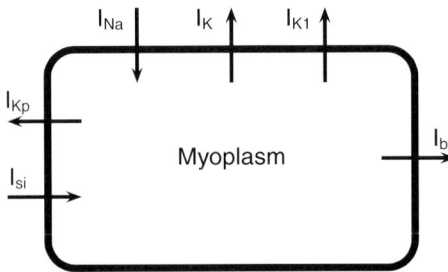

Fig. 2.11. The Luo-Rudy 1991 model for ventricular cells.

calcium. Furthermore, I_K is the time-dependent and I_{K1} the time-independent potassium current, I_{Kp} is the plateau potassium current, and I_b is the passive background current, which is similar to the leakage current in the Hodgkin-Huxley model. For a complete specification of model parameters, including expressions for ionic currents and rate functions, we refer the reader to the original publication [88].

2.4.9   Second Generation Models

The Beeler-Reuter model and the first Luo-Rudy model have a number of known limitations, related to their description of the physiology underlying the action potential. More recent models have been developed that give a more realistic description of the underlying details.

In 1994, Luo and Rudy [87] published a substantial upgrade of their 1991 model, often referred to as the Luo-Rudy phase two model. Compared to the first Luo-Rudy model, the new model gives a much more detailed description of the specific ionic currents across the membrane. The model describes a total of twelve membrane currents, including a number of important pumps and exchangers, and offers a more complete description of intracellular ionic concentrations. While the Beeler-Reuter model and the first Luo-Rudy model only describe the intracellular calcium concentration, the phase two model also models changes in intracellular sodium and potassium. From the Nernst equation (2.52) it is clear that changes in ionic concentrations will affect the equilibrium potential and hence the flux of the various ions, and this effect is included in the upgraded model.

In addition to a more detailed description of the membrane current and ionic concentrations, the Luo-Rudy phase two model also describes a number of internal fluxes. These include calcium flux in and out of the *sarcoplasmic reticulum* (SR), as well as buffering of calcium. The SR is a network inside the cell that takes up calcium from the myoplasm, and later releases it back to the myoplasm in response to an activation of the cell. The resulting variations in myoplasm calcium concentration are extremely important for the contractile ability of the muscle cells. The SR is divided into two separate compartments. The *network sarcoplasmic reticulum* (NSR) takes up the calcium from the myoplasm, while the *junctional sarcoplasmic reticulum* (JSR) is the cistarnae that releases the calcium back into the myoplasm. The two compartments of the SR, and the related calcium fluxes, are illustrated in Figure 2.12. The figure also shows the membrane currents included in the model, and illustrates the increased complexity compared to the earlier models. For an explanation of each current we refer to [87].

To describe the calcium dynamics of the SR, the Luo-Rudy phase two model describes three different intracellular calcium concentrations: the usual myoplasm calcium concentration $[Ca]_i$, and the two SR concentrations $[Ca]_{NSR}$ and $[Ca]_{JSR}$. The model describes four currents related to the uptake and release of calcium from the SR. The uptake of calcium to the NSR from the myoplasm, J_{up}, is a passive current that depends only on the myoplasm calcium concentration $[Ca]_i$. This uptake current is the primary current between the NSR and the myoplasm, but the model also includes a small leakage back from the NSR to the myoplasm, called J_{leak}.

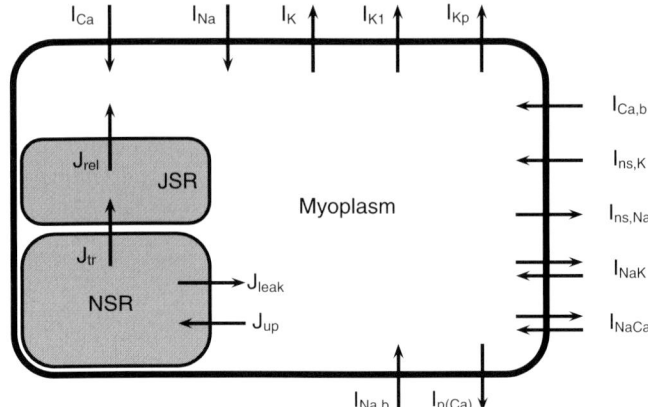

Fig. 2.12. A schematic view of the Luo-Rudy phase two model.

This current depends only on the NSR calcium concentration, and is described by a simple linear formulation. The third current directly related to the SR is a transfer current J_{tr}, which describes the transfer between the NSR and the JSR. This current is also described by a linear formulation, depending on the calcium concentrations in the two compartments. The final SR-related current is the release current J_{rel}, which releases calcium from the JSR back to the myoplasm. This calcium release is induced by an increase in the myoplasm calcium concentration, and is described by a complex relation related to the accumulation of calcium in the myoplasm. For a detailed specification of the model, the reader is again referred to the original publication [87].

Although the Luo-Rudy phase two model offers a very detailed description of the ventricular cell, a number of more recent models have been proposed, in which important processes in the cells are handled differently. One area where recent models differ substantially from the Luo-Rudy phase two model is the handling of calcium ions. Although the Luo-Rudy model includes both the sarcoplasmic reticulum and a number of calcium buffers, more recent experimental findings have led to the development of new models for calcium handling. A concept used in a model by Noble et al. [105] in 1998 is that of a restricted subspace, or diadic space, located between the cell membrane and the sarcoplasmic reticulum. Both the inward membrane current (L-type) of calcium, and the calcium release from the JSR end up in this restricted subspace, before the calcium ions diffuse into the bulk myoplasm. The resulting calcium concentration in the subspace is several orders of magnitude higher than in the myoplasm. The high calcium concentration is important both for triggering the calcium release from the JSR and for the calcium-induced inactivation of the L-type membrane current. The 1998 Noble model also includes effects related to the mechanical behaviour of the heart cells, through the addition of length-dependent and tension-dependent electrophysiological processes. Like the

Luo-Rudy phase two model, the 1998 Noble model is too complex to be described in full detail here, and the reader is referred to the original publication [105] for a complete specification of the model.

The restricted subspace in which the calcium channels empty is also included in a model by Jafri et al. [68]. The model uses a formulation that was originally proposed by Keizer and Levine [73] for the release of calcium from the SR. It also introduces a new model for the L-type calcium channels, featuring a new "mode switching" behaviour for the calcium-induced inactivation of this channel. Apart from the significantly different handling of calcium, most of the ionic currents of this model are similar to those of the Luo-Rudy phase two model.

Like the Luo-Rudy models and the Noble models described here, the 1998 Jafri model is for ventricular cells from guinea pigs. A later model by Winslow et al. [145] uses the same framework as the model by Jafri et al. for the calcium dynamics, but the model was adapted to fit experimental data from dogs. This adaptation involves adding more ionic currents, for instance the transient outward potassium current (I_{to1}), which is known to be significant in the dog but unimportant in the guinea pig. With more than 30 state variables to control 13 membrane current, the Winslow et al model is among the more advanced cardiac cell models available. Figure 2.13 shows the ionic currents included in the model, and the action potential is shown in Figure 2.14. For a complete specification of the model, the reader is referred to [145].

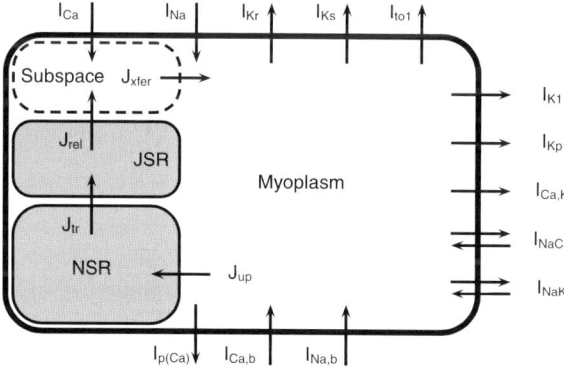

Fig. 2.13. A schematic view of the ionic currents in the 1999 model by Winslow et al.

2.5 Summary of the Mathematical Model

In this chapter we have presented a complete mathematical model for the electrical activity in the heart and the surrounding body. The body surrounding the heart is modelled as a passive volume conductor, while the model for the heart tissue was

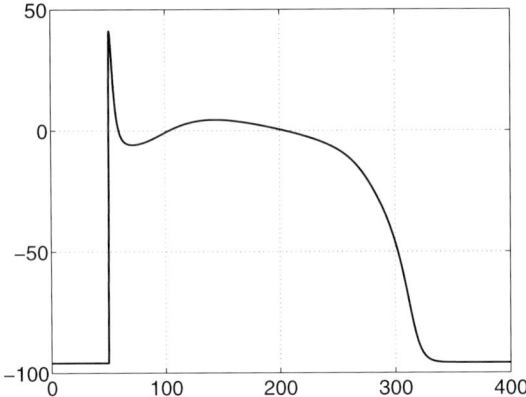

Fig. 2.14. Action potential for a dog ventricular cell, computed with the model by Winslow et al.

based on viewing the tissue as two continuous domains: intracellular and extracellular. For the heart tissue, the flux of ions across the cell membrane is essential for the activation of the cells, and for this process we have presented models based on the underlying physiology of the cell membrane. The complete model is given by

$$\frac{\partial s}{\partial t} = F(s, v, t) \qquad x \in H \qquad (2.65)$$

$$\nabla \cdot (M_i^* \nabla v) + \nabla \cdot (M_i^* \nabla u_e) = \frac{\partial v}{\partial t} + I_{ion}^* \qquad x \in H, \qquad (2.66)$$

$$\nabla \cdot (M_i^* \nabla v) + \nabla \cdot ((M_i^* + M_e^*) \nabla u_e) = 0 \qquad x \in H, \qquad (2.67)$$

$$\nabla \cdot (M_T^* \nabla u_T) = 0 \qquad x \in T, \qquad (2.68)$$

$$u_e = u_T \qquad x \in \partial H, \qquad (2.69)$$

$$n \cdot (M_i^* \nabla v + (M_i^* + M_e^*) \nabla u_e) = n \cdot (M_T^* \nabla u_T) x \in \partial H, \qquad (2.70)$$

$$n \cdot (M_i^* \nabla v + M_i^* \nabla u_e) = 0 \qquad x \in \partial H, \qquad (2.71)$$

$$n \cdot M_T^* \nabla u_T = 0 \qquad x \in \partial T. \qquad (2.72)$$

Equation (2.65) is a system of ordinary differential equations that describe the electrophysiological behaviour of the heart cells, while (2.66)–(2.67) describes the signal propagation in the heart tissue. Furthermore, (2.68) describes the potential distribution in the body surrounding the heart. The connection between the heart and the surrounding body is described by the boundary conditions (2.69)–(2.71), while (2.72) specifies the boundary conditions on the surface of the body.

Note that the system (2.66)–(2.72) has been scaled with the membrane capacitance C_m and the surface area to volume ratio χ. The resulting scaled quantities are given by

$$M_i^* = \frac{1}{C_m \chi} M_i, \quad M_e^* = \frac{1}{C_m \chi} M_e, \tag{2.73}$$

$$M_o^* = \frac{1}{C_m \chi} M_o, \quad I_{ion}^* = \frac{1}{C_m} I_{ion}. \tag{2.74}$$

For the remainder of this book we will use the scaled version of the equations, but for notational convenience we will omit the the stars on the scaled conductivities and ionic current. Values of the intra- and extracellular conductivities are found in Table 2.1, with the conductivity tensors computed from (2.27). Suitable values for the conductivity M_T in (2.68) may be found in e.g. [76] and [51]. For the experiments described later in this book, we have modelled the torso surrounding the heart as isotropic, so that M_T is a scalar. For most of the simulations we have used $M_T = 2.39$ mS/cm, but for the more realistic cases we distinguish between the lungs and the rest of the torso, with a conductivity value of $M_T = 0.96$ mS/cm in the lungs.

For the ionic current term I_{ion} and the description of the cells in (2.65), we have presented a number of widely used models, with large variations in complexity and physiological realism. However, we have only given a brief overview of a small selection of the available models for cardiac cells. Several other models exist, some of which may be at least as important as the models mentioned here. A large collection of available cell models may be found on the CellML web page [21], which uses the XML markup language to give a precise specification of the models. The CellML model repository is updated regularly with recent advances in both cellular electrophysiology, contractile mechanisms, and other physiological phenomena. The experimental techniques upon which the models are based are being developed continuously, and the ever-increasing sophistication allows greater physiological detail to be observed in the cells. The increasing amount of available data allows models of greater and greater complexity to be constructed, offering an accurate description of very small-scale details of the physiology of cardiac cells.

As noted in the discussion of the Beeler-Reuter model, a problem faced when using the most recent models for large-scale simulations is that they tend to be computationally demanding. The more accurate description requires an increasing number of state variables, such as ionic concentrations and gate variables. Adding further to the computational demand is that some of the processes underlying the action potential occur on a very small time scale. As will be discussed later in this book, the effect of this is that the equations describing the reactions become stiff, and therefore challenging to solve numerically. A consequence of the increasing complexity of the cell models is that the older models, such as the Beeler-Reuter model and the first Luo-Rudy model, are still widely used. In many cases, these models give a sufficiently realistic representation of the action potential, and the computational advantage over the more sophisticated models is significant. For some studies, even simple models of FitzHugh-Nagumo type may be sufficient, if the main interest is the qualitative behaviour of the tissue rather than a quantitatively accurate description.

Although simplified models in some form may be sufficient for many simulation purposes, an important motivation for this type of mathematical modelling is to integrate small-scale information up to the level of a complete organ. To fully accomplish this task, and investigate how small changes in cellular physiology affect the function of the complete heart, a fairly detailed description of the cells will, in most cases, be required. For full-scale simulations of this type the computational demand of solving the cell model equations will be significant, and using efficient numerical techniques becomes essential. This topic is discussed further in Chapter 5.

Chapter 3

Computational Models

The mathematical models derived in the previous chapter give a quantitative description of the electrical activity in the heart, from the level of electrochemical reactions in the cells to body surface potentials that may be recorded as ECGs. However, the models are formulated as systems of nonlinear partial and ordinary differential equations, for which analytical solutions are not available. To be of any practical use, the equations of the models must therefore be solved with numerical methods. The choice of numerical methods that may be applied to the equations is large, see e.g. [83], but we have chosen to focus entirely on finite element methods (FEM). One reason for this is that the geometries of the heart and the body are irregular, and this is more conveniently handled by FEM than, for instance, by finite difference methods.

3.1   The Finite Element Method for the Torso

To introduce the finite element method, we start with the simpler mathematical models derived in Chapter 2, which give the potential distribution in the body that results from a given source term. In contrast to the more advanced monodomain and bidomain models, these are stationary models, and the reduced complexity makes them suitable for introducing numerical techniques. However, to further ease the introduction of FEM computations, we first consider an even simpler model problem.

3.1.1   A Simplified Model Problem

In order to introduce the finite element method, we consider a simple model problem of the form

$$-\nabla^2 u = f(\mathbf{x}), \ \mathbf{x} \in \Omega, \tag{3.1}$$
$$u = 0 \qquad \mathbf{x} \in \partial\Omega. \tag{3.2}$$

We see that this problem is fairly similar to (2.5)–(2.6), discussed in Chapter 2, which describes the potential distribution that results from a dipole representation of the heart. The difference is that the Neumann boundary conditions along $\partial\Omega$ have been replaced by Dirichlet conditions for u, and we assume a constant, scalar conductivity $M = 1$ (isotropic and homogeneous material).

We now choose Ω to be the 2D unit square $\Omega = [0, 1] \times [0, 1]$, and

$$f(\mathbf{x}) = f(x, y) = -2x(x - 1) - 2y(y - 1).$$

The boundary value problem (3.1)–(3.2) then has the analytical solution

$$u = x(x-1)y(y-1). \tag{3.3}$$

Hence, for this simple equation, there is no need for numerical solution techniques, but to investigate the convergence properties of the numerical schemes it is convenient to study a problem with known analytical solution.

The foundation of the finite element method is the weak form, or variational formulation, of the PDE. To obtain a weak form of (3.1), we introduce a suitable function space V, in which we seek the solution u. The space V will be a Hilbert space; see, e.g., [69,38] for a more detailed discussion. We then multiply (3.1) with an arbitrary function $\psi \in V$, called a test function, and integrate over the domain Ω. This gives

$$-\int_\Omega \nabla^2 u\psi dx = \int_\Omega f\psi dx. \tag{3.4}$$

The variational problem is then to find a solution $u \in V$ that satisfies (3.4) for all choices of the test function $\psi \in V$. Now, applying Green's lemma (see, e.g., [69,79]) to the left hand side of (3.4) gives

$$\int_\Omega \nabla u \cdot \nabla \psi - \int_{\partial\Omega} n \cdot \nabla u\psi dx = \int_\Omega f\psi dx. \tag{3.5}$$

If we choose the function space V such that all functions $\psi \in V$ satisfy the boundary condition (3.2), we see that the boundary integral vanishes and we end up with the variational problem:

Find $u \in V$ so that

$$\int_\Omega \nabla u \cdot \nabla \psi dx = \int_\Omega f\psi dx \text{ for all } \psi \in V. \tag{3.6}$$

This is the weak form, or variational formulation, of (3.1)–(3.2).

While the original formulation required the PDE (3.1) to be satisfied at all points in Ω, the weak form only requires the integral equation (3.6) to be satisfied for all choices of $\psi \in V$. The integral represents a form of averaging, and so this formulation is, in some sense, weaker than the original formulation of the problem, hence the term "weak form". The solution of the weak form is called a weak solution of the PDE. That the variational formulation is weaker than the original PDE form may also be seen from the fact that while the weak formulation involves only first derivatives of u, the original PDE includes second derivatives, which imposes stricter smoothness requirements on the solution.

It is easy to see that any solution u satisfying (3.1)–(3.2) also satisfies (3.6). Based on the requirement that (3.6) is to be satisfied for all ψ, it can be proven that any solution of (3.6) that is sufficiently smooth, i.e. twice differentiable, is also a solution of (3.1). The details of this proof lie beyond the scope of this text, and the reader is referred to [69,38]

The weak formulation (3.6) is a continuous mathematical problem, and the solution space V has infinite dimension. To be suitable for solution on a computer, the problem needs to be discretized, i.e. we need to introduce a finite dimensional space within which to seek the solution. Any function belonging to this space may then be described by a finite number of parameters, which may be determined by solving a discretized version of the weak formulation. Formulating the weak form of the equation, defining the discrete subspace, and solving the resulting discretized weak form are the essential steps of the finite element method.

Several alternatives exist for the finite dimensional space $V_h \subset V$; see e.g. [79]. In this text we will restrict the discussion to classical finite element formulations, in which the solution domain Ω is divided into a number of polygonal domains, and V_h is defined in terms of piecewise polynomial functions over these domains. In general, any kind of polygonal shapes can be used for the partition of Ω, but we will here restrict our attention to triangulations of Ω, where the domain is partitioned into non-overlapping triangles. We will later consider three-dimensional problems, in which it is natural to extend the triangulations by partitioning the domain into tetrahedra. Other common choices are quadrilateral elements in 2D and hexahedra in 3D, see e.g. [69].

We assume that the domain Ω may be partitioned into a set of non-overlapping triangles $\tau_k, k = 1, \ldots, m$, which are oriented such that no corner of one triangle lies on the interior of a side of another triangle. The union of the triangles form a polygonal domain $\Omega_h \subset \Omega$, such that the boundary vertices of Ω_h lie on $\partial \Omega$. If Ω is a polygonal domain, as we have chosen for our simple model problem, we have $\Omega_h = \Omega$, but in the general case Ω_h is a polygonal approximation to Ω. The left panel of Figure 3.1 shows a possible triangulation of the unit square used for our model problem, and the right panel shows a triangulation of a cross section of the heart muscle.

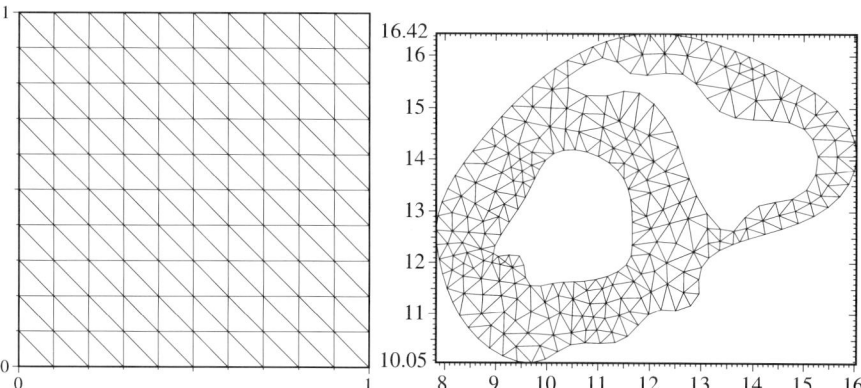

Fig. 3.1. Triangulations of two different computational domains: the unit square and a cross section of the heart muscle.

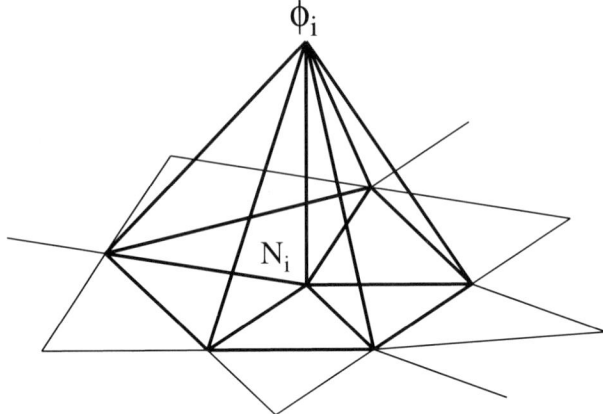

Fig. 3.2. The shape of the piecewise linear basis functions defined over the triangulation Ω_h.

We now want to use the triangulation Ω_h to define a finite dimensional function space $V_h \subset V$. Again, several choices exist, but one common choice is to define V_h as a space of piecewise polynomial functions defined over Ω_h. In principle, any order of polynomial may be used. For a given partition Ω_h, higher-order polynomials offer more degrees of freedom and yield generally higher accuracy[1]. The polynomial basis functions may also be used to describe the geometry of the elements (see, e.g., [69]), and high-order polynomials are, therefore, particularly suited for describing smooth, irregular geometries.

The basic steps in the finite element method are independent of the chosen partition of Ω and of the chosen order of polynomials. We will here focus on the simplest possible choice, which is to define the functions in V_h as piecewise linear functions. To give a precise definition of this choice of finite element space V_h, we define n to be the total number of vertices in the triangulation Ω_h, and denote the coordinates of the vertices by $\mathbf{x}_i, i = 1, \ldots n$. The discrete function space V_h is now defined as the space spanned by the basis functions $\phi_j, j = 1, \ldots n$, defined as piecewise linear functions satisfying

$$\phi_j(\mathbf{x}_i) = \begin{cases} 1 \text{ if } j = i \\ 0 \text{ otherwise} \end{cases}. \tag{3.7}$$

This definition gives basis functions with local support, restricted to a few triangles. The shape of a basis function is shown in Figure 3.2. The basis function ϕ_i depicted in the figure will be nonzero only over the triangles in which the node N_i is a corner. By definition, any function ψ_h in V_h may be written as a linear combination of the

---

[1] For instance, commonly used finite element formulations of the equations for the electrical activity in the heart are based on polynomials of orders two and three, defined over hexahedron shaped elements [10].

basis functions ϕ_j,

$$\psi_h = \sum_{j=1}^{n} \alpha_j \phi_j, \tag{3.8}$$

where $\alpha_j, j = 1, \ldots, n$ are scalars.

Formulated in the discrete space, the variational formulation (3.6) reads as follows:

Find $u_h \in V_h$ such that

$$\int_{\Omega_h} \nabla u_h \cdot \nabla \psi_h = \int_{\Omega_h} f \psi_h dx, \quad \text{for all } \psi_h \in V_h. \tag{3.9}$$

We require that (3.9) is fulfilled for all functions $\psi_h \in V_h$. However, if we insert the representation (3.8) into (3.9), it is easy to see that the equation is fulfilled for all $\psi_h \in V_h$ if it holds for all the basis functions $\{\phi_i\}$. Since the approximate solution u_h lies in V_h, the representation (3.8) may also be used for the solution,

$$u_h = \sum_{j=1}^{n} u_j \phi_j, \tag{3.10}$$

where $u_j, j = 1, \ldots, n$ are scalars. Inserting this representation of u_h into (3.9) gives

$$\int_{\Omega_h} \nabla \left(\sum_{j=1}^{n} u_j \phi_j \right) \cdot \nabla \phi_i = \int_{\Omega_h} f \phi_i dx, \quad i = 1, \ldots, n. \tag{3.11}$$

This type of finite element formulation, where the same basis functions are used both to approximate the solution and as test functions, is called a Galerkin method. Other choices of test function are possible, and may be more suitable for certain applications; see, e.g., [149] for details.

Using the linearity of the differential operators, (3.11) may be rewritten as

$$\sum_{j=1}^{n} u_j \int_{\Omega_h} \nabla \phi_j \cdot \nabla \phi_i = \int_{\Omega_h} f \phi_i dx, \quad i = 1, \ldots, n. \tag{3.12}$$

The basis functions ϕ_i, ϕ_j are known, so both integrals in this equation can be computed. The problem is reduced to finding the scalars u_j that satisfy this discrete variational formulation.

Equation (3.12) is a system of linear equations, which may be written in matrix form:

$$Au = f^n, \tag{3.13}$$

with

$$A_{ij} = \int_{\Omega_h} \nabla \phi_i \cdot \nabla \phi_j,$$

$$f_i^n = \int_{\Omega_h} f \phi_i dx.$$

The solution u of this linear system is the vector formed by the scalar coefficients $u_j, j = 1, \ldots, n$. The approximate solution u_h is given by (3.10), and will be a piecewise linear function that approximates the solution u of the continuous problem (3.6). In fact, it can be proven that u_h is the function in V_h that best approximates the solution u of the original weak form (3.6). This is a property of the Galerkin finite element method; see, e.g., [69,149] for details.

Handling essential boundary conditions. The discussion so far has been based on the assumption that the boundary condition (3.2) is automatically satisfied for all functions in the solution space V, and in the discrete subspace V_h. It is not particularly difficult to construct a solution space V_h that satisfies this requirement, but in practice a slightly different approach is normally adopted. The solution space V_h is constructed as piecewise linear functions with no attention to the Dirichlet boundary condition, and this condition is instead incorporated by adjusting the linear system (3.13). Following the definition of the basis functions $\{\phi_i\}$, the system (3.13) has one equation for each node in the grid. The Dirichlet boundary condition can be incorporated by deleting all equations that correspond to boundary nodes, and replacing them with explicit statements that enforce $u_j = 0$ if $\mathbf{x}_j$ is a boundary node. It might be feared that this approach is in conflict with the removal of the boundary integral in (3.5), which was based on the functions in V satisfying the Dirichlet boundary condition. However, since we use basis functions ϕ_i with local support, the boundary integral term in (3.5) will only contribute to those equations that will be replaced by statements to enforce the boundary condition. Hence, even with this more practical approach, it is appropriate to disregard the boundary integral in (3.5). The same approach can be used for nonhomogeneous Dirichlet conditions, i.e. if the solution takes given, nonzero values on the boundary. For practical applications, slightly more sophisticated methods for adjusting the linear system are often used, in order to conserve the symmetry of the matrix A in (3.13). A description of such a symmetric incorporation of Dirichlet boundary conditions can be found in [79]. The principle of this method is exactly the same as described here: the equations are adjusted to force the solution in the boundary nodes to the prescribed value.

An estimate of the error. It is possible to derive an estimate of the error $u_h - u$. We assume that the shapes of the triangles in Ω_h satisfy certain restrictions; see, e.g., [69] for details. We then define the quantity h to be the maximum length of a side in the triangulation. More precisely, if we define h_k to be the longest side in a triangle τ_k belonging to the triangulation Ω_h, we have

$$h = \max_{\tau_k \in \Omega_h} h_k.$$

It can then be proven that the L_2 norm of the error $u_h - u$ is proportional to h^2, i.e., we have

$$\|u_h - u\|_{L_2} = O(h^2), \tag{3.14}$$

where the L_2 norm is defined by

$$\|u\|_{L_2} = \left(\int_\Omega u^2 dx \right)^{1/2},$$

for a given function u defined over the domain Ω. The detailed proof of this error estimate lies beyond the scope of this text, and may be found in, e.g., [69,79].

Since we know the analytical solution of our model problem, we can easily perform numerical experiments to see whether the convergence is in accordance with (3.14). In Table 3.1 we present the L_2 norm of the error, $\|u - u_h\|$, with u given by (3.3), for different choices of the discretization parameter. The L_2 norm of the error is shown, and n is the number of nodes in the triangulations, i.e. the number of degrees of freedom in the numerical solution. In the rightmost column we have divided the error by the square of the maximum side length h, and we see that this number is approximately constant. Hence, the error is proportional to h^2, as predicted by the theoretical result.

Table 3.1. Convergence results for the finite element method applied to the test problem (3.1)–(3.2).

n	h	$\|u_h - u\|$	$\|u_h - u\|/h^2$
36	$2.828 \cdot 10^{-1}$	$3.454 \cdot 10^{-3}$	$4.317 \cdot 10^{-2}$
121	$1.414 \cdot 10^{-1}$	$8.964 \cdot 10^{-4}$	$4.482 \cdot 10^{-2}$
441	$7.071 \cdot 10^{-2}$	$2.262 \cdot 10^{-4}$	$4.524 \cdot 10^{-2}$
1681	$3.535 \cdot 10^{-2}$	$5.669 \cdot 10^{-5}$	$4.535 \cdot 10^{-2}$
6561	$1.767 \cdot 10^{-2}$	$1.418 \cdot 10^{-5}$	$4.538 \cdot 10^{-2}$
25921	$8.838 \cdot 10^{-3}$	$3.545 \cdot 10^{-6}$	$4.538 \cdot 10^{-2}$

Plots of the solution of (3.1)–(3.2) are shown in Figure 3.3. The upper left plot is computed with 36 nodes and 50 elements, and the upper right plot is computed with 121 nodes and 200 elements. The 200 element grid is shown in Figure 3.1. The solutions in the the two lower plots are computed with an 800 element grid with 441 nodes, and a 3200 element grid with 1681 nodes. We see that there is very little difference between these two solutions. The solution computed with the finest grid is visually almost identical to the analytical solution.

3.1.2 A Dipole Model of the Heart

In Chapter 2 we derived a simplified model for computing the electrical potential in the body, where the heart was represented by a source term in the form of a dipole. This model gives the potential distribution as the solution of (2.5)–(2.6). For convenience we repeat the equations here:

$$-\nabla \cdot (M\nabla u) = f, \quad x \in \Omega, \tag{3.15}$$

$$n \cdot (M\nabla u) = 0, \quad x \in \partial\Omega, \tag{3.16}$$

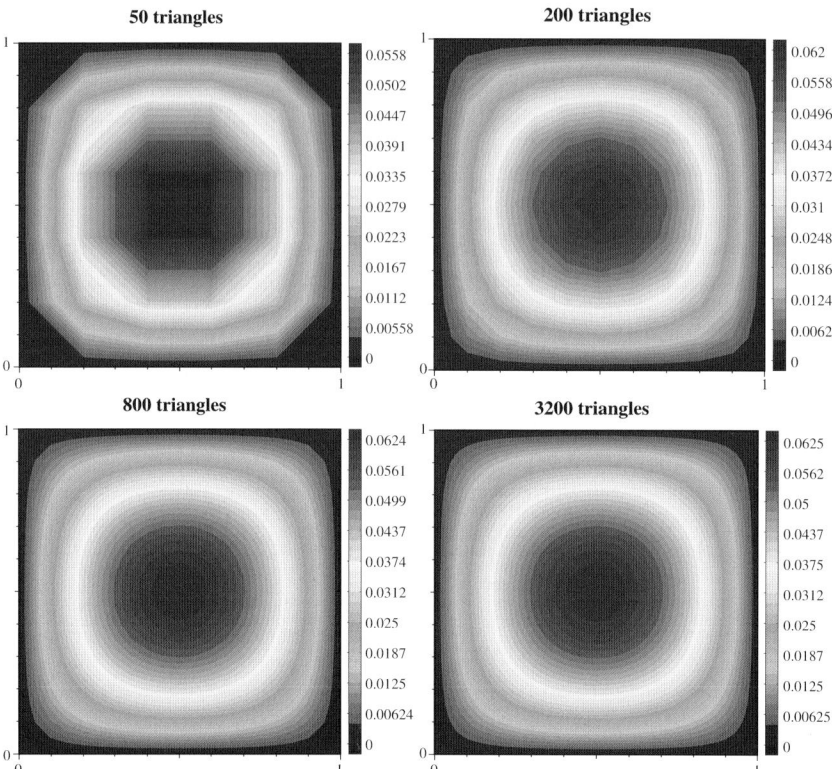

Fig. 3.3. Plots of the solution u of (3.1)–(3.2) for four different levels of grid refinement. (For the color version, see Figure A.1 on page 287).

where we again note that the source term f needs to satisfy

$$\int_\Omega f\,dx = 0,$$

for the boundary value problem to have a solution. As noted above, this problem is fairly similar to the model problem (3.1)–(3.2) considered in the previous section. The differences lie in the boundary conditions and the fact that we allow anisotropic conductivity properties, which implies that M is a tensor quantity. The source term f must also be adjusted to represent the dipole model of the heart. Many suitable choices exist for this function. One example is

$$f(x) = a(e^{-\|\mathbf{x}-\mathbf{x}_1\|/d} - e^{-\|\mathbf{x}-\mathbf{x}_2\|/d}) \qquad (3.17)$$

where $\mathbf{x}_1$ and $\mathbf{x}_2$ denote the position of the positive and negative poles, respectively; d is a parameter describing the size of each pole, and a describes the strength of the poles. Figure 3.4 shows a plot of this function for a domain with dimensions $[-20, 20] \times [-20, 20]$, and where $\mathbf{x}_1 = (-4, -4)$, $\mathbf{x}_2 = (4, 4)$, $d = 5$, and $a = 1$.

dipole

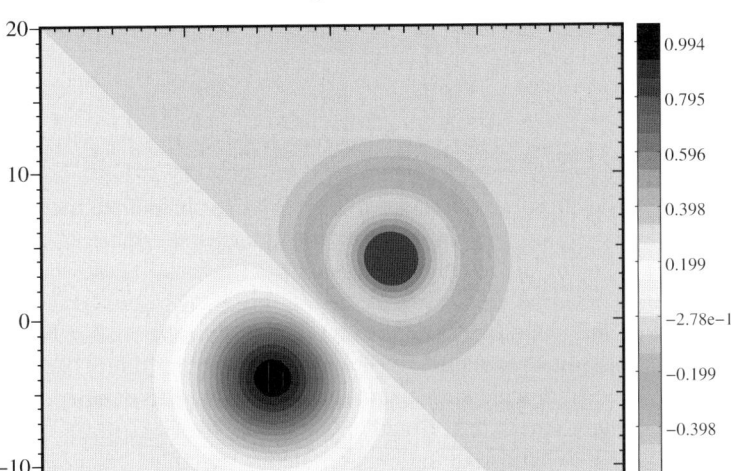

Fig. 3.4. An example of the function $f(\mathbf{x})$ in (3.15). (For the color version, see Figure A.2 on page 288).

Following the procedure introduced above, we derive a weak formulation of this problem by multiplying (3.15) by a test function ψ, and integrating over the domain Ω. We get

$$-\int_{\Omega} \nabla \cdot (M\nabla u)\psi dx = \int_{\Omega} f\psi dx, \tag{3.18}$$

which by application of Green's lemma gives

$$\int_{\Omega} M\nabla u \cdot \nabla \psi dx - \int_{\partial\Omega} n \cdot M\nabla u\psi dx = \int_{\Omega} f\psi dx. \tag{3.19}$$

We see that the boundary integral vanishes because of the boundary condition (3.16), and the result is the following weak formulation:

Find $u \in V$ such that

$$\int_{\Omega} M\nabla u \cdot \nabla \psi dx = \int_{\Omega} f\psi dx \quad \text{for all } \psi \in V. \tag{3.20}$$

Introducing a finite dimensional subspace $V_h \subset V$, and expanding the approximate solution u_h in terms of the basis functions in V_h, yields a system of linear equations

$$Au = b, \tag{3.21}$$

that must be solved for the unknown nodal values $u_j, j = 1, \ldots, n$. The entries in A and b are given by

$$A_{ij} = \int_\Omega M\nabla\phi_i \cdot \nabla\phi_j dx$$

$$b_i = \int_\Omega f\phi_i dx.$$

In the previous section, we discussed two alternative procedures for incorporating the Dirichlet boundary condition, either by adjusting the space V of possible solutions, or by adjusting the final linear system. For the present case, the boundary condition (3.16) is not explicitly present, either in the linear system or the spaces V and V_h. However, this boundary condition was naturally incorporated in the weak formulation, through the cancellation of the boundary integral in (3.19). Boundary conditions that enter the finite element equations in this implicit manner are referred to as natural boundary conditions. Boundary conditions such as (3.2), which must be explicitly fulfilled by adjusting the linear systems, are called essential boundary conditions.

It is important to note that the solution of (3.15)–(3.16) is only unique up to an additive constant. This is easy to see by a simple examination of the equations. For any function u that satisfies (3.15)–(3.16), we can add a constant, and the resulting function will still satisfy both the differential equation and the boundary condition. This is different from the problem (3.1)–(3.2), for which the specification of u on the boundary ensures a unique solution. For the linear system (3.21), the nonuniqueness of the solution has the effect that the matrix A is singular. This requires some attention when solving the equations, because direct methods, such as Gaussian elimination, will not work. One would assume that a mathematical model with a nonunique solution would also cause problems for the practical applications of the model. However, the main application of a model of this kind is to simulate ECG signals, and for this purpose we are only interested in the potential difference between different points on the surface of the body. These differences do not change when a constant is added to the solution.

Figure 3.5 shows the solution of (3.15)–(3.16), with the source term defined by (3.17), and the geometry and parameters specified above. The left panel shows the potential distribution u, while the right panel shows the magnitude of the resulting current. As described in Chapter 2, the current J is defined by

$$J = -M\nabla u.$$

For this example we have used a constant, scalar conductivity, corresponding to a homogeneous, isotropic medium. The current will, therefore, be directly proportional to the gradient of u, and in the figure it is easy to see that the magnitude of the current is largest between the two dipoles, where we also have the largest gradients in the potential field. Note also that although the boundary condition (3.16) states that no current exits the domain, the magnitude of the current is not zero at the boundary. It is sufficient that the current component normal to the boundary is zero, i.e. that the current flows parallel to the boundary.

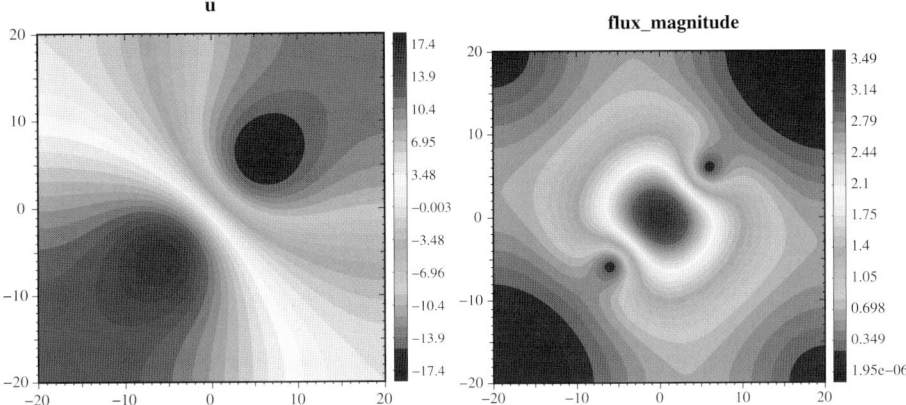

Fig. 3.5. The solution resulting from the source term function $f(\mathbf{x})$ given above. The left panel shows the potential, while the right panel shows the magnitude of the current. (For the color version, see Figure A.3 on page 288).

For practical applications, it is, of course, more relevant to consider (3.15)–(3.16) in three space dimensions. The finite element discretization procedure in 3D is completely analogous to the 2D case. For the purpose of using piecewise linear basis functions, the domain Ω is divided into nonoverlapping tetrahedra. The basis functions are then defined according to (3.7), as piecewise linear functions where basis function number i is equal to one in grid node i and zero in all other grid nodes. Both the weak forms and the resulting linear equations given above are independent of whether the geometry is two- or three-dimensional. Hence, the linear system we need to solve is identical to (3.21), but with the integration now performed over a three-dimensional domain, and the basis functions ϕ_i, ϕ_j defined over the tetrahedron-shaped elements.

An example of a 3D simulation is shown in Figures 3.6 and 3.7. The dipole is given by (3.17), with $a = 100$ and $d = 1$. The location of the two poles are shown in Figure 3.6, while Figure 3.7 shows the resulting potential distribution on the torso surface. The dipole is located closer to the front of the torso surface than to the back, which results in the potential gradients on the front surface being more visible.

3.1.3   Known Potential on the Heart Surface

In Chapter 2 we also derived an alternative approach for computing the electrical potential in the torso, based on the assumption that the potential distribution on the surface of the heart is known. The situation is described by (2.7)–(2.9). For convenience we repeat the equations here:

$$\nabla \cdot (M\nabla u) = 0 \qquad x \in T, \tag{3.22}$$

$$n \cdot M\nabla u = 0 \qquad x \in \partial T, \tag{3.23}$$

$$u = u_{\partial H} \quad x \in \partial H. \tag{3.24}$$

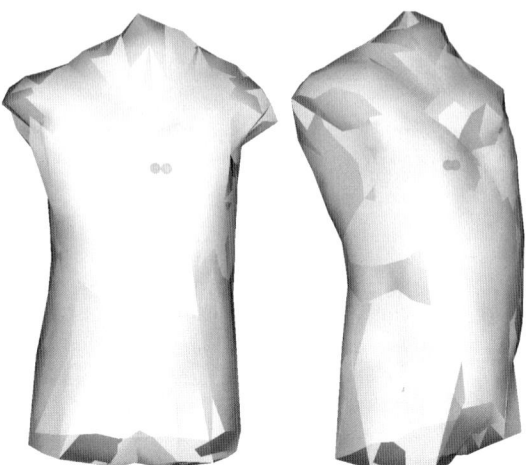

Fig. 3.6. The location of the poles are shown as red and blue spheres. (For the color version, see Figure A.4 on page 289).

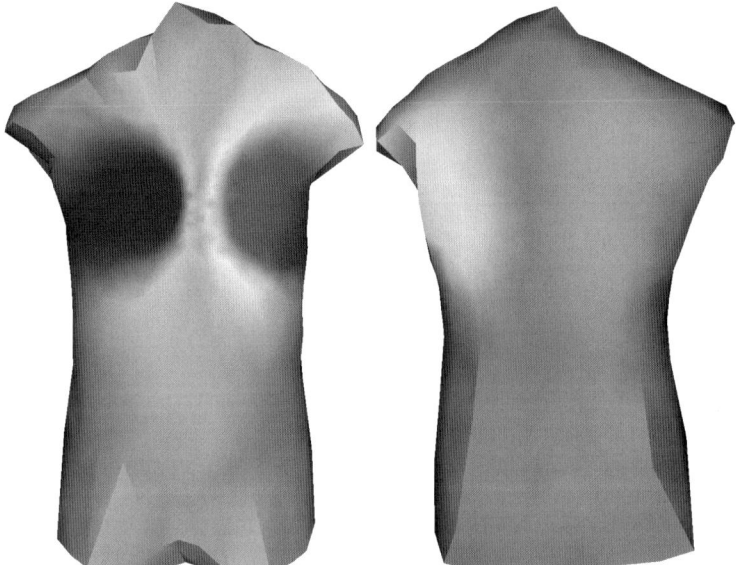

Fig. 3.7. The electrical potential set up by the dipole. The colour scale is from -4mV to 4mV. Note that the field is stronger on the front (left). (For the color version, see Figure A.5 on page 289).

A schematic view of the two boundaries ∂T and ∂H is shown in Figure 2.1 in Chapter 2. There are two important differences between this model and the previous ones: there is no source term on the right hand side and the model has two distinct boundaries, each with different boundary conditions. On the outer surface of the body we have a homogeneous Neumann condition similar to the dipole case, while on the inner boundary, corresponding to the surface of the heart, we have a nonhomogeneous Dirichlet condition.

A finite element discretization of this problem may be derived in the same way as for the model problems above. The weak form associated with (3.22) is

$$\int_T M\nabla u \cdot \nabla \psi dx - \int_{\partial H} (n \cdot M\nabla u)\psi dx - \int_{\partial T} (n \cdot M\nabla u)\psi dx = 0, \quad (3.25)$$

which is to be fulfilled for all choices of ψ within some suitable function space V. As usual, n is the outward unit normal vector on the boundary. On the inner boundary ∂H, n will point out of the body, and into the heart. We see that the Neumann boundary condition (3.23) on the outer boundary ∂T makes the surface integral vanish on this part of the boundary. This is hence a natural boundary condition, which is implicitly incorporated in the weak form.

Handling the essential boundary condition. For the Dirichlet condition on the inner boundary ∂H, we use a technique similar to that introduced in Section 3.1.1. We modify the technique slightly because for the present problem, the specified boundary values $u_{\partial H}$ are nonzero. As above, we first present a theoretical treatment of the boundary conditions, and then a slightly different procedure that is more suitable for practical use.

We first assume, as above, that all functions in V are zero on the inner boundary ∂H. We then construct a discrete subspace $V_h \subset V$, and approximate the solution u as

$$u = \xi(\mathbf{x}) + \sum_{j=1}^{n} u_j \phi_j, \quad (3.26)$$

where $\xi(\mathbf{x})$ is a known function that takes on the given values $u_{\partial H}$ on the inner boundary.

With this approximation of u, the problem is again reduced to determining the parameters $u_j, j = 1, \ldots, n$. Following the standard Galerkin method, we use the basis functions ϕ_i as test functions, and obtain

$$\int_T M\nabla u \cdot \nabla \phi_i dx - \int_{\partial H} n \cdot M\nabla u \phi_i dx = 0 \;\; i = 1, \ldots, n.$$

Since the basis functions ϕ_i vanish on ∂H, the boundary integral is zero, and we are left with

$$\int_T M\nabla u \cdot \nabla \phi_i dx = 0 \;\; i = 1, \ldots, n.$$

Inserting the approximation (3.26) for u, we get

$$\sum_{j=1}^{n} u_j \left(\int_T M \nabla \phi_j \cdot \nabla \phi_i dx \right) = - \int_T M \nabla \xi \cdot \nabla \phi_i dx \text{ for } i = 1, \ldots, n. \quad (3.27)$$

This is a linear system that can be written in matrix form:

$$Au = b,$$

where the coefficients u_j have been collected in the vector u. The coefficient matrix A is given by

$$A_{ij} = \int_\Omega M \nabla \phi_j \cdot \nabla \phi_i,$$

and the right hand side vector contains integrals of the function ξ.

Expanding the approximate solution u_h as (3.26) is suitable for a theoretical discussion of the finite element method, but for practical applications it is common to use a different technique. Instead of the expansion (3.26), we define the solution space V_h with no assumption that the functions are zero on the boundary, and expand the solution according to (3.10). As for the case described in Section 3.1.1, the essential boundary condition is then enforced by replacing the equations corresponding to the boundary nodes with explicit statements that ensure that $u_h = u_{\partial H}$ on the inner boundary. As described earlier, the equations that are adjusted in this way are precisely the the ones that receive a contribution from the surface integral over ∂H in (3.25). Hence, although this surface integral is not zero in this case, the enforcement of the essential boundary condition ensures that it will not be present in the final linear system. We can, therefore, safely disregard this boundary integral in the finite element formulation.

3.2   The Heart Equations

The models considered so far in this chapter all give a very simplified description of the electrical activity in the heart and the surrounding torso. They are all stationary models, which describe the potential distribution in the body resulting from a given source term. For a full description of the dynamical behavior of the potentials in the heart and the body, we need more sophisticated models. We recall from Chapter 2 that the most complete model we derived was the bidomain model, given by (2.65)–(2.72). To ease the discussion of computational techniques for this model, we repeat the equations here.

$$\frac{\partial s}{\partial t} = f(s, v, t) \qquad x \in H, \qquad (3.28)$$

$$\nabla \cdot (M_i \nabla v) + \nabla \cdot (M_i \nabla u_e) = \frac{\partial v}{\partial t} + I_{ion}(v, s) \quad x \in H, \qquad (3.29)$$

$$\nabla \cdot (M_i \nabla v) + \nabla \cdot ((M_i + M_e) \nabla u_e) = 0 \qquad x \in H, \qquad (3.30)$$

$$\nabla \cdot (M_o \nabla u_o) = 0 \qquad\qquad x \in T, \qquad (3.31)$$

$$u_e = u_o \qquad\qquad x \in \partial H, \qquad (3.32)$$

$$n \cdot (M_i \nabla v + (M_i + M_e) \nabla u_e) = n \cdot (M_o \nabla u_o) x \in \partial H, \qquad (3.33)$$

$$n \cdot (M_i \nabla v + M_i \nabla u_e) = 0 \qquad\qquad x \in \partial H, \qquad (3.34)$$

$$n \cdot M_o \nabla u_o = 0 \qquad\qquad x \in \partial T. \qquad (3.35)$$

In Section 2.2.3 we derived a simplified version called the monodomain model, given by (2.35) and (2.36). If we combine this model with one of the models derived in Section 2.4 for the ionic current term, we have a complete model given by

$$\frac{\partial s}{\partial t} = f(s, v, t) \qquad\qquad x \in H, \qquad (3.36)$$

$$\frac{\lambda}{1 + \lambda} \nabla \cdot (M_i \nabla v) = \frac{\partial v}{\partial t} + I_{ion}(v, s) \ \ x \in H, \qquad (3.37)$$

$$n \cdot (M_i \nabla v) = 0 \qquad\qquad x \in \partial H. \qquad (3.38)$$

Compared with the models discussed in Section 3.1, both the monodomain model and the bidomain model are significantly more complex. They are time-dependent systems, which must be discretized in time as well as in space. They also involve coupled systems of nonlinear ODEs and PDEs, which makes it much more challenging to derive efficient numerical methods.

In this section we will present numerical methods for solving both the monodomain model and the bidomain model. The methods are based on a technique known as operator splitting, in combination with a finite difference discretization in time. The spatial discretization of the equations will be based on a finite element procedure, similar to the one we used for the simpler equations considered above.

3.2.1   Operator Splitting

Operator splitting is an attractive technique for solving coupled systems of PDEs, since complex equation systems may be split into smaller parts that are easier to solve. Several operator splitting techniques exist, but we will apply a class of methods often referred to as fractional step methods; see, e.g., [82]. To introduce this technique, consider an initial value problem of the form

$$\frac{du}{dt} = (L_1 + L_2)u, \qquad (3.39)$$

$$u(0) = u_0, \qquad (3.40)$$

where L_1 and L_2 are operators on u, and u_0 is a given initial condition. If we choose a small time step Δt, an approximate solution at $t = \Delta t$ may be computed by first solving the problem

$$\frac{dv}{dt} = L_1(v) \qquad (3.41)$$

$$v(0) = u_0, \qquad (3.42)$$

for $t \in [0, \Delta t]$. Thereafter, we solve the problem

$$\frac{dw}{dt} = L_2(w), \tag{3.43}$$

$$w(0) = v(\Delta t), \tag{3.44}$$

for $t \in [0, \Delta t]$. Note that the final value of the solution of (3.41), $v(\Delta t)$, is used as an initial condition in this step.

Although it may now seem that we have found an approximate solution after a time interval $2\Delta t$, we have only included parts of the right hand side in each integration step. To see that the result $w(\Delta t)$ is in fact a consistent approximation to $u(\Delta t)$, we perform a Taylor series expansion of both the original solution u, and the approximation w obtained by the operator splitting. We have

$$u(\Delta t) = u_0 + \Delta t \left. \frac{du}{dt} \right|_0 + \frac{\Delta t^2}{2} \left. \frac{d^2 u}{dt^2} \right|_0 + O(\Delta t^3).$$

From (3.39) we have

$$\frac{du}{dt} = (L_1 + L_2)u,$$

and if L_1 and L_2 do not depend explicitly on t, we obtain by direct differentiation

$$\frac{d^2 u}{dt^2} = (L_1 + L_2)(L_1 + L_2)u,$$

for which we introduce the shorter notation

$$\frac{d^2 u}{dt^2} = (L_1 + L_2)^2 u.$$

Repeating these steps n times gives the general result

$$\frac{d^n u}{dt^n} = (L_1 + L_2)^n u,$$

where the notation $(L_1 + L_2)^n$ simply means that the operator $(L_1 + L_2)$ is applied n times to u.

Inserted into the Taylor series, this gives

$$u(\Delta t) = u_0 + \Delta t(L_1 + L_2)u_0 + \frac{\Delta t^2}{2}(L_1 + L_2)^2 u_0 + O(\Delta t^3). \tag{3.45}$$

A similar Taylor expansion can be made for the solution v of the simplified equation (3.41). We get

$$v(\Delta t) = u_0 + \Delta t L_1 u_0 + \Delta t^2 L_1^2 u_0 + O(\Delta t^3).$$

We now use the same series expansion for the solution of (3.43), with $v(\Delta t)$ as the initial condition. We get

$$w(\Delta t) = v(\Delta t) + \Delta t L_2 v(\Delta t) + \Delta t^2 L_2^2 v(\Delta t) + O(\Delta t^3),$$

and inserting the series expansion for $v(\Delta t)$ gives

$$w(\Delta t) = u_0 + \Delta t (L_1 + L_2) u_0 + \frac{\Delta t^2}{2} (L_1^2 + 2L_2 L_1 + L_2^2) u_0 + O(\Delta t^3). \quad (3.46)$$

The splitting error at $t = \Delta t$ is the difference between the operator splitting solution $w(\Delta t)$ and the solution $u(\Delta t)$ of the original problem. Inserting the series expansions (3.45) and (3.46), we get

$$w(\Delta t) - u(\Delta t) = \frac{\Delta t^2}{2} (L_1 L_2 - L_2 L_1) u_0 + O(\Delta t^3).$$

We see that the error after one time step is proportional to Δt^2, and we expect this error to accumulate to $n \Delta t^2$ after n time steps. When solving the equations over a fixed time interval, e.g. $t \in [0, b]$, the number of time steps n is proportional to Δt^{-1}, and the error at $t = b$ is therefore proportional to Δt. The splitting method, commonly referred to as Godunov splitting, is hence a first-order method.

By a small modification it is possible to make the splitting algorithm second-order accurate. The idea is that instead of first solving (3.41) for a full time step of length Δt, we solve the problem for a time step of length $\Delta t/2$. We then solve the problem (3.43) for a full step of length Δt, and finally (3.41) once more, again for a time interval of length $\Delta t/2$. The details of the three steps of the algorithm are as follows. We first solve the problem

$$\frac{dv}{dt} = L_1(v), \quad (3.47)$$

$$v(0) = u_0, \quad (3.48)$$

for $t \in [0, \Delta t/2]$. Then the problem

$$\frac{dw}{dt} = L_2(w), \quad (3.49)$$

$$w(0) = v(\Delta t/2), \quad (3.50)$$

is solved for $t \in [0, \Delta t]$. Finally, we solve

$$\frac{dv}{dt} = L_1(v), \quad (3.51)$$

$$v(\Delta t/2) = w(\Delta t), \quad (3.52)$$

for $t \in [\Delta t/2, \Delta t]$. This three-step algorithm is called Strang splitting.

To see that the Strang splitting gives second-order accuracy, we compute the Taylor series expansion of the approximate solution v at $t = \Delta t$. Following the steps outlined above for the Godunov splitting, we first find a Taylor expansion of the solution v of (3.47)–(3.48), at $t = \Delta t/2$;

$$v(\Delta t/2) = u_0 + \frac{\Delta t}{2} L_1 u_0 + \frac{\Delta t^2}{4} L_1^2 u_0 + O(\Delta t^3).$$

Using this as the initial condition for a Taylor expansion of the solution $w(\Delta t)$ from the second step, we get

$$w(\Delta t) = u_0 + \frac{\Delta t}{2}L_1 u_0 + \Delta t L_2 u_0 + \frac{\Delta t^2}{8}L_1^2 u_0 + \frac{\Delta t^2}{2}L_2 L_1 u_0 + \frac{\Delta t^2}{2}L_2^2 u_0 + O(\Delta t^3).$$

And finally, by a Taylor expansion of the third step, we find

$$v(\Delta t) = u_0 + \Delta t(L_1 + L_2)u_0 + \frac{\Delta t^2}{2}(L_1^2 + L_1 L_2 + L_2 L_1 + L_2^2)u_0 + O(\Delta t^3).$$

Comparing this with the Taylor expansion (3.45) of the solution of (3.39), we see that the second-order terms in the local splitting error cancel, and we have

$$v(\Delta t) - u(\Delta t) = O(\Delta t^3).$$

With the local error proportional to Δt^3, the accumulated error after $n \sim \Delta t^{-1}$ steps is hence proportional to Δt^2.

We can derive a more compact notation for the operator splitting methods by introducing the concept of solution operators. We introduce operators V and W, defined so that for a given time step Δt, $V(\Delta t)$ applied to u_0 gives the solution of (3.41) at $t = \Delta t$, with u_0 as initial condition. Similarly, $W(\Delta t)u_0$ gives the solution of (3.43) at $t = \Delta t$, with u_0 as initial condition. With this notation, the solution u_n obtained by Godunov splitting at $t = t_n = n\Delta t$ is given by

$$u_n = (W(\Delta t)V(\Delta t))^n u_0.$$

As above, this is just a short notation for applying the combined operator $(W(\Delta t)V(\Delta t))$ to u_0 n times. The solution obtained with Strang splitting may be written as

$$u_n = (V(\Delta t/2)W(\Delta t)V(\Delta t/2))^n u_0.$$

By introducing a parameter $\theta \in [0, 1]$, we can formulate a more general operator-splitting algorithm, for which Godunov and Strang splitting are obtained as special cases. One step of the general formulation may be written as

$$u(\Delta t) = V((1 - \theta)\Delta t)W(\Delta t)V(\theta\Delta t)u_0.$$

Noting that $V(0)$ is simply the identity operator, we see that choosing $\theta = 0$ or $\theta = 1$ gives two different versions of the Godunov splitting. The Strang splitting is obtained by choosing $\theta = 1/2$. By a Taylor series expansion similar to that above, it can be shown that all choices of $\theta \neq 1/2$ give a first-order splitting scheme. We are mostly interested in choosing θ so that we get either Strang splitting or Godunov splitting, but other choices are possible. The greatest advantage of the general θ-formulation of the operator splitting is that it leads to a very convenient implementation of the algorithm, where the splitting method used can be varied by adjusting a single parameter.

3.2.2 Operator Splitting for the Monodomain Model

So far, we have introduced the ideas of Godunov and Strang splitting for the general initial value problem (3.39)–(3.40). To concretize the ideas, let us now apply the algorithm to the monodomain model (3.37)–(3.38), as proposed by Qu and Garfinkel [117].

Equation (3.37) can be written in the form (3.39), with the two operators defined by

$$L_1 v = -I_{ion}(v, s),$$

$$L_2 v = \frac{\lambda}{1 + \lambda} \nabla \cdot (M_i \nabla v).$$

With this definition of the two operators, the two subproblems in the splitting method defined above become

$$\frac{\partial v}{\partial t} = -I_{ion}(v, s), \tag{3.53}$$

$$\frac{\partial v}{\partial t} = \frac{\lambda}{1 + \lambda} \nabla \cdot (M_i \nabla v). \tag{3.54}$$

We see that the nonlinear PDE is reduced to a linear PDE and a nonlinear ODE. Recall that in order to solve the monodomain model we are already required to solve the ODE system (3.36), and this system may now be solved simultaneously with the ODE (3.53) resulting from the splitting of the PDE. Assuming that $v^n = v(t_n)$ and $s^n = s(t_n)$ are known, one time step of the splitting algorithm consists of the following three operations:

1. Solve the system

$$\frac{\partial v}{\partial t} = -I_{ion}(v, s), \quad v(t_n) = v^n,$$

$$\frac{\partial s}{\partial t} = f(v, s), \qquad s(t_n) = s^n,$$

for $t_n < t \le t_n + \theta \Delta t$. The solutions v and s at $t = t_n + \theta \Delta t$ are denoted by v_θ^n and s_θ^n, respectively.

2. Solve the linear PDE

$$\frac{\partial v}{\partial t} = \frac{\lambda}{1 + \lambda} \nabla \cdot (M_i \nabla v), \quad v(t_n) = v_\theta^n, \tag{3.55}$$

for $t_n < t \le t_n + \Delta t$, with the boundary condition (3.38). The resulting solution $v(t_n + \Delta t)$ is denoted by v_θ^{n+1}.

3. Solve the system

$$\frac{\partial v}{\partial t} = -I_{ion}(v, s), \quad v(t_n + \theta \Delta t) = v_\theta^{n+1},$$

$$\frac{\partial s}{\partial t} = f(v, s), \qquad s(t_n + \theta \Delta t) = s_\theta^n,$$

for $t_n + \theta \Delta t < t \leq t_n + \Delta t$, to obtain the approximate solutions v^{n+1} and s^{n+1} at $t = t_n + \Delta t$.

We are primarily interested in choosing $\theta = 1/2$ and $\theta = 1$, corresponding to the Strang and Godunov splitting.

In the general discussion of the fractional step methods we did not comment on the solution of the subproblems, implicitly assuming that these could be solved analytically. For some problems this may be the case, but for all reasonably realistic choices of cell models and geometry it is impossible to find analytical solutions of the subproblems that result from the monodomain model. We will, therefore, need to apply some numerical method for solving these problems. To obtain the overall second-order accuracy of the Strang splitting, we need to solve both the ODE systems in Steps 1 and 3 and the PDE in Step 2 with at least second-order accuracy. Solving the subproblems with an accuracy much higher than second order will yield no benefit, since the overall accuracy will still be limited by the splitting error. Similarly, if Godunov splitting is used, it is of no benefit to solve the subproblem equations with greater than first-order accuracy.

A large variety of methods exist for solving nonlinear ODE systems; see, e.g., [58,59]. Realistic ODE models for the cellular processes normally describe chemical reactions with very different time scales. As will be discussed in more detail in Chapter 5, this makes the equations stiff, and therefore challenging to solve with numerical methods. Restrictions imposed by the coupling to the PDEs, as well as the overall splitting error, must also be considered when choosing an ODE solver. Among the methods that have been found suitable are various implicit Runge-Kutta methods, preferably with a fairly low order of accuracy to match the accuracy of the operator splitting. These methods will be described in more detail in Chapter 5. Here, we focus on the discretization of the PDEs.

For solving the PDE in Step 2, we use a time discretization based on a θ-rule; see, e.g., [79]. The θ-rule is a commonly used technique for the time discretization of PDEs, in which the terms in the equation are computed as weighted averages of the values from the current and the next time step. As for the operator splitting method described above, θ is a parameter in $[0, 1]$. For different choices of the parameter θ we get methods with different accuracy and characteristics. In our case, the use of the θ-rule enables the accuracy of the PDE discretization to be matched easily to that of the operator splitting scheme. Assuming that the current value $v^n = v(t_n)$ is known, we want to find the unknown field v^{n+1} at the next time step. A time discretization based on the θ-rule approximates v^{n+1} from

$$
\frac{v^{n+1} - v^n}{\Delta t} = \theta \left(\frac{\lambda}{1 + \lambda} \nabla \cdot (M_i \nabla v^{n+1}) \right)
$$
$$
+ \left((1 - \theta) \frac{\lambda}{1 + \lambda} \nabla \cdot (M_i \nabla v^n) \right). \tag{3.56}
$$

We see that a simple finite difference approximation is used for the time derivative, and the right hand side is a weighted average of values from the current and the next time step. For the choice $\theta = 1$, the diffusion term on the right hand side is approximated at time t_{n+1}, giving the implicit backward Euler method. For $\theta = 0$ the scheme is easily recognized as the forward Euler scheme, which gives an explicit formula for the unknown field v^{n+1}. Because of its poor stability properties (see, e.g., [139]) or the discussion of ODE solvers in Chapter 5, this method is not very suitable for our application. Setting $\theta = 1/2$ gives the Crank-Nicolson scheme, which is second-order accurate with respect to time[2]. It can be shown that for all choices of $\theta \neq 1/2$, the resulting scheme is first-order accurate. By using the same value of θ for the PDE discretization and the operator splitting, we see that the accuracy matches well.

The time-discrete PDE (3.56) may be discretized in space with a finite element method similar to the ones introduced for the simple model problems. As above, we multiply the equation with a test function and integrate over the entire domain H. Applying Green's lemma, we find that the weak formulation corresponding to (3.56) is

$$\int_H v^{n+1}\psi dx + \theta\gamma \int_H M_i \nabla v^{n+1} \cdot \nabla \psi dx - \theta\gamma \int_{\partial H} n \cdot M_i \nabla v^{n+1}\psi ds$$
$$= \int_H v^n \psi dx + (1-\theta)\gamma \int_{\partial H} n \cdot M_i \nabla v^n \psi ds$$
$$- (1-\theta)\gamma \int_H M_i \nabla v^n \cdot \nabla \psi dx \quad \text{for all } \psi \in V.$$

For convenience of notation, we have introduced

$$\gamma = \frac{\Delta t \lambda}{1 + \lambda}.$$

We see that the boundary integral terms vanish because of the boundary condition (3.38), and we obtain the following weak form:
Find $v^{n+1} \in V$ such that

$$\int_H v^{n+1}\psi dx + \theta\gamma \int_H M_i \nabla v^{n+1} \cdot \nabla \psi dx$$
$$= \int_H v^n \psi dx - (1-\theta)\gamma \int_H M_i \nabla v^n \cdot \nabla \psi dx \quad \text{for all } \psi \in V,$$

$$(3.57)$$

for some suitable function space V.

As above, we introduce a finite dimensional subspace $V_h \subset V$, and approximate v^{n+1} as a linear combination of basis functions,

$$v^{n+1} = \sum_{j=1}^{n} v_j \phi_j,$$

---

[2] Recall from above that the finite element method for the simple stationary problems was second-order accurate with respect to space. This will also be the case when the finite element method is applied to the present problem, so that the spatial discretization is always second-order accurate.

where $\phi_j, j = 1, \ldots, n$ are basis functions spanning V_h, and v_j are scalars. We insert this approximation into (3.57), and use the basis functions ϕ_i as weight functions. This gives the finite element equations

$$\sum_{j=1}^{n} v_j \int_H \phi_j \phi_i dx + \theta \gamma \sum_{j=1}^{n} v_j \int_H M_i \nabla \phi_j \cdot \nabla \phi_i dx$$

$$= \int_H v^n \phi_i dx - (1 - \theta) \gamma \int_H M_i \nabla v^n \cdot \nabla \phi_i dx, i = 1, \ldots, n$$

$$(3.58)$$

which is a system of linear equations that can be written in the form

$$Av = b.$$

The matrix and the right side vector are given by

$$A_{ij} = \int_H \phi_j \phi_i dx + \theta \gamma \int_H M_i \nabla \phi_j \cdot \nabla \phi_i dx,$$

and

$$b_i = \int_H v^n \phi_i dx - (1 - \theta) \gamma \int_H M_i \nabla v^n \cdot \nabla \phi_i dx.$$

To summarize, we have seen that we can solve the monodomain model, given by (3.36)–(3.38), in a sequential manner, by alternately solving a nonlinear system of ODEs and a linear PDE. The PDE may be discretized in time with a θ-rule, and in space with a finite element method. Solving the original problem has thus been reduced to solving nonlinear systems of ODEs, and solving the discretized PDE in the form of a system of linear algebraic equations. Methods for solving the ODE systems will be presented in Chapter 5, while solvers for linear systems are discussed in Chapter 4.

3.2.3   Operator Splitting for the Bidomain Model

The splitting algorithm outlined above for the monodomain model may also be applied to the full bidomain model, as described in [132]. Assuming now that the operators L_1 and L_2 in (3.39) can also be functions of the extracellular potential u_e, we write (3.29) on the form (3.39), with

$$L_1 v = -I_{ion}(v, s),$$
$$L_2 v = \nabla \cdot (M_i \nabla v) + \nabla \cdot (M_i \nabla u_e).$$

Applying the θ-formulation of the operator splitting introduced above, the result is a three-step algorithm very similar to the one used for the monodomain model. Again, we assume that $v^n = v(t_n)$ and $s^n = s(t_n)$ are known, and perform the following steps to determine v^{n+1} and s^{n+1}.

1. Solve the system

$$\frac{\partial v}{\partial t} = -I_{ion}(v, s), \; v(t_n) = v^n,$$

$$\frac{\partial s}{\partial t} = f(v, s), \qquad s(t_n) = s^n,$$

for $t_n < t \leq t_n + \theta \Delta t$. As above, the solutions v and s at $t = t_n + \theta \Delta t$ are denoted by v_θ^n and s_θ^n, respectively.

2. Solve the linear PDE system

$$\nabla \cdot (M_i \nabla v) + \nabla \cdot (M_i \nabla u_e) = \frac{\partial v}{\partial t}, \text{ in } H \qquad (3.59)$$

$$\nabla \cdot (M_i \nabla v) + \nabla \cdot ((M_i + M_e) \nabla u_e) = 0, \quad \text{in } H \qquad (3.60)$$

$$\nabla \cdot (M_o \nabla u_o) = 0, \quad \text{in } T \qquad (3.61)$$

with $v(t_n) = v_\theta^n$, for $t_n < t \leq t_n + \Delta t$, with the boundary conditions (3.32)–(3.35). The resulting solution $v(t_n + \Delta t)$ is denoted by v_θ^{n+1}.

3. Solve the system

$$\frac{\partial v}{\partial t} = -I_{ion}(v, s), \; v(t_n + \theta \Delta t) = v_\theta^{n+1},$$

$$\frac{\partial s}{\partial t} = f(v, s), \qquad s(t_n + \theta \Delta t) = s_\theta^n,$$

for $t_n + \theta \Delta t < t \leq t_n + \Delta t$, to obtain the final approximate solutions v^{n+1} and s^{n+1} at $t = t_n + \Delta t$.

We see that Step 1 and Step 3 in this algorithm are equal to the steps described for the monodomain model. However, Step 2 is considerably more complex, because we now have to solve three coupled PDEs, defined over two different domains.

For simplicity, we first consider the simpler case of a heart that is surrounded by an insulating material. In this case (3.61) does not apply, and Step 2 in the algorithm reduces to solving the PDE system (3.59)–(3.60), with boundary conditions

$$n \cdot (M_i \nabla v + M_i \nabla u_e) = 0, \text{ on } \partial H, \qquad (3.62)$$

$$n \cdot (M_e \nabla u_e) = 0, \text{ on } \partial H. \qquad (3.63)$$

We want to derive a discrete version of these equations. We use an approach similar to the one used for the monodomain model in Section 3.2.3, with a θ-rule for time discretization and a finite element method in space. Applying this scheme to (3.59)–(3.60) gives the time discrete PDE system

$$\frac{v_\theta^{n+1} - v^n}{\Delta t} = \theta \nabla \cdot (M_i \nabla v_\theta^{n+1}) + (1 - \theta) \nabla \cdot (M_i \nabla v_\theta^n)$$

$$+ \nabla \cdot (M_i \nabla u_e^{n+\theta}), \qquad (3.64)$$

$$\theta \nabla \cdot (M_i \nabla v_\theta^{n+1}) + \nabla \cdot ((M_i + M_e) \nabla u_e^{n+\theta}) = -(1 - \theta) \nabla \cdot (M_i \nabla v_\theta^n), \quad (3.65)$$

which must be solved to find the unknown potential fields v_θ^{n+1} and $u_e^{n+\theta}$. Here $u_e^{n+\theta}$ denotes an approximation to u_e at time $t_n + \theta \Delta t$. For instance, u_e will be approximated at the midpoint of each time step if we choose $\theta = 1/2$.

The time-discrete system (3.64)–(3.65) must also be discretized in space. As for the monodomain model, we use a finite element method, for which the starting point is the weak formulation of the equations. To simplify the notation, we introduce the inner products

$$(\varphi, \phi) = \int_H \varphi \phi \, dx, \qquad\qquad \text{for } \varphi, \phi \in V,$$

$$a_I(\varphi, \phi) = \int_H M_i \nabla \varphi \cdot \nabla \phi \, dx, \qquad \text{for } \varphi, \phi \in V,$$

$$a_{I+E}(\varphi, \phi) = \int_H (M_i + M_e) \nabla \varphi \cdot \nabla \phi \, dx, \text{ for } \varphi, \phi \in V.$$

As above, V denotes the function space in which we seek the solution. The upper-case subscripts on the inner products are used to avoid confusion with the use of the index i for the basis functions.

Multiplying (3.64)–(3.65) with a test function ϕ and integrating over the domain gives the following weak formulation:

Find $v_\theta^{n+1}, u_e^{n+\theta} \in V$ such that

$$(v_\theta^{n+1}, \psi) + \theta \Delta t a_I(v_\theta^{n+1}, \psi) + \Delta t a_I(u_e^{n+\theta}, \psi)$$

$$- \theta \Delta t \int_{\partial H} \psi [(M_i \nabla v_\theta^{n+1}) \cdot n] ds - \int_{\partial H} \psi [(M_i \nabla u_e^{n+\theta}) \cdot n] ds$$

$$= (v_\theta^n, \psi) - (1-\theta) \Delta t a_I(v_\theta^n, \psi)$$

$$+ (1-\theta) \Delta t \int_{\partial H} \psi [(M_i \nabla v_\theta^n) \cdot n] ds \qquad \text{for all } \psi \in V, \quad (3.66)$$

$$- \theta a_I(v_\theta^{n+1}, \psi) - a_{I+E}(u_e^{n+\theta}, \psi) + \theta \int_{\partial H} \psi [(M_i \nabla v_\theta^{n+1}) \cdot n] ds$$

$$+ (1-\theta) \int_{\partial H} \psi [(M_i \nabla v_\theta^n) \cdot n] ds + \int_{\partial H} \psi [(M_i \nabla u_e^{n+\theta}) \cdot n] ds$$

$$+ \int_{\partial H} \psi [(M_e \nabla u_e^{n+\theta}) \cdot n] ds = 0 \qquad \text{for all } \psi \in V. \quad (3.67)$$

The boundary condition (3.63) must be fulfilled for all time steps, and this causes the last boundary integral term in (3.67) to vanish. The other boundary condition, (3.62), is slightly more complicated to include because our numerical scheme will, in general, approximate v and u_e at different time points. We therefore introduce a time-discrete version of this condition, given by

$$n \cdot (\theta M_i \nabla v_\theta^{n+1} + (1-\theta) M_i \nabla v_\theta^n + M_i \nabla u_e^{n+\theta}) = 0. \qquad (3.68)$$

This condition causes the three boundary integrals in (3.66) to cancel, as well as the remaining three boundary terms in (3.67). The equations are hence reduced to

$$(v_\theta^{n+1}, \psi) + \theta \Delta t a_I(v_\theta^{n+1}, \psi) + \Delta t a_I(u_e^{n+\theta}, \psi)$$
$$= (v_\theta^n, \psi) - (1 - \theta) \Delta t a_I(v_\theta^n, \psi) \quad \text{for all } \psi \in V, \quad (3.69)$$

$$\Delta t a_I(v_\theta^{n+1}, \psi) + \frac{\Delta t}{\theta} a_{I+E}(u_e^{n+1}, \psi) = -\Delta t \frac{1 - \theta}{\theta} a_I(v_\theta^n, \psi) \quad \text{for all } \psi \in V.$$
$$(3.70)$$

We multiplied the second equation by $\Delta t / \theta$ in order to obtain a symmetric coefficient matrix in the final linear system. Recall that choosing $\theta = 0$ corresponds to the forward Euler method, which, as noted above, is not of interest for this application. Hence, dividing the equation by θ does not cause any problems.

Introducing a finite element representation of the heart domain, with basis functions $\phi_j, j = 1, \ldots, n$, the solution fields u and v can be approximated by

$$v_\theta^{n+1} \approx v_h = \sum_{j=1}^n v_j \phi_j,$$

$$u^{n+\theta} \approx u_h = \sum_{j=1}^n u_j \phi_j.$$

Inserting these expressions into (3.69)–(3.70), we get finite element equations of the form

$$\sum_{j=1}^n v_j(\phi_j, \phi_i) + \theta \Delta t \sum_{j=1}^n v_j a_I(\phi_j, \phi_i) + \Delta t \sum_{j=1}^n u_j a_I(\phi_j, \phi_i)$$
$$= (v_\theta^n, \phi_i) - (1 - \theta) \Delta t a_I(v_\theta^n, \phi_i) \quad i = 1, \ldots, n, \quad (3.71)$$

$$\Delta t \sum_{j=1}^n v_j a_I(\phi_j, \phi_i) + \frac{\Delta t}{\theta} \sum_{j=1}^n u_j a_{I+E}(\phi_j, \phi_i)$$
$$= -\Delta t \frac{1 - \theta}{\theta} a_I(v_\theta^n, \phi_i) \quad i = 1, \ldots, n. \quad (3.72)$$

This is a linear system of equations with unknowns $v_j, u_j, j = 1, \ldots, n$. Collecting the unknowns into vectors v and u, the linear system may be written in matrix form:

$$\begin{bmatrix} A & B \\ B^T & C \end{bmatrix} \begin{bmatrix} v \\ u \end{bmatrix} = \begin{bmatrix} a \\ b \end{bmatrix} \quad (3.73)$$

The block matrices in this linear system arise from the terms in (3.71)–(3.72), and are given by

$$A_{ij} = (\phi_j, \phi_i) + \theta \Delta t a_I(\phi_j, \phi_i),$$
$$B_{ij} = \Delta t a_I(\phi_j, \phi_i),$$
$$C_{ij} = \frac{\Delta t}{\theta} a_{I+E}(\phi_j, \phi_i),$$
$$a_i = (v_\theta^n, \phi_i) - (1 - \theta) \Delta t a_I(v_\theta^n, \phi_i),$$
$$b_i = -\Delta t \frac{1 - \theta}{\theta} a_I(v_\theta^n, \phi_i).$$

We see that the result of applying the operator-splitting algorithm to the bidomain model is very similar to that which we observed above for the monodomain model. The original system of nonlinear PDEs and ODEs is reduced to systems of nonlinear ODEs and linear PDEs, which are solved sequentially to obtain an approximate solution. The difference from the monodomain model is that we have to solve a system of PDEs in Step 2 of the algorithm. Using a θ-rule for the time discretization and a Galerkin finite element method in space, we end up with a block structured linear system of the form (3.73) to be solved for each time step. As will be described in Chapter 4, the structure of the linear system may be utilized to construct efficient solvers for this system.

3.3 Coupling the Heart and the Torso

In the previous section we studied the monodomain model and the bidomain model for a heart surrounded by a nonconductive material. While this may be relevant for various experimental settings, we are more interested in the case where the heart is surrounded by a conducting body. In this case, we need to solve the full system (3.59)–(3.61), with boundary conditions (3.32)–(3.35), in Step 2 of the operator-splitting algorithm. To derive the weak form in this case we first introduce two function spaces for the solutions v, u_e, and u_o. As above, v and u_e will be found in a function space V, defined over the domain H. We now denote this function space by $V(H)$. The extracardiac potential field u_o is found in a similar function space defined over the torso domain T, which we denote by $V(T)$. Again we refer to Figure 2.1 in Chapter 2 for a schematic view of the two domains. To simplify the notation, we define an additional inner product:

$$a_T(\varphi, \phi) = \int_T M_o \nabla \varphi \cdot \nabla \phi dx, \text{ for } \varphi, \phi \in V(T).$$

Multiplying (3.59)–(3.60) with a test function $\psi \in V(H)$, and (3.61) with a test function $\eta \in V(T)$, and integrating over the respective domains, gives the weak formulations

$$(v_\theta^{n+1}, \psi) + \theta \Delta t a_I(v_\theta^{n+1}, \psi) + \Delta t a_I(u_e^{n+\theta}, \psi)$$

$$- \theta \Delta t \int_{\partial H} \psi[(M_i \nabla v_\theta^{n+1}) \cdot n]ds - \theta \int_{\partial H} \psi[(M_i \nabla u_e^{n+\theta}) \cdot n]ds$$

$$= (v_\theta^n, \psi) - (1 - \theta)\Delta t a_I(v_\theta^n, \psi)$$

$$+ (1 - \theta)\Delta t \int_{\partial H} \psi[(M_i \nabla v_\theta^n) \cdot n]ds \quad \text{for all } \psi \in V(H), \quad (3.74)$$

$$- \theta a_I(v_\theta^{n+1}, \psi) - a_{I+E}(u_e^{n+\theta}, \psi) + \theta \int_{\partial H} \psi[(M_i \nabla v_\theta^{n+1}) \cdot n]ds$$

$$+ (1 - \theta) \int_{\partial H} \psi[(M_i \nabla v_\theta^n) \cdot n]ds + \int_{\partial H} \psi[(M_i \nabla u_e^{n+\theta}) \cdot n]ds$$

$$+ \int_{\partial H} \psi[(M_e \nabla u_e^{n+\theta}) \cdot n]ds = 0 \quad \text{for all } \psi \in V(H), \quad (3.75)$$

$$- a_T(u_o^{n+1}, \eta) + \int_{\partial H} \phi[(M_o \nabla u_o^{n+\theta}) \cdot n_T]ds$$

$$+ \int_{\partial T} \eta[(M_o \nabla u_o^{n+\theta}) \cdot n_T]ds = 0 \quad \text{for all } \phi \in V(T). \quad (3.76)$$

Note that u_o is approximated at time points $t_n + \theta \Delta t$, indicated by the notation $u_o^{n+\theta}$. Approximating this field at the same time points as u_e is natural because of the boundary conditions along ∂H, which give a strong coupling between these two fields. The boundary condition (3.34) is identical to the case of an insulated heart, and the time-discrete version (3.68) causes the boundary integrals in (3.74), as well as the first two boundary integrals in (3.75), to vanish. Furthermore, the condition (3.35) forces the integral over the outer boundary in (3.76) to zero. The system is reduced to

$$(v_\theta^{n+1}, \psi) + \theta \Delta t a_I(v_\theta^{n+1}, \psi) + \Delta t a_I(u_e^{n+\theta}, \psi)$$
$$= (v_\theta^n, \psi) - (1 - \theta)\Delta t a_I(v_\theta^n, \psi), \quad \text{for all } \psi \in V(H), \quad (3.77)$$

$$- \theta a_I(v_\theta^{n+1}, \psi) - a_{I+E}(u_e^{n+\theta}, \psi)$$
$$+ \int_{\partial H} \psi[(M_e \nabla u_e^{n+\theta}) \cdot n]ds = 0 \quad \text{for all } \psi \in V(H), \quad (3.78)$$

$$-a_T(u_o^{n+\theta}, \eta) + \int_{\partial H} \eta[(M_o \nabla u_o^{n+\theta}) \cdot n_T]ds = 0 \quad \text{for all } \eta \in V(T). \quad (3.79)$$

We can use the continuity condition (3.32) to construct a new field u, defined over the complete domain $H \cup T$,

$$u^{n+\theta} = \begin{cases} u_e^{n+\theta} & \text{for } \mathbf{x} \in H \\ u_o^{n+\theta} & \text{for } \mathbf{x} \in T \end{cases}.$$

We also define a function space $V(H \cup T)$, where this new field belongs. Because all functions $u \in V(H \cup T)$ must be continuous over ∂H, it is natural to define the space $V(H \cup T)$ to be the set of functions defined over $H \cup T$ that belong to both $V(H)$ and $V(T)$, and are continuous over ∂H. We also assume that all functions in $V(H)$ and $V(T)$ can be obtained as restrictions of functions in $V(H \cup T)$. Following this definition, the test functions ψ and η used in (3.78) and (3.79) may be replaced by a single test function $\varphi \in V(H \cup T)$. The two weakly formulated equations can now be written

$$
- \theta a_I(v_\theta^{n+1}, \varphi) - a_{I+E}(u_e^{n+\theta}, \varphi)
$$
$$
+ \int_{\partial H} \varphi[(M_e \nabla u_e^{n+\theta}) \cdot n] ds = 0 \quad \text{for all } \varphi \in V(H \cup T),
$$

$$
- a_T(u_o^{n+\theta}, \varphi) - \int_{\partial H} \varphi[(M_o \nabla u_o^{n+\theta}) \cdot n] ds = 0 \quad \text{for all } \varphi \in V(H \cup T).
$$

Note here that the negative sign in front of the boundary integral in the second equation occurs because n is a unit normal vector pointing *into* the domain T, instead of the outward normal vector that is normally used. The definition of the space $V(H \cup T)$, and the requirement that all functions in $V(H)$ and $V(T)$, can be seen as restrictions of functions in $V(H \cup T)$, ensures that these equations are equivalent to (3.78)–(3.79). If we add these two equations, we see that the boundary integrals cancel. If we also replace the fields u_e and u_o with the combined field u, we get

$$
- \theta a_I(v_\theta^{n+1}, \varphi) - a_{I+E}(u^{n+\theta}, \varphi) - a_T(u^{n+\theta}, \varphi) = 0. \tag{3.80}
$$

Again, the definition of the function space $V(H \cup T)$, and the fact that the inner products $a_{I+E}(\cdot, \cdot)$ and $a_T(\cdot, \cdot)$ are defined over nonoverlapping domains, ensures that we can add the equation without any loss of information.

With (3.78) and (3.79) replaced by (3.80), the complete weak formulation can be written as:

Find $v^{n+1} \in V(H)$ and $u^{n+\theta} \in V(H \cup T)$ such that

$$
(v_\theta^{n+1}, \psi) + \theta \Delta t a_I(v_\theta^{n+1}, \psi) + \Delta t a_I(u^{n+\theta}, \psi)
$$
$$
= (v_\theta^n, \psi) - (1 - \theta) \Delta t a_I(v^n, \psi) \quad \text{for all } \psi \in V(H), \tag{3.81}
$$
$$
\Delta t a_I(v^{n+1}, \varphi) + \frac{\Delta t}{\theta} a_{I+E}(u^{n+\theta}, \varphi) + \frac{\Delta t}{\theta} a_T(u^{n+\theta}, \varphi)
$$
$$
= - \frac{1 - \theta}{\theta} a_I(v_\theta^n, \varphi) \quad \text{for all } \varphi \in V(H \cup T). \tag{3.82}
$$

As above, we discretize the weak formulation by defining finite element grids for the domains H and $H \cup T$, and use these grids to define discrete subspaces $V_h(H)$ and $V_h(H \cup T)$. Since the domain $H \cup T$ is an extension of the H domain, the discrete space $V_h(H \cup T)$ will normally have more basis functions than the space $V_h(H)$. As usual, the two solutions are approximated by the sums

$$v_\theta^{n+1} \approx v_h = \sum_{j=1}^{n} v_j \phi_j,$$

$$u^{n+\theta} \approx u_h = \sum_{j=1}^{m} u_j \phi_j,$$

where n and m are the number of basis functions in $V(H)$ and $V(H \cup T)$, respectively. We will normally have $m > n$. Inserting these approximations into the weak form (3.81)–(3.82), and using the basis functions as test functions, we obtain a system of linear equations that can be written in matrix-vector form:

$$\begin{bmatrix} A & B \\ B^T & C \end{bmatrix} \begin{bmatrix} v \\ u \end{bmatrix} = \begin{bmatrix} a \\ b \end{bmatrix}. \tag{3.83}$$

We see that the structure of this system is identical to the one we obtained for the insulated heart, but the definition of the blocks is slightly different:

$$A_{ij} = (\phi_j, \phi_i) + \theta \Delta t a_I(\phi_j, \phi_i), \qquad i, j = 1, \ldots n,$$
$$B_{ij} = \Delta t a_I(\phi_j, \phi_i), \qquad i = 1, \ldots n, j = 1, \ldots, m,$$
$$C_{ij} = \frac{\Delta t}{\theta} a_{I+E}(\phi_j, \phi_i) + \frac{\Delta t}{\theta} a_T(u^{n+\theta}, \phi), i, j = 1, \ldots m,$$
$$a_i = (v_\theta^n, \phi_i) - (1-\theta) \Delta t a_I(v^n, \phi_i), \qquad i = 1, \ldots, n,$$
$$b_i = -\Delta t \frac{1-\theta}{\theta} a_I(v_\theta^n, \phi_i), \qquad i = 1, \ldots, m.$$

Comparing with the block matrices resulting from the insulated heart model, we see that there is an additional term in block C, involving an integral over the torso domain T. Another important difference is the fact noted above that in general $n \neq m$, which implies that the dimensions of the matrices are not equal. Specifically, blocks A and C will be square matrices with different sizes, i.e.,

$$A \in \mathbb{R}^{n,n},$$
$$C \in \mathbb{R}^{m,m},$$

and B will be a rectangular matrix,

$$B \in \mathbb{R}^{n,m}.$$

However, in spite of these differences the method for solving this system will be the same as that used for the system that results from the insulated heart model. Algorithms for solving such systems of linear equations will be presented in Chapter 4.

3.4 Numerical Experiments

3.4.1 Convergence Tests

For the simple test cases in Section 3.1, we found the convergence rate by comparing our numerical results to the analytical solution. For the bidomain model we do not know the analytical solution, and this complicates the process of determining convergence rates for the numerical method.

When analytical solutions are not available, an alternative approach for computing convergence rates is to use a numerical solution as the reference. For the error estimates to be reliable it is, of course, essential that this numerical solution is very accurate, which places strict demands on the spatial and temporal resolution. For problems in two and three spatial dimensions, the resolution requirements may easily lead to a problem that is almost impossible to solve, because of the huge number of unknowns.

A very accurate numerical solution may be computed if we restrict ourselves to the consideration of problems with circular or spherical geometries. For a suitable choice of initial conditions and conductivity values the solution will, in this case, always be symmetric, with the spatial variations depending only on the distance r from the centre of the domain. Therefore, if formulated in polar coordinates, the equations may be solved as a one-dimensional problem, which allows a very high spatial resolution. The 1D solution may then be converted back to the two- or three-dimensional case, and used as a reference solution for checking numerical errors and convergence rates.

A potential problem with using a numerical solution as reference, regardless of whether a radial formulation or a more general numerical solution is used, is that the results may converge towards the wrong solution. It is convenient to use the same software for computing the reference solution that we use for the normal solutions, by simply increasing the spatial and temporal resolutions. Then, if there is something wrong in the implementation, the convergence tests may very well give good results, but the convergence will be towards the wrong solution. However, for complicated problems such as the present one, there are few options other than to use a numerical solution as the reference. Still, the problem needs to be considered when the tests are performed, so that care is taken to avoid this type of error.

In the two-dimensional case, the relations between Cartesian and polar coordinates are given by

$$x = r \cos \phi, \quad y = r \sin \phi,$$

with the inverse relations

$$r = \sqrt{x^2 + y^2}, \quad \phi = \arctan\left(\frac{y}{x}\right). \tag{3.84}$$

From this, we get

$$\frac{\partial u}{\partial x} = \frac{\partial u}{\partial r}\frac{\partial r}{\partial x} + \frac{\partial u}{\partial \phi}\frac{\partial \phi}{\partial x},$$

and since we have assumed circular symmetry, $\partial u/\partial \phi = 0$. From (3.84) we have $\partial r/\partial x = \cos \phi$, and we get

$$\frac{\partial u}{\partial x} = \frac{\partial u}{\partial r}\frac{\partial r}{\partial x} = (\cos \phi)\frac{\partial u}{\partial r}.$$

We differentiate this expression once more to obtain

$$\frac{\partial^2 u}{\partial x^2} = \frac{\partial}{\partial x}\left((\cos \phi)\frac{\partial u}{\partial r}\right) = \frac{\partial(\cos \phi)}{\partial x}\frac{\partial u}{\partial r} + (\cos \phi)\frac{\partial}{\partial x}\frac{\partial u}{\partial r},$$

and using the chain rule gives

$$\frac{\partial^2 u}{\partial x^2} = \frac{\partial(\cos\phi)}{\partial\phi}\frac{\partial\phi}{\partial x}\frac{\partial u}{\partial r} + (\cos\phi)\frac{\partial}{\partial r}\frac{\partial u}{\partial r}\frac{\partial r}{\partial x}.$$

Inserting

$$\frac{\partial\phi}{\partial x} = -\frac{(\sin\phi)}{r},$$

which we get from (3.84), gives

$$\frac{\partial^2 u}{\partial x^2} = \frac{\sin^2\phi}{r}\frac{\partial u}{\partial r} + (\cos^2\phi)\frac{\partial^2 u}{\partial r^2}.$$

Similar calculations give

$$\frac{\partial u}{\partial y} = (\sin\phi)\frac{\partial u}{\partial r},$$

and

$$\frac{\partial^2 u}{\partial y^2} = \frac{\cos^2\phi}{r}\frac{\partial u}{\partial r} + (\sin^2\phi)\frac{\partial^2 u}{\partial r^2}.$$

A radial formulation of the monodomain model. For simplicity, we assume constant, scalar conductivities. This does not lead to a serious loss of generality, since we already know that the conductivity must be constant in the ϕ direction to get a symmetric solution. In addition, it can be shown for this circular symmetric case that the only significant value in the conductivity tensor is the conductivity in the r direction. Denoting the scalar intracellular conductivity by m_i, the monodomain model can be written as

$$\frac{\partial v}{\partial t} + I_{ion} = \frac{m_i\lambda}{1+\lambda}\left(\frac{\partial^2 v}{\partial x^2} + \frac{\partial^2 v}{\partial y^2}\right).$$

Inserting the expressions above for the spatial derivatives, we get

$$\frac{\partial v}{\partial t} + I_{ion} = \frac{m_i\lambda}{1+\lambda}\left((\cos^2\phi)\frac{\partial^2 v}{\partial r^2} + \frac{\sin^2\phi}{r}\frac{\partial v}{\partial r} + (\sin^2\phi)\frac{\partial^2 v}{\partial r^2} + \frac{\cos^2\phi}{r}\frac{\partial v}{\partial r}\right),$$

which reduces to

$$\frac{\partial v}{\partial t} + I_{ion} = \frac{m_i\lambda}{1+\lambda}\left(\frac{\partial^2 v}{\partial r^2} + \frac{1}{r}\frac{\partial v}{\partial r}\right).$$

After multiplication with r, this equation can be written

$$r\frac{\partial v}{\partial t} + rI_{ion} = \frac{m_i\lambda}{1+\lambda}\frac{\partial}{\partial r}\left(r\frac{\partial v}{\partial r}\right). \qquad (3.85)$$

This equation can be discretized with the technique used for the original monodomain model. An operator splitting technique is used to separate the nonlinear term from the PDEs, resulting in a three-step algorithm as presented in Section

3.2.2. The nonlinear ODE system in Steps 1 and 3 is identical to the one for the original model. The linear PDE, now defined over a one-dimensional domain, is slightly different from the Cartesian formulation, but the steps in the discretization of the equation are still exactly the same. We use the θ-rule in time, and introduce a 1D finite element space for the solution. Deriving a weak form and following the steps described above then gives a system of linear equations:

$$Av = b.$$

The unknown vector v contains the nodal values $v_j, j = 1, \ldots, n$, and A and b are defined as

$$A_{ij} = \int_H r\phi_i\phi_j dx + \theta\gamma \int_H rm_i\nabla\phi_i \cdot \nabla\phi_j,$$

$$b_i = \int_H rv^n\phi_i dx - (1-\theta)\gamma \int_H rm_i\nabla v^n \cdot \nabla\phi_i dx.$$

The left panel of Figure 3.8 shows a circular grid used for convergence tests with the monodomain model. The right panel shows the transmembrane potential at $t = 40$ ms, computed with this fairly coarse grid. The dimension of the grid is given in centimeters, which gives a node spacing of approximately 2 mm, and the size of the time step is 1 ms. The FitzHugh-Nagumo model is used for the cellular reactions, and the initial condition is

$$v = \begin{cases} -20\,\text{mV for } r \leq 0.2\,\text{cm}, \\ -80\,\text{mV for } r > 0.2\,\text{cm}. \end{cases}$$

This initial condition produces a circular excitation wave that travels from the centre of the circle towards the boundary. We see that the contours shown in the right panel

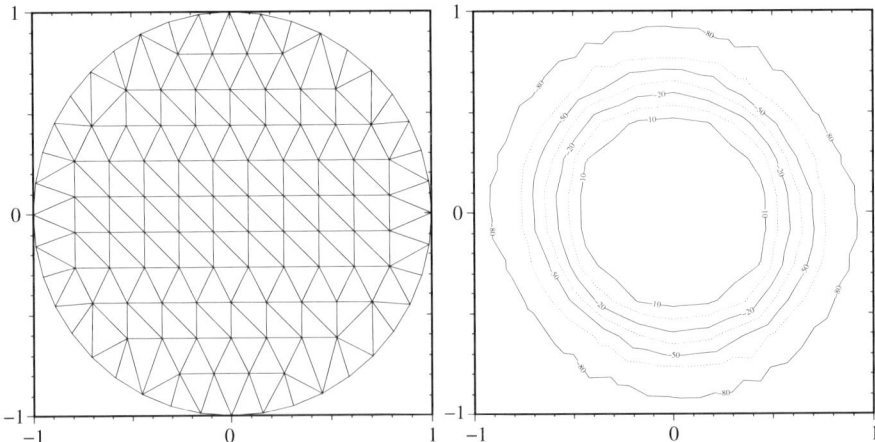

Fig. 3.8. The left panel shows the circular grid used in convergence tests for the monodomain model, while the right panel is a coarse numerical solution at $t = 40$ ms.

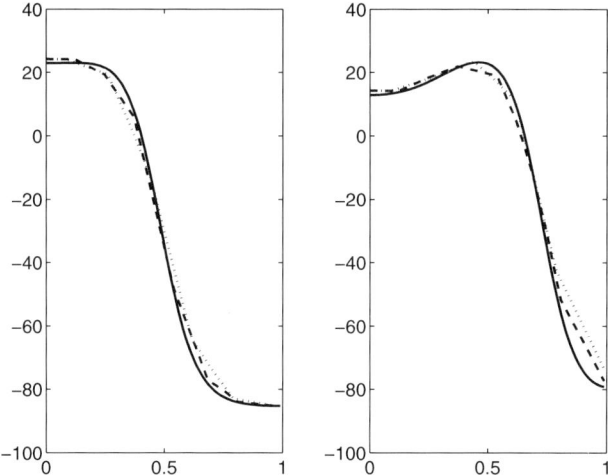

Fig. 3.9. The solution of the monodomain model for a circular domain. A 2D solution plotted along two different lines is compared to a reference solution computed with the radial formulation of the equations. The left panel shows solutions for $t = 20$ ms, and the right panel at $t = 50$ ms.

of the figure are not completely circular, which indicates that the grid used for this computation is too coarse to produce satisfactory results.

Figure 3.9 shows solutions from the coarse circular grid compared to a reference solution computed with the radial formulation of the equations. The solid lines show the reference solution at $t = 20$ ms and $t = 50$ ms, the dotted curves show the 2D solutions plotted along the line from (0,0) to (1,0), while the dashed curves are plotted along the line from (0,0) to ($\sqrt{2}, \sqrt{2}$). We see that there is a significant difference between the 2D solutions and the reference solution. Because the 2D solution is not completely circular, there is also a visible difference between the curves plotted along the two lines.

Convergence results for the monodomain model coupled to the FitzHugh-Nagumo equations are shown in Table 3.2. The table shows the L_2 norm of the error, which is computed by comparing circular 2D solutions to the reference solution obtained using the radial formulation of the model. The reference solution is computed with $\theta = 1/2$, $\Delta t = 0.001$ ms and $h = 0.00625$ mm. The errors are computed at $t = 10$ ms, and results are presented for both Strang splitting ($\theta = 1/2$) and Godunov splitting ($\theta = 1$). We see that the solutions converge to the reference solution with the expected convergence rates. For Godunov splitting, the convergence is approximately first-order, and close to second-order convergence rates are observed for Strang splitting.

A radial formulation of the bidomain equations. Although more complicated, formulating the bidomain equations in polar coordinates involves exactly the same steps as the transformation of the monodomain model. Again, we assume scalar,

Table 3.2. Convergence results for the monodomain model, with the FitzHugh-Nagumo cell model. The columns marked $e_{\theta=1}$ and $\alpha_{\theta=1}$ show the L_2 error and the estimated convergence rate for Godunov splitting, while the columns marked $e_{\theta=1/2}$ and $\alpha_{\theta=1/2}$ show similar results for Strang splitting. The unit for the time step size is milliseconds, while the spatial resolution h is given in millimeters.

Δt	h	$e_{\theta=1}$	$\alpha_{\theta=1}$	$e_{\theta=1/2}$	$\alpha_{\theta=1/2}$
0.25	0.5	0.18979	-	0.160208	-
0.125	0.25	0.08159	1.217	0.04121	1.958
0.0625	0.125	0.04066	1.004	0.01044	1.980
0.03125	0.0625	0.02085	0.963	0.00261	1.999

constant conductivities, and each term involving spatial second derivatives is handled exactly as for the monodomain model. We end up with the system

$$\frac{\partial}{\partial r}\left(m_i r \frac{\partial v}{\partial r}\right) + \frac{\partial}{\partial r}\left(m_i r \frac{\partial u_e}{\partial r}\right) = r\frac{\partial v}{\partial t} + r I_{ion} \quad r \in H, \qquad (3.86)$$

$$\frac{\partial}{\partial r}\left(m_i r \frac{\partial v}{\partial r}\right) + \frac{\partial}{\partial r}\left((m_i + m_e) r \frac{\partial u_e}{\partial r}\right) = 0 \qquad\qquad r \in H, \qquad (3.87)$$

$$\frac{\partial}{\partial r}\left(m_T r \frac{\partial u_o}{\partial r}\right) = 0 \qquad\qquad r \in T, \qquad (3.88)$$

with boundary conditions

$$\frac{\partial v}{\partial r} + \frac{\partial u_e}{\partial r} = 0 \qquad\qquad r \in \partial H, \qquad\qquad (3.89)$$

$$u_e = u_o \qquad\qquad r \in \partial H, \qquad\qquad (3.90)$$

$$\frac{\partial}{\partial r}(m_e u_e) = \frac{\partial}{\partial r}(m_T u_o) \quad r \in \partial H, \qquad\qquad (3.91)$$

$$\frac{\partial}{\partial r}(m_T u_o) = 0 \qquad\qquad r \in \partial T. \qquad\qquad (3.92)$$

These equations can be discretized and solved in exactly the same way as for the standard, Cartesian formulation of the bidomain equations. The result is a block-structured linear system of the form (3.83). With a redefinition of the inner products defined above, to

$$(v, \phi) = \int_H r v \phi \, dx,$$

$$a_I(v, \phi) = \int_H r m_i \nabla v \cdot \nabla \phi \, dx,$$

$$a_{I+E}(v, \phi) = \int_H r(m_i + m_e) \nabla v \cdot \nabla \phi \, dx,$$

$$a_T(v, \phi) = \int_H r m_o \nabla v \cdot \nabla \phi \, dx,$$

the definition of the matrix blocks is also the same as for the system (3.83).

Table 3.3. Convergence results for the bidomain model, with the cell model by Winslow et al. [145]. The columns marked $e_{\theta=1}$ and $\alpha_{\theta=1}$ show the L_2 error and the estimated convergence rate for Godunov splitting, while the columns marked $e_{\theta=1/2}$ and $\alpha_{\theta=1/2}$ show similar results for Strang splitting.

Δt	h	$e_{\theta=1}$	$\alpha_{\theta=1}$	$e_{\theta=1/2}$	$\alpha_{\theta=1/2}$
0.25	0.5	10.9339	-	7.65882	-
0.125	0.25	4.69081	1.2209	2.7012	1.5035
0.0625	0.125	2.79365	0.7477	2.16758	0.3175
0.03125	0.0625	1.84566	0.5980	0.745513	1.5398
0.015625	0.03125	1.09955	0.7472	0.196549	1.9233

The radial formulation of the bidomain model has been used to compute a reference solution, on a circular domain where the heart H occupies the region $0 < r \leq 1.0$ and the surrounding torso T is $1.0 < r \leq 1.4$. The reference solution was computed on a grid with a total of 2241 nodes, which gives a nodal distance h of 0.00625 mm. The time step used is $\Delta t = 0.001$ ms. It is clear that this resolution level would be difficult to achieve on a two- or three-dimensional domain, because the number of unknowns would become too large to handle. Table 3.3 shows convergence results for the bidomain model, using the cell model by Winslow et al. [145]. Only the errors in the transmembrane potential v are reported, and the error is measured at $t = 3$ ms. As above, the error is measured in the L_2 norm. We see that the results are similar to those for the monodomain model, although the convergence results are slightly less favourable. For the choice $\theta = 1$, which combines a Godunov splitting with the backward Euler method, we see less than first-order accuracy, while the algorithm that combines Strang splitting with the Crank-Nicolson discretization gives an accuracy close to order two on the finer grids. The more variable results seen in this computation compared to the previous one may be caused by the very rapid dynamics of the applied cell model. Compared to the FitzHugh-Nagumo model, the model by Winslow et al. has a very rapid upstroke, which leads to sharp gradients in the solution. This places stricter demands on the spatial resolution required to observe consistent convergence results.

The limitations of the present approach, where a numerical solution is used as the reference for convergence estimates, were discussed above. Although the convergence of the algorithms is confirmed, our confidence in the results depends on our confidence in the correctness of the numerical reference solution. Furthermore, the tests are performed on highly idealized geometries, and we can only assume that the demonstrated convergence properties hold for more realistic experiments. Still, for a complex model like the bidomain model, it is difficult to come up with good alternatives to this type of convergence experiment. Analytical solutions can only be obtained after substantial simplifications of the equation, and the value of using these solutions for convergence tests is, therefore, also very limited. Although the results should be used with some care, the convergence tests presented here are, therefore, valuable for examining the accuracy of the complex numerical schemes that are employed.

The formulation of the equations in polar coordinates can also be performed in 3D, using the spherical coordinates (r, θ, φ). By making the same assumptions as for the 2D case, i.e. spherical symmetry and scalar, constant conductivities, we obtain systems of equations very similar to those presented above. The equations can be solved as 1D problems depending only on the radius r, and used as a reference for checking the convergence of 3D solutions. However, the convergence tests in 3D are not as easy to perform as in the 2D case. To get a picture of the convergence the experiments must be performed for a number of different grid refinement levels, and in 3D this will quickly lead to a very large number of nodes, which easily exceeds the capacity of the available computers. Therefore, until computer speed and memory have increased sufficiently, it is easier to base convergence tests on 2D experiments.

3.4.2 Simulation on a 2D Slice

The results shown in previous sections were only suitable for showing the convergence properties of the numerical methods. The practical value of simulations with perfect circular symmetry is, of course, very limited. To illustrate how the numerical simulations can be used for practical purposes, we will now present a different set of experiments. The geometry used for the experiments is a two-dimensional slice through the heart and torso. The advantages of doing 2D studies, rather than 3D, are that the computational burden is much smaller and it is easier to visualize the results. The disadvantage is that the results may be unrealistic in many respects; particularly, in our case, regarding the propagation pattern. Forcing the signal propagation to occur in only one plane will not give a realistic activation pattern. However, 2D simulations can still provide insight into several important mechanisms of the heart. The goal of the present study is to investigate the relation between ischemic heart disease and shifts in the ST segment of the ECG. ST segment deflection is an important indicator for ischemic heart disease, and so the relation is important from a clinical point of view.

Geometry and data. The geometry used for the simulations is shown in Figure 3.10. The torso boundary is based upon a cross section from the Visible Human dataset [141]. This dataset consists of photographs of horizontal cross sections of a male torso body. The chosen section cuts through the middle of the ventricles. The Visible Human dataset was not used for the heart boundary, because the heart cavities are not so clear in the photographs. The boundary was traced from a different set of photographs [84].

The tissue is stimulated at three different locations, as indicated in Figure 3.10 (a). The locations represent the endings of fibres in the Purkinje network. The left ventricular sites are triggered 20 ms before the right ventricular site, the difference in timing being based on the activation map of Dürrer [33].

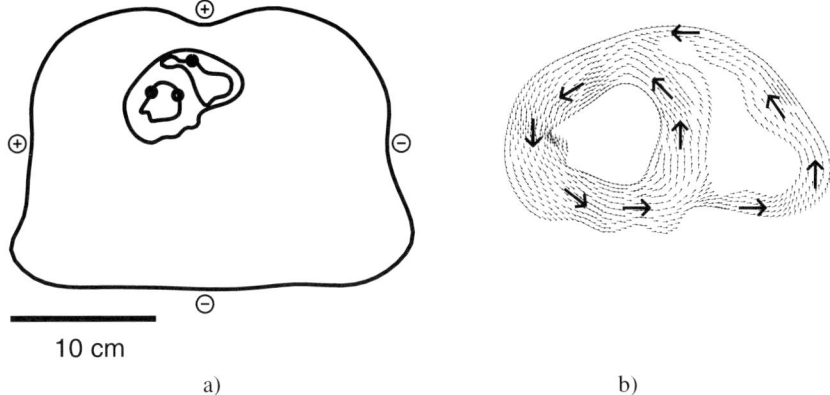

10 cm

a)   b)

Fig. 3.10. a) The heart and torso boundaries. The black dots on the endocardial surface indicate stimulation points. The plus and minus poles on the torso surface show the location where the surface potential is recorded. **b)** The fibre directions in the heart. The large arrows are the selected directions and the small arrows are interpolations.

The directions of the fibres are shown in Fig. 3.10b). They run tangentially to the heart surfaces. The directions are not based directly on measurements, but are selected to give a reasonable representation of the orientation in the two-dimensional slice. The out-of-plane component of the fibres is not taken into account.

3.4.3 Normal Propagation

For the simulation under consideration, there were 61167 nodes in the heart and a total of 292417 in the body. The average node density was about 0.2 mm in the heart and 0.4 mm in the body. The time step was 0.125 ms and the simulation time was 500 ms, giving a total of 4000 steps. The computational loads of the ODE and the PDE steps of the operator-splitting algorithm were approximately equal.

Figure 3.11 shows four snapshots of the simulation. After 25 ms, we see that the depolarization front has spread out from the two stimulation points on the left endocardial surface and we also see the early contribution from the third stimulation point. Notice from the position of the wave front that the propagation velocity is larger in the direction along the fibres than across them. This reflects the larger conductivity in the direction of the fibres. After 75ms, most of the tissue has been depolarized. The third image shows the situation after 220 ms. The whole heart has now depolarized and we see the beginnings of repolarization in some areas. These areas coincide with the initial depolarization. The reason that the repolarization follows the same pattern as the depolarization is that the action potential duration is the same in all locations This is not realistic, since it has been reported that the endocardial cells, and in particular the cells near the middle of the heart wall, have longer action potential durations than epicardial cells. The effect of this is that although the depolarization of the tissue starts at the endocardium, the repolarization

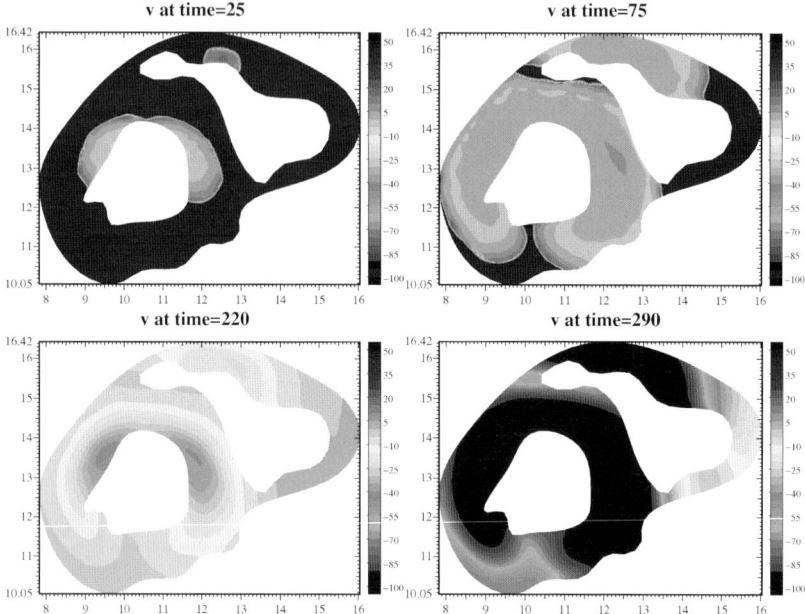

Fig. 3.11. The transmembrane potential (mV) at four stages during normal propagation. (For the color version, see Figure A.6 on page 290).

wave starts at the epicardium. In our model we use identical cells throughout the myocardium, which results in both the depolarization wave and the repolarization wave starting from the endocardium. In the last snapshot the repolarization is further advanced. Notice that the repolarization front is less steep than the depolarization front. This reflects the shape of the action potential, which has a fast upstroke and a much slower repolarization.

3.4.4   Ischemia

In this section we look at the effect of changing the ionic model in parts of the tissue. Figure 3.12 shows the area where the model has been modified. The dark area represents ischemic tissue, and the ionic model is modified to reflect this. Figure 3.13 shows the normal action potential used in the healthy part of the tissue, along with the modified action potential used in the ischemic region.

Four changes have been made to the model, to reflect the changes occurring in the cells during ischemia. The extracellular potassium concentration has been increased from 4 mM to 9 mM; the maximum conductance of the fast sodium channels is reduced by 25%; the maximum conductances for I_{to}, I_{Kr}, and I_{Ks} have been reduced by 50%; and conductance of the time independent potassium current I_{K_1} has been reduced by 60%. These changes may not correspond exactly to the changes taking place during ischemia, but the resulting action potential is reasonable, which

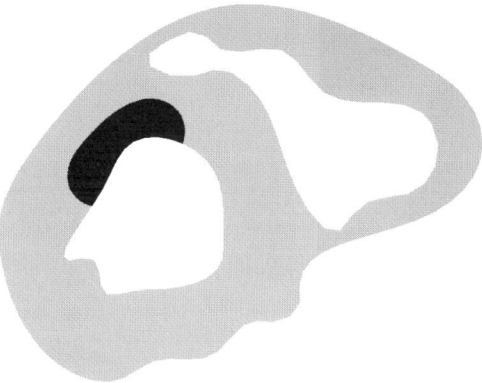

Fig. 3.12. In the dark area the ODE model is modified to represent ischemic tissue.

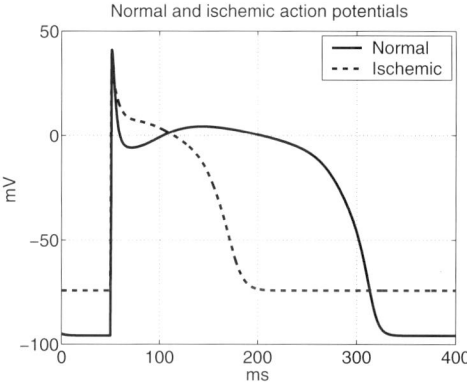

Fig. 3.13. The normal and modified action potential of the Winslow model.

is sufficient for our present purpose. The resting potential is elevated, the action potential duration is reduced and there is no notch after the upstroke. The upstroke velocity is also reduced, although that is not visible on the figure.

The simulation from Section 3.4.3 was repeated with the modified model. Figure 3.14 shows four snapshots of the transmembrane potential. After 40 ms we see that the propagation has spread out through a large part of the myocardium. Notice that the potential around the stimulation site located in the ischemic area is larger than the potential around the second stimulation point in the left ventricle. The reason is that the action potentials for the ischemic cells have a higher value during the early part of the plateau phase; see Figure 3.13. The scale in this panel has been magnified to make this easier to see. The second snapshot shows that when the complete heart has been depolarized, the ischemic heart, in contrast to a normal heart, is not iso-electric. The time interval when the whole muscle is depolarized corresponds to the ST segment in the ECG, and this difference is one of the causes

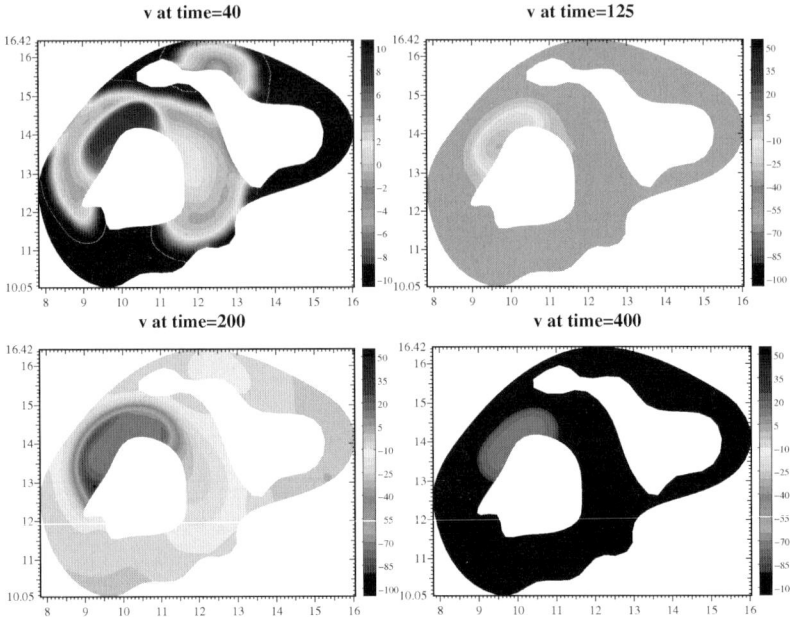

Fig. 3.14. Four stages during ischemic propagation. (For the color version, see Figure A.7 on page 290).

for the ST segment shifts observed during ischemic heart disease. The third image illustrates the effect of the difference in action potential duration between the healthy and ischemic area; the repolarization is completed in the ischemic area but has barely started in the other parts of the tissue. The last snapshot shows that the heart is not iso-electric in the TP interval either; the ischemic cells have a higher resting potential. This effect also contributes to the ECG changes during ischemia.

It is interesting to look at the extracellular potential during the ST phase (e.g. at 125 ms) and the TP phase (e.g. at 400 ms). Figure 3.15 shows a magnified image of part of the ischemic area. We see that a gradient in the transmembrane potential also gives a gradient in the extracellular potential. This is in contrast to the normal, nonischemic case, in which the potential in the area will be comparatively flat both in the ST and TP phases. There is hence a local difference in the extracellular potential between the healthy and the ischemic case, and this difference will generate changes in body surface potential.

Note in Figure 3.15 that the polarity during the ST segment is opposite to that during the TP segment. This is the underlying mechanism for the ST changes seen during ischemia. What is typically observed in patients suffering from ischemia is either an elevation or a depression in the ST segment. These changes are partly due to the gradient observed in Figure 3.15a), and partly due to the opposite polarity of the field in Figure 3.15b). Whether we see an elevation or a depression depends on the location of the electrodes and of the ischemic area. If we have an ST elevation the

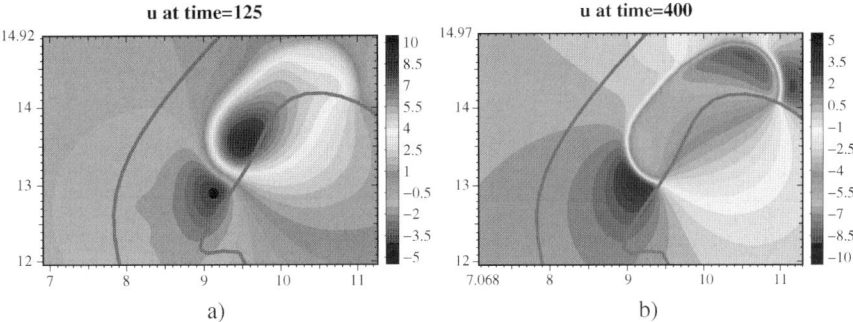

Fig. 3.15. The extracellular potential around the ischemic area during **a)** the ST segment **b)** the TP segment. The heart boundary is indicated by the solid line. (For the color version, see Figure A.8 on page 291).

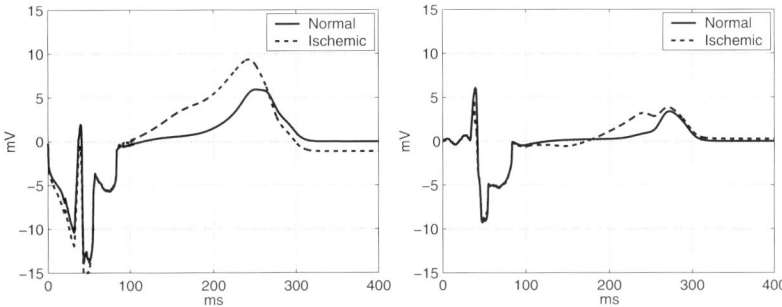

Fig. 3.16. The surface potential recorded at the two pairs indicated in Figure 3.10. The left panel shows the front-to-back electrodes and the right panel shows the left-to-right electrodes.

TP segment will be depressed, and vice versa. In the normal ECG, the TP segment is normally taken as the baseline, and consequently both contributions will add up to an apparent ST change only. If one is only interested in the ST segment shift, this can also be studied with a simpler stationary version of the bidomain model, as described in [92].

Figure 3.16 shows the simulated ECGs, corresponding to the electrode positions shown in Figure 3.10. For both electrode pairs we see changes between the normal and the ischemic conditions. There is relatively little change during the depolarization phase (the first 100 ms), but obvious changes in the ST segment. The discrepancies do not have the same magnitudes in the different directions. This is as expected, because the orientation of the potential gradients will influence where the signal is strongest. It is well known that pathological condition can be visible in some leads, but not in others. Indeed, this is what makes it possible to determine where an infarction or ischemic region is located.

Chapter 4

Solving Linear Systems

The physical relevance of computations based on the model problems arising from the electrical activity in the heart depends on high accuracy of the solution. High accuracy requires the solution of large linear or nonlinear systems of ODEs and PDEs. This chapter deals with solution algorithms for the discretization of (linear) PDEs, which is a huge research field around the world. Much of the research in this field has been centred around simple model problems such as the Poisson problem, where a solid theoretical framework has been developed. We will briefly review this theory in the simplest possible manner. Then, at the end of the chapter, we explain how the powerful concept of (block) preconditioning extends these algorithms to systems of PDEs that arise from the discretization of the Bidomain model.

4.1 Overview

As discussed in the previous chapter, the computation of the electrical activity in the heart and body requires the solution of large linear systems. With today's extremely fast computers, when even desktop computers are capable of performing more than 10^9 floating point operations per second, this might not seem to be a problem; but it is. In our quest for more and more accurate solutions to models of physical problems, we tend to solve larger and larger linear systems. This trend is unavoidable, because the accuracy of the solution is proportional to some power of the number of unknowns.

Let us, for the moment, consider a linear system arising from a finite element discretization of a partial differential equation on the form

$$-\nabla^2 u = f, \tag{4.1}$$

equipped with suitable boundary conditions. The linear system then becomes

$$Au_h = b, \tag{4.2}$$

where A typically is large and very sparse. We have used the subscript h to emphasize that u_h is an approximation of u and that the accuracy depends on h, which is the typical distance between the vertices in the grid. Because u_h approaches u as h approaches zero, we want to be able to choose as small a value of h as possible. The smallest possible h is dictated by the largest number of unknowns that can be handled by the computer. On a modern desktop computer we can solve (4.2) with $N \sim 10^7$ unknowns, provided that we use an order-optimal solution algorithm.

However, we may want even larger values of N and, as we have already seen, we want to solve systems of partial differential equations leading to even larger linear systems. In addition, the linear systems arising from the Bidomain equations are more challenging to solve than the system (4.2) generated by the Poisson equation (4.1).

The matrices that arise from low-order discretization of PDEs are typically very sparse, that is, relatively few entries are non-zeros. In fact, the number of non-zeros in each row is usually a fixed small number, e.g., between five and ten, and the matrices therefore require $\mathcal{O}(N)$ floating point numbers of storage. For such matrices, it is essential that only the nonzero entries are stored. The matrix may be banded or completely unstructured, depending on the structure of the underlying grid. For banded matrices it is possible to develop an unpivoted factorization procedure using $\mathcal{O}(Nb^2)$ floating point operations, where b is the bandwidth of the system. However, the bandwidth is usually at least $\mathcal{O}(N^{1/2})$, which entails that the total solution algorithm will require $\mathcal{O}(N^2)$ floating point operations. The storage requirement is usually somewhat smaller, $\mathcal{O}(N^{3/2})$ floating points. The case is even worse for unstructured matrices.

Improvements in computing power has, for the last 30 years, been predicted remarkably well by Moore's law, which states that the number of transistors per area will double every 18 months[1]. A consequence of this is that the speed of the CPU and the amount of RAM double in the same period. In fact, CPU speed has improved roughly by a factor 10^5 over the last 30 years. In recent years, the amount of RAM has increased at a similar rate. Assuming that Moore's law will continue to hold for the next 10 years, we can expect computers with CPUs that are 50 times faster, and with 50 times more RAM. Still, it may very well be that the accuracy is questionable and that we need as many unknowns as possible. Hence, for simplicity we assume that it is possible to generate the matrix with $50\,N$ unknowns. It would then not be possible to solve the system, because the banded Gaussian elimination is $\mathcal{O}(N^{3/2})$) in storage, which in this concrete case means $\approx 350\,N$. However, the situation is even worse for the number of floating point operations, which is $\mathcal{O}(N^2)$, or $2500\,N$, in this concrete case. Because the CPU is able to perform $50\,N$ operations per second, we would then need to wait 50 times longer for the answer than today. This is not acceptable and we must, therefore, search for faster methods that require less storage. More precisely, we shall seek order–optimal methods that require $\mathcal{O}(N)$ operations and $\mathcal{O}(N)$ memory allocations for a system with N unknowns.

4.2   Iterative Methods

As discussed above, direct methods have two main problems:

– the required storage is $\mathcal{O}(N^\alpha)$ floating point numbers, and
– the required floating points operations are $\mathcal{O}(N^\beta)$,

---

[1] Moore never formulated the law clearly, but this is commonly known as Moore's law. The doubling period may vary.

Table 4.1. The CPU time (in seconds) required to solve a Poisson problem in 2D, using different types of solution algorithms, with respect to the number of unknowns. For the iterative algorithms, the stopping criterion requires that the discrete L_2-norm of the residual is reduced by a factor of 10^8. The measurements were obtained on an Itanium2 1.3 GHz processor.

Unknowns	Gauss Elim.	CG	CG/MILU	MG	CG/MG
65^2	0.29	0.12	0.04	0.04	0.04
129^2	4.30	1.07	0.28	0.15	0.18
257^2	68.49	12.30	2.77	0.64	0.92
513^2	-	123.06	18.65	2.89	4.08
1025^2	-	969.21	111.31	12.07	16.90

where $\alpha, \beta > 1$ and the matrix to be solved is very sparse; in fact, the required storage is $\mathcal{O}(N)$. These observations have led to the development of algorithms that take advantage of the particular structures of the matrices. This field of research is huge and successful. As seen in Table 4.1, the computational time needed to solve the problem with a highly efficient multigrid algorithm is about 1% of the banded Gaussian elimination, when using 257^2 unknowns. Moreover, banded Gaussian elimination runs out of memory at this point and cannot be compared with the iterative methods. Still, it is worth noting that the multigrid algorithms using 1025^2 unknowns outperforms banded Gaussian elimination with 257^2 unknowns.

4.2.1   The Richardson Iteration

Here we will briefly introduce iterative methods designed to solve linear systems. We will begin with classical schemes and then move on to more advanced methods. Our ultimate goal is to derive an order-optimal method for the linear system arising from discretizations of the Bidomain model. We want to solve the linear system,

$$Au = b.$$

The matrix, A, is sparse, but A^{-1} is, in general, full and requires the storage of N^2 floating points. When $N \sim 10^7$, which is quite common, the inverse can neither be computed nor stored. However, the matrix-vector product, $w = Av$, requires only $\mathcal{O}(N)$ operations. Therefore, a first attempt for a more memory-efficient algorithm might be a fixed-point iteration,

$$u^n = u^{n-1} - \tau(Au^{n-1} - b). \tag{4.3}$$

This algorithm is commonly called the *Richardson iteration* or the *simple iteration*. The obvious question is whether the iteration converges to the solution or not. A standard approach to analyze iterative methods is to assume that the solution u is known, so that we can investigate how the error behaves. The error in the n'th iteration is defined as

$$e^n = u^n - u. \tag{4.4}$$

Hence, by subtracting u from both sides of (4.3) and using the relation, $Au = b$, we obtain a recursion for the error,

$$e^n = e^{n-1} - \tau A e^{n-1}. \tag{4.5}$$

The error at the n'th iteration, e^n, should then be smaller, in some sense, than the previous ones. To quantify the error behaviour in "some sense" we introduce the discrete L_2-norm,

$$\|e\|_{L_2} = \left(\frac{1}{N} \sum_{i=1}^{N} e_i^2 \right)^{1/2}, \tag{4.6}$$

and the discrete L_2-inner product,

$$(e, f)_{L_2} = \left(\frac{1}{N} \sum_{i=1}^{N} e_i f_i \right). \tag{4.7}$$

The corresponding matrix norm is defined by

$$\|A\| = \sup_v \frac{(Av, v)_{L_2}}{(v, v)_{L_2}}.$$

We will, in what follows, drop the subscript L_2.

A necessary and sufficient condition for convergence is that the error decreases during the iteration

$$\|e^n\| \le \rho \|e^{n-1}\|,$$

where $0 \le \rho < 1$. This implies that the Richardson iteration is a contraction and convergent if and only if

$$\|I - \tau A\| \le \rho < 1, \tag{4.8}$$

or

$$0 < (1 - \rho) \le \|\tau A\| \le (1 + \rho) < 2, \tag{4.9}$$

From (4.5) and (4.8) the error will decrease

$$\|e^n\| = \|(I - \tau A)e^{n-1}\| \le \|I - \tau A\| \|e^{n-1}\| \le \rho \|e^{n-1}\|.$$

A more detailed mathematical description of this method and the corresponding convergence proofs can be found in Chapter 3 in [57].

Let the parameter τ be $\tau = \frac{c}{\|A\|}$, with $c < 2$, then (4.8) (or (4.9)) is satisfied and the Richardson iteration will converge. However, $\|A\|$ has to be computed or estimated. In the general case, the computation of $\|A\|$ is not easy, but for the applications considered in this chapter it is sufficient to assume that it is on the form $\tau_2 N^{2/d}$, where d is the number of space dimensions, and manually tune τ_2. The estimate does not need to be very accurate, but too small a τ_2 will lead to divergence.

This iteration has some of the basic characteristics we are seeking. The matrix requires $\mathcal{O}(N)$ floating points in storage, which imply that

 – Each iteration involves $\mathcal{O}(N)$ floating point operations.

 – The entire method requires only the storage of $\mathcal{O}(N)$ floating points.

This is a potential advantage over direct methods, but we will see that it is not enough. The Richardson iteration performs poorly when the matrices come from the discretization of PDEs. To explain this we introduce a FDM discretization of the Poisson problem in 1D. This example is very simple, but it still has the basic features of the problems that we want to address.

4.2.2 The FDM Discretization Poisson Equation in 1D

Efficient algorithms take advantage of particular properties of the matrix. The classical iterations; Jacobi, Gauss–Seidel, SOR, and SSOR are usually more efficient than the Richardson iteration, but they are not general-purpose algorithms. Some properties are required. This also apply to the Conjugate Gradient method. Order-optimal solution algorithms, such as multigrid and domain decomposition, need even more properties to ensure convergence. Since multigrid was introduced in 1960 (domain decomposition having been introduced as early as 1870), these algorithms have been extensively studied by many researchers and the theoretical framework has reached a high degree of sophistication and abstraction [13] and [57]. We will start by explaining the underlying ideas in the simplest possible fashion. In this respect, the exposition is similar to that in [15].

First, the basic properties of the FDM discretization of the Poisson problem in 1D are reviewed. This model problem reads

$$-u''(x) = f(x), \ x = (0,1), \ u(0) = 0, \ u(1) = 0. \tag{4.10}$$

We are not interested in any particular solution, but rather the class of problems on this form. We therefore introduce the differential operator $L = -\frac{\partial^2}{\partial x^2}$, such that the problem can be written as

$$Lu = -u''(x) = f(x), \ x = (0,1), \ u(0) = 0, \ u(1) = 0. \tag{4.11}$$

The operator L has the following properties:

 – It is *positive definite*; this means that $(Lu, u) > 0$ for all relevant functions u.

 – It is *symmetric*; this means that $(Lu, v) = (u, Lv)$ for all relevant functions u.

 – It is *invertible* for any $f \in C(0,1)$; this means that the problem (4.10) has a unique solution for $f \in C(0,1)$.

These properties can be derived directly from the definition of L; see, e.g., [139].

Much of the intuition concerning the solution of PDEs and the corresponding solution algorithms is closely connected to eigenvalues and eigenfunctions for the differential operators. The definition of eigenvalues and eigenfunctions for differential operators is equivalent to the definition for matrices in linear algebra, i.e., λ is an eigenvalue of L if

$$Lw = \lambda w, \quad \lambda \in \mathbb{C}, \tag{4.12}$$

and w is the corresponding eigenfunction. Because the operator is symmetric and positive definite, the eigenvalues are real and positive, and the eigenfunctions are orthogonal. In fact, it is known that the eigenvalues for L in (4.11) are

$$\mu_k = \pi^2 k^2, \quad \text{for } k = 1, \ldots, \infty,$$

and the eigenfunctions are

$$w_k(x) = \sin(k\pi x), \quad \text{for } k = 1, \ldots, \infty.$$

These properties can be derived directly from the definition of L; see, e.g., [139].

The first four eigenfunctions are shown in Figure 4.1. Notice that for small k, the eigenvalues are small and the corresponding eigenfunctions are low-frequency functions. However, as $k \to \infty$, $\mu_k \to \infty$ and the eigenfunctions oscillates between 1 and -1 with a period $\frac{\pi}{k}$ that approaches zero. This relation between the smoothness of the eigenfunctions and the magnitude of the eigenvalues is typical for elliptic PDEs, regardless of dimensions, domains and boundary conditions.

This 1D problem can be solved analytically, but this is not feasible for the Poisson problem on general domains in 2D or 3D. However, discretization techniques,

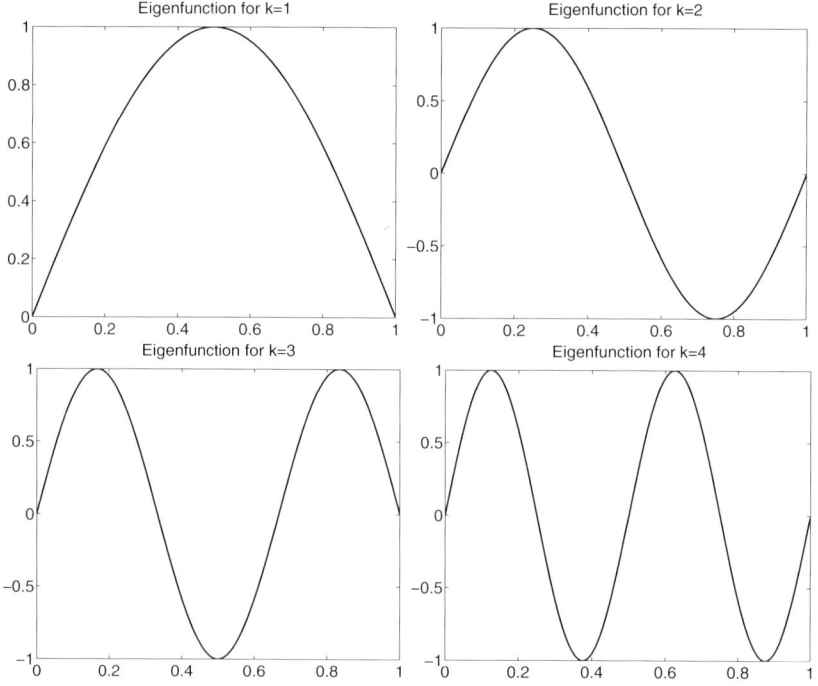

Fig. 4.1. The first four eigenfunctions.

such as the finite element method (FEM)[2] or the finite difference method, (FDM)[3] are general-purpose strategies. The next step is to make a discrete approximation of the problem such that we arrive at a linear system that can be solved. Let the grid consist of grid points $x_i = ih$. A FDM discretization of (4.10) reads

$$-\frac{u_{i-1} - 2u_i + u_{i+1}}{h^2} = f_i, \quad i = 1, \ldots N, \tag{4.13}$$

$$u_0 = 0, \tag{4.14}$$

$$u_{N+1} = 0. \tag{4.15}$$

These equations can be written as the linear system

$$A u_h = f, \tag{4.16}$$

where

$$A = \frac{1}{h^2} \begin{pmatrix} 2 & -1 & & & \\ -1 & 2 & -1 & & \\ & \ddots & \ddots & \ddots & \\ & & -1 & 2 & -1 \\ & & & -1 & 2 \end{pmatrix}, \quad u = \begin{pmatrix} u_1 \\ \cdot \\ \cdot \\ \cdot \\ u_N \end{pmatrix}, \quad f = \begin{pmatrix} f_1 \\ \cdot \\ \cdot \\ \cdot \\ f_N \end{pmatrix}. \tag{4.17}$$

The unknowns u_i are pointwise approximations of u, $u_i \approx u(x_i)$ and $f_i = f(x_i)$. This matrix will be used extensively throughout this chapter to explain the behaviour of the various algorithms. Although this matrix is very simple, it has the characteristic properties that will be studied and used throughout most of this chapter.

The eigenvalues of this system are

$$\mu_k = \frac{4}{h^2} \sin^2 \left(\frac{k\pi h}{2} \right), \quad k = 1, \ldots, N, \tag{4.18}$$

and the corresponding eigenfunctions are

$$v_{k,i} = v_k(x_i) = \sin(k\pi x_i), \quad k = 1, \ldots, N, \tag{4.19}$$

where $v_{k,i}$ is the value associated with the node i at the point x_i and the k'th eigenfunction. Similar to the continuous case, large eigenvalues correspond to high-frequency eigenvectors, whereas smooth eigenfunctions are associated with small eigenvalues. This property is exactly what is exploited in the multigrid algorithm that we will discuss later.

A more detailed description of the continuous and discrete Poisson problem can be found in [139].

---

[2] FEM requires that it is possible to generate a reasonable grid.
[3] FDM is difficult to apply in general geometries.

4.2.3   The Richardson Iteration Revisited

We saw in Section 4.2.1 that the Richardson iteration is a memory-efficient method for solving general matrix equations, given a reasonable estimate on $\|A\|$. In order to estimate the computational work needed by the method, it is necessary to estimate the number of floating point operations the CPU has to perform. To this end, it is necessary to estimate the number of iterations that will be needed.

The first thing to consider is a *stopping criterion*, designed to prevent endless iterations. We are seeking a numerical approximation u_h of the actual unknown u and have introduced an error, e_h, inherited from the numerical method. In the previous Chapter 3 we saw that the error could be estimated:

$$\|e_h\| = \|u - u_h\| \leq ch^\alpha,$$

where h is a characteristic grid size that depends directly on the number of unknowns. The discretization error e_h determines the level of accuracy needed by the iterative method. Let e_h^n be the error at the n'th iteration. We can split this error into two parts: $e_h^n = e^n + e_h$, where e^n is the part induced by the iterative method, and e_h, which is the discretization error. It is then reasonable to require that both contributions are equally sized, $e^n \approx e_h$. Hence, we will stop the iteration when $\|e^n\| \leq \|e_h\|$. The concretization of such a stopping criterion is not an easy task. The discretization error, e_h, is of course, in general not available. However, the residual can be computed:

$$r^n = b - Au^n.$$

From the residual-error equation

$$Ae^n = r^n,$$

we obtain

$$\|e^n\| \leq \|A^{-1}\|\|r^n\|.$$

Another possible technique is to check $u^n - u^{n-1}$. We will not go deeply into a discussion about various stopping criteria here. However, a frequently used criterion stops the sequence of iterations when

$$\|r^n\| < \gamma,$$

where γ is a small number that is usually found by numerical experiments.

The second parameter that determines the number of iterations is the *convergence rate* or minimal error reduction per iteration, which will be explained below. We remember from Section 4.2.1 that the error at the n'th iteration is governed by

$$e^n = (I - \tau A)e^{n-1}. \tag{4.20}$$

Given that $\|I - \tau A\| < 1$, we saw that the iteration was convergent, but we have still not estimated the number of iterations needed to reach a given stopping criterion. Let the convergence rate ρ be defined as

$$\rho = \|I - \tau A\|.$$

Then

$$\|e^n\| \leq \rho\|e^{n-1}\|. \tag{4.21}$$

Moreover,

$$\|e^n\| \leq \rho^n\|e^0\|. \tag{4.22}$$

Assuming that ϵ is the discretization error, we can tolerate an iteration error:

$$\|e^n\| \leq \epsilon. \tag{4.23}$$

If we assume equality in (4.22) and (4.23) we get

$$\|e^n\| = \rho^n\|e^0\| = \epsilon. \tag{4.24}$$

Hence, the number of iterations can be estimated as

$$n = \frac{\log \frac{\epsilon}{\|e^0\|}}{\log \rho}. \tag{4.25}$$

The goal of this chapter is to present algorithms where $\rho \leq c$, where $c < 1$ is a constant independent of the grid size. If the convergence criterion ϵ is fixed, independent of the grid size, n will be bounded independent of the grid size. This is referred to as an *order-optimal algorithm*.

Finally, we will consider some additional properties that are typical for the type of matrix we are working with. The matrices are symmetric positive definite and for these matrices it is known that

$$\|A\| = \lambda_{\max}(A), \quad \|A^{-1}\| = \lambda_{\min}^{-1}(A),$$

where $\lambda_{\max}$ and $\lambda_{\min}$ are the largest and smallest eigenvalues of A, respectively. From the eigenvalues for the 1D discretization of the Poisson equation (4.18) we have

$$\lambda_{\max}(A) \approx \frac{4}{h^2}, \quad \lambda_{\min}(A) \approx \pi^2.$$

The ratio between the largest and the smallest eigenvalues is commonly called the condition number of the matrix, $\kappa(A)$:

$$\kappa(A) = \frac{\lambda_{\max}(A)}{\lambda_{\min}(A)}. \tag{4.26}$$

Hence, the condition number of the matrix A in (4.17) is $\mathcal{O}(h^{-2})$. Notice that this condition number is typical for elliptic PDEs, regardless of the dimension, the boundary conditions and the domain.

These eigenvalues are now used to gain more insight into what will happen during the iteration. Let the initial error be expanded in terms of eigenvectors:

$$e_0 = \sum_{k=1}^{N} c_k w_k,$$

where, by orthonormality, the coefficients are determined by

$$c_k = (e_0, w_k).$$

Let $\tau = \frac{c}{\lambda_{\max}}$. Then, the error at the n'th iteration is from (4.20):

$$e^n = \sum_{k=1}^{N} \left(1 - \frac{c\lambda_k}{\lambda_{\max}}\right)^n c_k w_k.$$

If $0 < c < 2$, the iteration is convergent in the sense that

$$\|e^n\| \to 0, \qquad \text{as } n \to \infty.$$

To see this, we note that

$$\|e^n\| = (e^n, e^n)^{1/2} = \left(\frac{1}{N} \sum_{k=1}^{N} \left(1 - \frac{c\lambda_k}{\lambda_{\max}}\right)^n c_k w_k, \left(1 - \frac{c\lambda_k}{\lambda_{\max}}\right)^n c_k w_k\right)^{1/2}$$

$$= \left(\frac{1}{N} \sum_{k=1}^{N} \left(1 - \frac{c\lambda_k}{\lambda_{\max}}\right)^{2n} c_k^2\right)^{1/2} \leq (1 - c)^n \|e^0\|,$$

where we have used that $\{w_k\}_{k=1}^{N}$ are orthonormal.

Still, different parts of the error will decrease at different speeds. This is made clear in what follows, by considering the error components that correspond to the high and low eigenvalues. Let the initial error, e^0, be dw_N, where d is a constant and w_N is the N'th eigenvector, corresponding to the largest eigenvalue. The error after the first iteration will then be

$$e^1 = \left(1 - \frac{c\lambda_{\max}}{\lambda_{\max}}\right) w_N = (1 - c)e^0.$$

For instance, the choice $c = 0.9$ implies an error reduction by a factor 0.1 for the error component associated with w_N. We also notice that choosing a smaller c leads to slower convergence. Similarly, all the high-frequency parts of the error are removed rather efficiently.

However, the situation is quite different for the low-frequency parts of the error. Let e^0 be dw_1, where d is a constant and w_1 is the eigenvector corresponding to the lowest eigenvalue. Then we have

$$e^1 = \left(1 - \frac{c\lambda_{\min}}{\lambda_{\max}}\right) w_1 = \left(1 - \frac{c}{K(A)}\right) w_1 \approx e^0,$$

where $\kappa(A)$ is the condition number of the matrix. As observed for the 1D discretization of the Poisson equation, the condition number is $\sim h^{-2}$, which is very bad. Moreover, reducing c only makes things worse.

The second disadvantage of using the Richardson iteration is that it relies on the estimation of the largest eigenvalue. The iteration will diverge if τ is not chosen properly.

However, despite the fact that this method is primitive and cannot be used for large linear systems, it captures the basics of the classical iterations. Some parts of the error are dealt with efficiently, while others remain essentially unchanged. In particular, the smooth components are troublesome. This property will be exploited later, when we consider the (relaxed) Jacobi and Gauss–Seidel methods.

4.2.4 Preconditioning

An "obvious" generalization of the Richardson iteration is to include a matrix B, usually called a preconditioner, in the following way:

$$u_i^n = u^{n-1} - \tau B(Au^{n-1} - b). \tag{4.27}$$

The error iteration will then be

$$e^n = e^{n-1} - \tau BAe^{n-1}, \tag{4.28}$$

which is convergent if and only if

$$\|I - \tau BA\| < 1. \tag{4.29}$$

Although this idea seems quite simple, it is remarkably powerful. In fact, nearly all the methods we will consider in what follows, multigrid and domain decomposition as well as Jacobi and Gauss–Seidel, fit into this framework. It is the general form of any *linear iteration*. The only exception in this chapter is the Conjugate Gradient method.

Notice that we will refer to B as both a linear operator and a matrix. The reason is that if B represents a linear operator it can, in principle, always be represented as a matrix. Still, it is usually easier and more efficient to implement B as an algorithm. To be more specific, only the action of B on a vector, $u = Bv$, has to be implemented.

We will now briefly go through the classical iterations. These methods cannot, in general, be used to solve matrices with more than 1000 unknowns. Still, they deserve attention, at least for the following four reasons:

– They are used as smoothers in multigrid algorithms.

– They serve to illustrate important aspects of the multigrid method.

– Domain decomposition algorithms are generalizations of the block versions of these algorithms.

– Our final order-optimal block preconditioner is an inexact variant of the block Jacobi method.

4.2.5 The Jacobi Method

The simplest algebraic operator-splitting technique is the Jacobi method. This algorithm is easily explained by considering the i'th equation in the matrix equation

$$\sum_{j=1}^{N} a_{ij} u_j = b_i.$$

Rearranging the i'th equation, we get

$$u_i = \frac{1}{a_{ii}} \left(b_i - \sum_{i \neq j} a_{ij} u_j \right).$$

The problem is, of course, that the other unknowns, u_j, have not yet been computed. However, we can assume that we have either initial guesses or the values from the previous iteration, which suggests the Jacobi iteration:

$$u_i^n = \frac{1}{a_{ii}} \left(b_i - \sum_{j \neq i} a_{ij} u_j^{n-1} \right), \quad \text{for } i = 1, \ldots N. \tag{4.30}$$

Notice that all the variables in (4.30) can be updated independently. This is important when parallel computers are used, because on such computers the unknowns may be distributed on different processors and the update can then be done in parallel. This issue is discussed in Chapter 6.

The strength of the Richardson iteration (4.27) now becomes apparent because the Jacobi iteration can be seen as a preconditioned Richardson iteration. To see this, consider

$$u^n = u^{n-1} - D^{-1}(Au^{n-1} - b),$$

with $B = D^{-1}$, $D = \text{diag}(A)$ and $\tau = 1$. Hence, the convergence properties can be studied in the same way as the convergence of the Richardson iteration.

Properties and Problems. To understand when the Jacobi method is convergent we consider the error iteration,

$$e^n = e^{n-1} - D^{-1} A e^{n-1}. \tag{4.31}$$

Hence, the iteration is convergent if and only if

$$\rho = \|I - D^{-1}A\| < 1,$$

or

$$0 < \|D^{-1}A\| < 2.$$

It is not easy to understand the behaviour of this algorithm directly from these inequalities. Hence, we will, briefly, go through the 1D discretization of the Poisson equation. The Jacobi iteration matrix associated with (4.17) is

$$J = I - D^{-1}A = \frac{1}{2}\begin{pmatrix} 0 & 1 & & & \\ 1 & 0 & 1 & & \\ & \ddots & \ddots & \ddots & \\ & & 1 & 0 & 1 \\ & & & 1 & 0 \end{pmatrix}.$$

Inserting the eigenfunctions of (4.17),

$$w_k = \sin(k\pi x_j), \quad k = 1, \dots N. \tag{4.32}$$

where $x_j = jh$, gives

$$(I - D^{-1}A)w_k = w_k - D^{-1}\mu_k w_k = \left(1 - \frac{h^2}{2}\mu_k\right)w_k,$$

where μ_k is given in (4.18). Therefore, w_k is an eigenfunction of J, with the corresponding eigenvalue

$$\lambda_k = 1 - \frac{h^2}{2}\mu_k = 1 - \frac{h^2}{2}\frac{4}{h^2}\sin^2\left(\frac{k\pi h}{2}\right) = \cos(k\pi h), \quad k = 1, \dots, N. \tag{4.33}$$

The eigenvalues are shown in Figure 4.2.

Notice that from Figure 4.2 it seems that the eigenvalues $\lambda_k \in [-1, 1]$. In fact, $\lambda_k \in (-1, 1)$ and the iteration is therefore convergent. However, the convergence

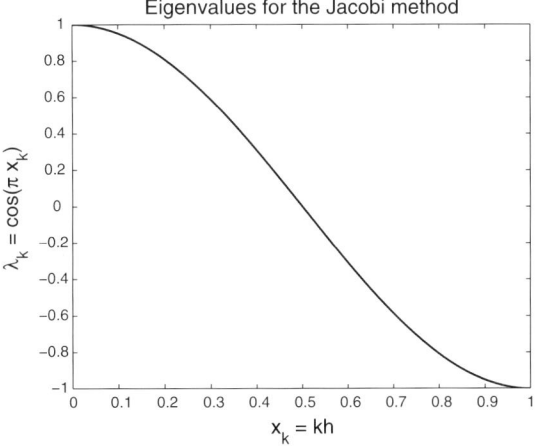

Fig. 4.2. Eigenvalues of the Jacobi Matrix.

Table 4.2. Number of iterations for the Jacobi method to achieve an error reduction by a factor of 10^{-4} for the 1D discretization of the Poisson problem.

Unknowns	Convergence rate	Iterations
10	$1 - 8.2e^{-2}$	108
100	$1 - 9.7e^{-4}$	9514
1000	$1 - 9.8e^{-6}$	$9.4 \cdot 10^5$
10000	$1 - 9.9e^{-8}$	$9.3 \cdot 10^7$

rate, determined by $\lambda_{\max}$ and $\lambda_{\min}$, depends strongly on the number of unknowns. To see this, we use the Taylor expansion of $\cos(x)$ around $x = \pi$,

$$\cos(x) = \cos(\pi) - \sin(\pi)(x - \pi) - \cos(\pi)(x - \pi)^2 + \ldots \qquad (4.34)$$

Letting $x = \frac{N\pi}{N+1}$ in (4.34), corresponding to the N'th eigenvector in (4.33), we get

$$\lambda_N = \cos\left(\frac{N\pi}{N+1}\right) \approx \cos(\pi) - \cos(\pi)\left(\frac{\pi N}{N+1} - \pi\right)^2 = 1 - \left(\frac{\pi}{N+1}\right)^2 = 1 - (\pi h)^2.$$

A similar estimate can be obtained for the lowest eigenvalue, with $N = 1$, using the Taylor expansion around $x = 0$. This means that the error corresponding to the most high frequent eigenfunction, $k = N$, and the most low frequent eigenfunction, $k = 1$, only decreased by a factor $1 - (\pi h)^2$. The number of Jacobi iterations needed to reduce the error by a factor of 10^{-4} is shown in Table 4.2.

Still, for a subset of the eigenfunctions, $\frac{N}{4} \leq k \leq \frac{3N}{4}$, the convergence is fast, in the sense that $\rho < 0.71$. Hence, the Jacobi method performs quite well for at least half of the error components. The iteration would, therefore, be efficient if the initial error only contained these parts. Naturally, it is difficult to make such an initial guess.

4.2.6 The Relaxed Jacobi Method

By looking at the Jacobi iteration (4.30) we see that u_i is computed based on u_j for $j \neq i$. A natural extension is to include information from the previous iteration,

$$u_i^n = (1 - \omega)u_i^n + \frac{\omega}{a_{ii}}\left(b_i - \sum_{j \neq i} a_{ij} u_j^{n-1}\right), \qquad \text{for } i = 1, \ldots N, \qquad (4.35)$$

where ω can be chosen. The inclusion of the data from the previous iteration is commonly called relaxation. As before, we can express this iteration in terms of the Richardson iteration (4.3)

$$u^n = u^{n-1} - \omega D^{-1}(Au^{n-1} - b),$$

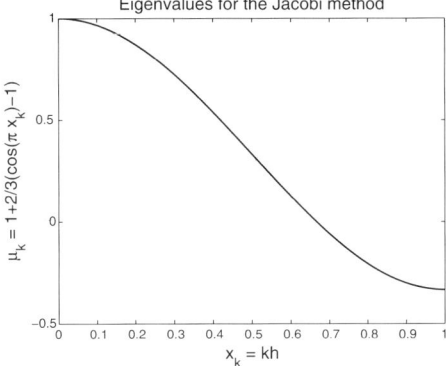

Fig. 4.3. Eigenvalues of the Relaxed Jacobi iteration Matrix.

where $B = D = \text{diag}(A)$ and $\omega = \tau$. The necessary condition for convergence is that

$$0 < \|\omega D^{-1} A\| < 2.$$

Again, we briefly review the properties of this iteration in terms of eigenvalues and eigenfunctions. The eigenfunctions are the same as for Jacobi (4.32), and a simple calculation shows that the eigenvalues are

$$\lambda_k(\omega) = 1 + \omega(\cos(k\pi h) - 1), \quad k = 1, \ldots, n. \tag{4.36}$$

The eigenvalues for $\omega = 2/3$ are shown in Figure 4.3. The spectrum has changed from $(-1, 1)$ for $\omega = 1$ to $(-\frac{1}{3}, 1)$ for $\omega = 2/3$. Hence, changing the parameter ω dramatically changes the behaviour of the relaxed Jacobi for the high frequency parts of the error. However, low frequency or smooth parts of the error are inefficiently handled for any ω. Even so, the news is good, because it is the high frequency parts of the error that are most problematic. A simple idea might be to compute the solution on a coarser grid and use this as the initial guess. Multigrid methods are generalizations of this idea.

4.2.7 The Exact and the Inexact Block Jacobi Methods

Another natural extension of the pointwise Jacobi algorithm expressed above constitutes what we will call the block Jacobi method. This iteration arise when replacing the numbers a_{ij}, u_j, and b_j in (4.30) with the matrices A_{ij}, and the vectors u_j and b_j, respectively. Consider the linear system

$$Au = b,$$

where

$$
A = \begin{pmatrix} A_{11} & A_{12} & \cdots & A_{1N} \\ A_{21} & A_{22} & & \\ & \ddots & \ddots & \ddots \\ A_{N1} & & & A_{NN} \end{pmatrix}, \quad u = \begin{pmatrix} u_1 \\ \cdot \\ \cdot \\ \cdot \\ u_N \end{pmatrix}, \quad b = \begin{pmatrix} b_1 \\ \cdot \\ \cdot \\ \cdot \\ b_N \end{pmatrix}. \tag{4.37}
$$

Here, A_{ij} are matrices, and u_i and b_i are vectors. The algorithm takes the form

$$
u_i^n = A_{ii}^{-1} \left(b_i - \sum_{j \neq i} A_{ij} u_j^{n-1} \right). \tag{4.38}
$$

In order for this method to be applicable we have to assume that the block matrices, A_{ii}, are invertible.

As for the previous Jacobi variants, this iteration can be written in terms of the Richardson iteration (4.3):

$$
u^n = u^{n-1} - D^{-1}(Au^{n-1} - b),
$$

where D contains the diagonal blocks matrices A_{ii}. The iteration is convergent if

$$
0 < \|D^{-1}A\| < 2.
$$

Of course, A_{ii} may be large blocks and in such cases we do not necessarily want to invert A_{ii}, but rather make an approximation of D^{-1}, i.e.,

$$
u^n = u^{n-1} - \hat{D}^{-1}(Au^{n-1} - b).
$$

The convergence is determined by $\|\hat{D}^{-1}A\|$.

We will postpone the discussion of the merits of this algorithm. However, notice that this framework covers both the order-optimal preconditioner for the Bidomain model and the domain decomposition method of additive Schwarz type.

4.2.8 The Gauss–Seidel Method

Looking at the Jacobi iteration, we observe that when computing the unknown u_i^n the unknowns u_j^n for $j < i$ have already been computed. Moreover, u_j^n should be "closer" to the actual solution than u_j^{n-1}. Therefore, it seems natural to modify the Jacobi iteration to use the new values instead. This is the Gauss–Seidel iteration,

$$
u_i^n = \frac{1}{a_{ii}} \left(b_i - \sum_{j<i} a_{ij} u_j^n - \sum_{j>i} a_{ij} u_j^{n-1} \right), \quad \text{for } i = 1, \ldots N. \tag{4.39}
$$

Let

$$
A = D + U + L,
$$

where D is the diagonal, and U and L are the strictly upper and lower diagonal parts of A, respectively. Then the Gauss–Seidel iteration can be written in terms of the preconditioned Richardson iteration (4.27),

$$u^n = u^{n-1} - (D+L)^{-1}(Au^{n-1} - b),$$

where $B = (D+L)^{-1}$ and $\tau = 1$.

4.2.9   The Relaxed Gauss–Seidel Method

The Gauss–Seidel method can be relaxed in the same way as the Jacobi method:

$$u_i^n = (1-\omega)u^{n-1} + \omega\left(\frac{1}{a_{ii}}\left(b_i - \sum_{j<i} a_{ij}u_j^n - \sum_{j>i} a_{ij}u_j^{n-1}\right)\right), \text{ for } i = 1,\dots N.$$
(4.40)

Written in terms of the preconditioned Richardson iteration, we obtain

$$u^n = u^{n-1} - \omega(D + wL)^{-1}(Au^{n-1} - b).$$

4.2.10   The Symmetric Gauss–Seidel Method

The Gauss–Seidel method is not symmetric (unless A is a diagonal matrix). However, symmetry is an important property for some solution algorithms, such as the Conjugate Gradient method described later. If the matrix is symmetric, it is fairly straightforward to derive a symmetric version of the Gauss–Seidel method. The symmetric Gauss–Seidel simply consists of one standard Gauss–Seidel sweep followed by an additional sweep using the unknowns numbered backwards.

$$u_i^{n-1/2} = \frac{1}{a_{ii}}\left(b_i - \sum_{j<i} a_{ij}u_j^{n-1/2} - \sum_{j>i} a_{ij}u_j^{n-1}\right), \text{ for } i = 1,\dots N,$$

$$u_i^n = \frac{1}{a_{ii}}\left(b_i - \sum_{j>i} a_{ij}u_j^n - \sum_{j<i} a_{ij}u_j^{n-1/2}\right), \text{ for } i = N,\dots 1.$$

Written in terms of the preconditioned Richardson iteration,

$$u^n = u^{n-1} - (D+L)^{-1}D(D+U)^{-1}(Au^{n-1} - b).$$

4.2.11   The Exact and the Inexact Block Gauss–Seidel Method

The block version of Gauss–Seidel method is a straightforward extension of the pointwise Gauss–Seidel. This is similar to the extension we performed with the

Jacobi method. The algorithm is as follows:

$$u_i^n = A_{ii}^{-1} \left(b_i - \sum_{j<i} A_{ij} u_j^n - \sum_{j>i} A_{ij} u_j^{n-1} \right). \tag{4.41}$$

A necessary condition is that A_{ii} is invertible, which was also the case with the block Jacobi.

For the inexact block Gauss–Seidel method, we may use an approximate inverse $\hat{A}_{ii}^{-1}$, instead of A_{ii}^{-1}. This leads to the iteration

$$u_i^n = \hat{A}_{ii}^{-1} \left(b_i - \sum_{j<i} A_{ij} u_j^n - \sum_{j>i} A_{ij} u_j^{n-1} \right). \tag{4.42}$$

The Multiplicative Schwarz algorithm is a variant of this iteration.

4.3   The Conjugate Gradient Method

So far, we have considered classical iterative methods for solving linear systems on the form

$$Au = f. \tag{4.43}$$

The classical schemes are easily derived and their motivation is rather simple. In 1952, Hestenes and Stiefel [63] broke this tradition and published a completely different algorithm. Initially, they viewed their scheme as an alternative to Gaussian elimination; i.e. they derived a direct method that would, in exact arithmetics, give the exact solution in at most N iterations. Here, N is the number of unknowns in (4.43) and the matrix A is supposed to be symmetric and positive definite. Later on, it was realized that the approximate solution obtained after much fewer than N iterations was actually quite good. Soon, it became common to use the Conjugate Gradient (CG) method of Hestenes and Stiefel as an iterative method rather than as a direct method.

One of the most important ideas in numerical analysis is that of "best approximation". Given a large space of functions V and a subspace with fewer functions V_N; how is it possible to find the best approximations $v_N \in V_N$ of the "true" solution $v \in V$. This approach is the basis of the finite element method and, as we shall see, the fundamental idea of the CG method as well. In this text, we shall merely sketch the development of the method. For a full story we refer the reader to, e.g., Golub and van Loan [49] or to Stoer and Bulirsch [129].

4.3.1   The CG Algorithm

As mentioned above, the CG method seeks the best possible approximation in a certain subspace. To this end, the method consists of two basic ingredients: computing a proper subspace and computing the best approximation in this subspace.

In the CG method, we use two inner products: the standard one

$$(u, v) = \frac{1}{N} \sum_{i=1}^{N} u_i v_i, \tag{4.44}$$

and the A-inner product

$$(u, v)_A = (Au, v). \tag{4.45}$$

Since A is symmetric and positive definite, (4.45) defines an inner product. Similarly, we define the associated norms:

$$\|u\| = (u, u)^{1/2}, \tag{4.46}$$

and

$$\|u\|_A = (Au, u)^{1/2}. \tag{4.47}$$

The algorithm computes the best solution measured in the A-norm. Furthermore, we will derive a set of search vectors that are A-orthogonal, thus spanning the subset in which we seek an approximate solution. Suppose the subspace of $\mathbb{R}^N$ is denoted by W, and let w be the best approximation of u measured in the A-norm, i.e.

$$\|u - w\|_A \le \|u - v\|_A, \quad \forall\, v \in W. \tag{4.48}$$

Then, it is generally known that

$$(u - w, v)_A = 0, \quad \forall\, v \in W, \tag{4.49}$$

i.e. the error is orthogonal to the subspace. This result is fundamental in the derivation of the CG-method.

Let us now assume that we have already computed k search vectors

$$p_0, p_1, \ldots, p_{k-1},$$

that are mutually A-orthogonal, i.e.

$$(p_i, p_j)_A = 0 \quad i \ne j. \tag{4.50}$$

Let

$$W_k = \mathrm{span}\{p_0, \ldots, p_{k-1}\}, \tag{4.51}$$

and note that

$$\dim(W_k) = k. \tag{4.52}$$

We assume that

$$u_k \in W_k, \tag{4.53}$$

satisfies

$$\|u - u_k\|_A \le \|u - v\|_A, \quad \forall\, v \in W_k, \tag{4.54}$$

such that $u_k \in W_k$ is the best approximation of the exact solution u measured in the A-norm. It is fairly obvious that if W_k spans all of $\mathbb{R}^N$, then $u_k = u$ so we will have the exact solution in at most N iterations.

Let us also define the residual

$$r_k = f - Au_k, \tag{4.55}$$

which, of course, is zero if $u_k = u$. The residual has a very interesting property that can be derived from the best approximation property. It follows from (4.49) and (4.54) that

$$(u - u_k, v)_A = 0, \quad \forall\, v \in W_k. \tag{4.56}$$

By using the definition of the A-norm, we get

$$(A(u - u_k), v) = 0, \quad \forall\, v \in W_k, \tag{4.57}$$

and since $Au = f$, we have

$$(r_k, v) = 0, \quad \forall\, v \in W_k. \tag{4.58}$$

So r_k is orthogonal to all the vectors in W_k, and in particular

$$(r_k, p_j) = 0, \quad j = 0, 1, \ldots, k - 1. \tag{4.59}$$

We now want to step from iteration k to iteration $k + 1$, and in order to do so, we need to increase the dimension of $W_k = \text{span}\{p_0, \ldots, p_{k-1}\}$ and then compute the best approximation in the new and larger subspace. In order to increase the dimension of W_k, we will apply the Gram–Schmidt algorithm. This is an algorithm that computes an orthogonal basis based on linearly independent vectors. Now, we already have k linearly independent, and in fact also A-orthogonal, vectors. In order to apply the Gram–Schmidt algorithm, we will need a vector that is linearly independent of all vectors in W_k. Since we have already seen that r_k is orthogonal to all vectors in W_k we can use this vector to increase the dimension to $k + 1$, and then use the Gram–Schmidt orthogonalization process to generate an orthogonal basis. Using this algorithm, we find that

$$p_k = r_k + \beta_{k-1} p_{k-1}, \tag{4.60}$$

where

$$\beta_{k-1} = -\frac{(r_k, p_{k-1})_A}{(p_{k-1}, p_{k-1})_A}. \tag{4.61}$$

Now we have $W_{k+1} = \text{span}\{p_0, p_1, \ldots, p_k\}$ and we want to find the best approximation $u_{k+1} \in W_{k+1}$. We seek $u_{k+1} \in W_{k+1}$ of the form

$$u_{k+1} = u_k + \alpha_k p_k, \tag{4.62}$$

and the task is to determine α_k such that

$$\|u - u_{k+1}\|_A \leq \|u - v\|_A, \quad \forall\, v \in W_{k+1}. \tag{4.63}$$

Because of (4.49), we require that

$$(u - u_{k+1}, v)_A = 0, \quad \forall\, v \in W_{k+1}, \tag{4.64}$$

such that, in particular, we have

$$(u - u_{k+1}, p_k)_A = 0. \tag{4.65}$$

Using (4.62), we have

$$\alpha_k = \frac{(u - u_k, p_k)_A}{(p_k, p_k)_A}. \tag{4.66}$$

Here we notice that

$$(u - u_k, p_k)_A = (A(u - u_k), p_k)_A = (r_k, p_k)_A,$$

such that

$$\alpha_k = \frac{(r_k, p_k)_A}{(p_k, p_k)_A}. \tag{4.67}$$

By (4.62) we have

$$u_{k+1} = u_k + \alpha_k p_k, \tag{4.68}$$

and the derivation is complete.

The CG-method can be formulated in many variants using the orthogonality properties discussed above. All these variants are mathematically equivalent, but may behave differently on a computer due to round-off issues. The version we give here has been applied successfully in many practical computations.

The Conjugate Gradient Algorithm

Let $A \in \mathbb{R}^{N,N}$, $f \in \mathbb{R}^N$, $u^0 \in \mathbb{R}^N$, and $0 < \varepsilon < 1$ be given. The matrix A is supposed to be symmetric and positive definite.

$u = u^0$
$r = f - Au$
$p = r$
$\rho_0 = (r, r)$
$k = 0$
While $\rho_k / \rho_0 > \varepsilon$ **do**

z	$= Ap$	(4.69)
γ	$= (p, z)$	(4.70)
α	$= \rho_k / \gamma$	(4.71)
u	$= u + \alpha p$	(4.72)
r	$= r - \alpha z$	(4.73)
ρ_{k+1}	$= (r, r)$	(4.74)
β	$= \rho_{k+1} / \rho_k$	(4.75)
p	$= r + \beta p$	(4.76)
k	$= k + 1$	(4.77)

end

The update of r_k is worth noting. Recall that

$$r_k = f - Au_k, \tag{4.78}$$

since, according to (4.68),

$$u_{k+1} = u_k + \alpha_k p_k, \tag{4.79}$$

and we have

$$\begin{aligned}
r_{k+1} &= f - Au_{k+1} \\
&= f - A(u_k + \alpha_k p_k) \\
&= f - Au_k - \alpha_k Ap_k \\
&= r_k - \alpha_k Ap_k.
\end{aligned} \tag{4.80}$$

Since we have already computed Ap_k (see (4.69)), we can avoid an extra matrix-vector multiplication by using (4.80).

4.3.2   Convergence Theory

As mentioned above, the CG method is now considered to be an iterative scheme and it is important to study the convergence behaviour of the method. It transpires that the convergence can be studied in terms of the condition number of the matrix A in (4.43). Recall that

$$K = K(A) = \frac{\lambda_{\max}}{\lambda_{\min}}, \tag{4.81}$$

where $\lambda_{\max}$ and $\lambda_{\min}$ are the largest and smallest eigenvalues of A, respectively. It is well known (see, e.g., Knabner and Angermann [77]), that the error after k iterations with the CG method can be bounded as follows:

$$\frac{\|e_k\|_A}{\|e_0\|_A} \leq 2 \left(\frac{\sqrt{K} - 1}{\sqrt{K} + 1} \right)^k, \tag{4.82}$$

where the error is given by

$$e_k = u - u_k. \tag{4.83}$$

We observe from (4.82) that if K is small (close to one), the convergence is very fast, and if K is large the convergence may be very slow. As discussed above, linear systems on the form (4.43) that arise from the discretization of partial difference equations are often poorly conditioned, i.e. K is very large.

Suppose K is large and we want to compute the number of iterations k such that

$$\frac{\|e_k\|_A}{\|e_0\|_A} \leq \varepsilon, \tag{4.84}$$

for a given ε, $0 < \varepsilon < 1$. Then, from (4.82) we need

$$k \geq \frac{\ln(\varepsilon/2)}{\ln\left(\frac{\sqrt{K}-1}{\sqrt{K}+1}\right)} \tag{4.85}$$

iterations. Since K is large, we have

$$\frac{\sqrt{K}-1}{\sqrt{K}+1} \approx 1 - \frac{1}{\sqrt{K}},$$

and since

$$\ln(1+x) = x + O(x^2),$$

we have

$$\ln\left(\frac{\sqrt{K}-1}{\sqrt{K}+1}\right) \approx -1/\sqrt{K},$$

and thus

$$k \gtrsim \ln(2/\varepsilon)\sqrt{K}. \tag{4.86}$$

This formula explains the importance of the condition number for the convergence of the CG method. We return to this issue on several occasions below. It is clear that if (4.86) is a sharp estimate, then the number of iterations is $O(\sqrt{K})$ and thus increases quite rapidly as the condition number increases.

4.3.3 Numerical Experiments

In this section, we will present some numerical experiments using the CG method. To this end, we consider the following two problems:

$$-\Delta u_2 = f_2 \quad (x,y) \in \Omega = [0,1]^2, \quad u_2 = 0 \text{ at } \partial\Omega, \tag{4.87}$$
$$-\Delta u_3 = f_3 \quad (x,y) \in \Omega = [0,1]^3, \quad u_3 = 0 \text{ at } \partial\Omega. \tag{4.88}$$

We use

$$f_2 = e^{xy}, \tag{4.89}$$
$$f_3 = e^{xyz}, \tag{4.90}$$

and discretize (4.87) and (4.88) using straightforward finite differences. In a 1D finite difference discretization we use n internal nodes leading to a grid spacing of

$$h = \frac{1}{1+n}. \tag{4.91}$$

Table 4.3. The table shows the number of nodes (N), the number of iterations (k_2), and $c_2 = k_2/N^{1/2}$.

N	k_2	$c_2 = k_2/N^{1/2}$
10 404	287	2.81
40 804	579	2.87
161 604	1167	2.90
643 204	2361	2.94
2 566 404	4770	2.98

Table 4.4. The table shows the number of nodes (N), the number of iterations (k_3), and $c_3 = k_3/N^{1/3}$.

N	k_3	$c_3 = k_3/N^{1/3}$
1 728	34	2.83
10 648	69	3.14
74 088	138	3.29
551 368	278	3.39
4 251 528	558	3.44

In 2D we use n^2 internal nodes and n^3 internal nodes in 3D. Based on the PDEs above, this leads to linear systems on the form

$$A_2 u_2 = g_2, \tag{4.92}$$
$$A_3 u_3 = g_3. \tag{4.93}$$

It can be shown that the condition numbers of A_2 and A_3 are $O(h^{-2})$; see, e.g., [77]. Since the number of iterations is given by (4.86), we have $k = O(\sqrt{K}) = O(h^{-1})$. In 2D, the number of nodes is $N \approx 1/h^2$ and in 3D, we have $N \approx 1/h^3$, hence $k_2 \approx c_2 N^{1/2}$ in 2D, and $k_3 \approx c_3 N^{1/3}$ in 3D. Using $\frac{\|r_k\|}{\|r_0\|} \leq 10^{-7}$, we have applied the CG-method to the systems (4.92) and (4.93). The number of iterations are given in Table 4.3 and Table 4.4.

We observe from the tables that the number of iterations is about

$$k_2 \approx 3\,N^{1/2}$$

in 2D and

$$k_3 \approx 3.5\,N^{1/3}$$

in 3D. Since the amount of work in each iteration is $O(N)$, we have that the solution process requires $O(N^{3/2})$ in 2D and $O(N^{4/3})$ in 3D. The order-optimal result would be $O(N)$ and we will derive methods that are order-optimal in this sense later in the text.

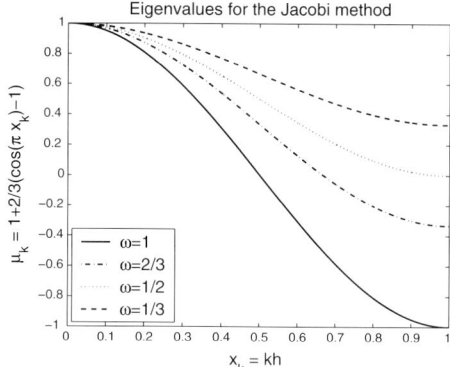

Fig. 4.4. Eigenvalues of the Jacobi iteration Matrix.

4.4 Multigrid

4.4.1 Idea

In this section, we will first try to motivate multigrid methods with the intuition derived from the classical iterations in the previous section, before we present the polished and abstract framework. Readers interested in the theory of multigrid methods can consult [13] and [57]. More practical introductions to multigrid methods are given in [15], [98], and [135]. An overview of multigrid methods and generalizations thereof can be found in [14] and the references therein.

In the previous section we observed that both the standard and the relaxed Jacobi iterations had problems with some parts of the error (associated with certain eigenfunctions), but were very good for other parts of the error. The Jacobi method showed poor performance for errors associated with eigenfunctions corresponding to either the larger or the smaller eigenvalues, whereas the relaxed Jacobi with $\omega = \frac{2}{3}$ was rather effective for all the higher frequencies. We recall that the error iteration associated with the relaxed Jacobi method can be written

$$e^n = e^{n-1} - \omega D^{-1} A e^{n-1}. \tag{4.94}$$

Using the eigenfunction expansion of e^n and (4.94), we arrived at an estimate for the error after n iterations, e^n, in terms of the eigenvalues for the relaxed Jacobi method in (4.33). We restate the eigenvalues here for convenience:

$$\lambda_k(\omega) = 1 + \omega(\cos(k\pi h) - 1), \quad k = 1, \ldots, N. \tag{4.95}$$

For the high frequencies, this results in

$$\max_{k \geq \frac{n}{2}} \left| \lambda \left(\frac{2}{3} \right) \right| \leq \frac{1}{3}. \tag{4.96}$$

However, the performance is bad for the smooth components. A natural idea for improving the performance is to combine the strengths of Jacobi with some other method. Earlier, we suggested that a coarse grid solution could be used as an initial guess. The initial error would then only contain the high frequency components and the relaxation would be highly efficient. In this section we will develop this idea further, and eventually arrive at the multigrid algorithm.

Let us assume that we have performed a number, m, of relaxed Jacobi iterations. The error is then smooth and the continuing iterations barely alter u^m. It is clear that we need another strategy. The first step is to observe that instead of solving

$$Au = b.$$

We can solve

$$Ae_m = r_m,$$

where

$$e_m = u - u_m,$$

and

$$r_m = Au_m - b.$$

The solution u is then obtained as

$$u = u_m + e_m.$$

The multigrid idea is based on the observation that while u can be any function, the error is smooth after a number of relaxations. Hence, the error can be represented well on a coarse grid. Moreover, the computations on a coarse grid is, of course, cheaper than on a fine grid. To clarify this idea, we again consider the 1D discretization of the Poisson problem. The generalization is performed afterwards.

In our 1D example it is easy to define a coarse grid. Let the fine grid be

$$\Omega_h = \{x_j = jh, j = 0, 1, \ldots, N + 1\}.$$

Then a coarser grid can be defined analogously,

$$\Omega_H = \{x_j = jH, j = 0, 1, \ldots, M + 1\}.$$

An obvious choice is $H = 2h$, which for simplicity is assumed in what follows. Hence, instead of solving

$$A_h e_h = r_h,$$

we solve

$$A_H e_H = r_H,$$

because we know that the Jacobi iterations have removed the high frequency error. Given the grid Ω_H we are able to define the coarse grid matrix A_H. However, a coarse-grid representation of the residual on the fine grid, r_h, is also needed. Such a representation may be constructed by using a restriction operator such that

$$r_H = I_h^H r_h.$$

The notation indicates that I_h^H generates an approximation r_H on the coarse scale, based on the fine-scale version represented by r_h. A suitable choice, in this case, is the weighted restriction operator defined by

$$u_j^H = (I_h^H u^h)_j = \frac{1}{4}(u_{2j-1} + 2u_{2j} + u_{2j+1}), \quad j = 1, \dots, M.$$

With this restriction operator we are able to compute the coarse error

$$Ae_H = r_H = I_h^H r_h.$$

Finally, it is necessary to represent the coarse error on the fine grid. A suitable interpolation is

$$I_H^h = 2(I_h^H)^T,$$

which, componentwise, reads

$$u_{2j}^h = u_j^H, \quad j = 1, \dots, M,$$
$$u_{2j+1}^h = \frac{1}{2}(u_j^H + u_{j+1}^H) \quad j = 0, \dots, M.$$

4.4.2 Theoretical Framework

It is now time to generalize the ideas and present a more polished theoretical framework of multigrid methods; see also [13] and [146]. Let $\Omega_0 \subset \Omega_1 \subset \dots \subset \Omega_L$ be a nested sequence of quasi-uniform grids, where Ω_J is the finest grid and Ω_0 the coarsest. We assume that the coarsest grid is significantly coarser than the finest, such that the cost of the solution of the coarse-grid problem can be neglected when compared to a smoothing operation on the finest grid. Furthermore, let V_J be a finite element space on Ω_J that consists of piecewise continuous polynomials of degree r; typically r is 1 or 2. Then $V_0 \subset V_1 \subset \dots \subset V_L \subset H_0^1$. Notice that this assumption about nested grids and finite element spaces is not essential, but that it simplifies the analysis significantly.

The next thing to notice is that the restriction and the interpolation operators are defined naturally by the finite element spaces. The natural restriction operator is the L_2 projection, $Q_J : L_2 \rightarrow V_J$, defined by

$$(Q_J f, N) = (f, N), \quad \forall N \in V_J \text{ and } f \in L_2,$$

where $(\cdot, \cdot)$ is the continuous L_2 inner product defined by

$$(f, g) = \int_\Omega fg \, \partial\Omega.$$

This inner product is the natural extension of the discrete L_2 inner product in (4.7). The interpolation operator is defined implicitly because $V_{J-1} \subset V_J$. In other words, a basis function N_j^{J-1} on the coarse grid Ω_{J-1} can be expressed by a sum of basis functions $\{N_i^J\}$ on the fine grid Ω_J,

$$N_j^{J-1} = \sum_i \alpha_{ij} N_i^J.$$

Hence, if the solution on grid $J-1$ is $u_{J-1} = \sum_j u_j^{J-1} N_j$, then the interpolation operator I_{J-1}^J, mapping u^{J-1} to u^J, is defined by:

$$u^J = I_{J-1}^J u_{J-1} = \sum_j u_j^{J-1} \sum_i \alpha_{ij} N_i^J.$$

Next, we must define the linear systems on the different grids. In the previous Chapter 2 we defined the weak formulation of the Poisson problem,

$$-\Delta u = f, \text{ in } \Omega, \tag{4.97}$$
$$u = 0, \text{ on } \partial\Omega. \tag{4.98}$$

The weak formulation, introduced in Chapter 2 of (4.97)–(4.98) is:
Find $u \in V$ such that

$$(Au, v) = (\nabla u, \nabla v), \quad \forall u, v \in V,$$

where V was some Hilbert space.

We are now able to define the linear systems to be solved approximately on the different grids by using the weak formulation on the different finite element spaces V_J. The linear systems are

$$A_J u_J = f_J, \quad J \in [0, L],$$

where

$$f_J = Q_J f, \quad A_J = Q_J A.$$

The explicit expressions for the matrices and right-hand sides are as follows. Let $\{N_j^J\}$ be the finite element basis functions that span V_J. Then

$$f_j = \int_\Omega f N_j^J dx, \quad A_{ij}^J = \int_\Omega \nabla N_i^J \nabla N_j^J dx.$$

Finally, we need to specify sufficient conditions for the approximate solvers (smoothers) on the different grids to ensure convergence. Let S_J denote a smoother. As an example, the relaxed Jacobi method would correspond to $S_J = \omega D_J^{-1}$, with D_J being the diagonal of A_J. As before, these smoothers need to be convergent; that is

$$\rho(S_J A_J) \le \sigma, \quad J \in [0, L], \tag{4.99}$$

where $\sigma \subset (0, 2)$. The other condition basically says that the smoother has to be close to A_J^{-1} for the high frequency components. This can be stated as

$$(S_J^{-1} v, v) \le \alpha(A_J v, v), \quad \forall v \in (I - Q_{J-1}) V_J, \quad J \in [1, L] \tag{4.100}$$

where $\alpha > 0$. This condition can be seen as a generalization of the eigenvalue result for the relaxed Jacobi (4.96).

Finally, we state the V-cycle multigrid algorithm. Notice that there are a number of generalizations of this algorithm, e.g., W-cycle and full multigrid; see e.g., [15] and [135].

The Multigrid V-cycle Algorithm

I: $B_1 = A_1^{-1}$
II: $B_j g = v^3$ where
$\qquad v^0 = 0$
$\qquad v^1 = v^0 - S_j(A_j v^0 - f)$
$\qquad v^2 = v^1 - B_{j-1} Q_{j-1}(A_j v^1 - f)$
$\qquad v^3 = v^2 - S_j(A_j v^2 - f)$

4.4.3 Convergence Theory

In Section 4.2.3 we described how the efficiency of iterative methods can be stated in terms of the convergence rate

$$\rho = \|I - \tau B A\|,$$

where,

$$\|e^n\| \le \rho \|e^{n-1}\|.$$

For order-optimal algorithms $\rho \le c < 1$ independent of h and this leads to a bounded number of iterations independent of h. In the following we set $\tau = 1$ for simplicity.

We used the norm $\| \cdot \|$ for convenience, but in fact, any norm can be used[4]; see [57], Chapter 3. Another common norm in the convergence analysis of multigrid methods (and domain decomposition methods) is the $\| \cdot \|_A$-norm. The A-norm of v is defined as

$$\|v\|_A = (Av, v)^{1/2},$$

and the corresponding A-norm of a matrix C is

$$\|C\|_A = \sup_v \frac{(Cv, v)_A}{(v, v)_A}.$$

With these norms the convergence rate is

$$\rho_A = \|I - BA\|_A,$$

and the error estimate is

$$\|e^n\|_A \le \rho_A \|e^{n-1}\|_A.$$

---

[4] As long as the matrix norm corresponds to the vector norm.

It is shown in [4] and [146] that given that the smoothers S_J satisfy the assumptions (4.99) and (4.100), we have $\rho_A < 1$ independent of h. In fact,

$$\rho_A \leq 1 - \frac{2 - \sigma}{\alpha} < 1.$$

As a rule of the thumb, the convergence rate $\rho_A \leq \frac{1}{10}$ in the case of a discretization of the Poisson equation on a simple geometry.

4.4.4 Numerical Experiments

We now present an experiment with multigrid. Let the model problem be

$$-\Delta u = f, \quad \text{in } \Omega,$$
$$u = e^{xy}, \quad \text{on } \partial\Omega,$$

where the solution is $u = e^{xy}$ and $\Omega = [0,1] \times [0,1]$. The grid hierarchies are constructed as successive refinements of a 2×2 partition of the unit square. The initial guess is a highly oscillating random function. This random function is used to make the stress test for multigrid as difficult as possible. It should contain "every possible" error. We use $\frac{\|r_k\|}{\|r_0\|} \leq 10^{-8}$ as the stopping criterion. The number of iterations needed to achieve convergence is shown in Table 4.5. The number of iterations seems to be independent of h, which is as it should be .

Table 4.5. The number of iterations, n, to achieve convergence with for a Poisson problem with respect to the grid size h.

h	2^{-2}	2^{-3}	2^{-4}	2^{-5}	2^{-6}	2^{-7}
n	5	6	6	6	6	6

In Figure 4.5 the dramatic improvement of the solution during one V-cycle is displayed. It should be clear that the combination of smoothing and coarse grid correction is powerful.

4.5 Domain Decomposition

Due to the development of parallel computers, domain decomposition methods have become very popular, particularly over the last two decades. Many researchers have worked on such methods and the result is an abstract and mature theoretical foundation, at the same level as multigrid methods. The idea was introduced as early as 1870 by Schwarz, who made an algorithm for computing the solution on a compound domain $\Omega = \Omega_1 \cup \Omega_2$, by successively solving similar problems on the

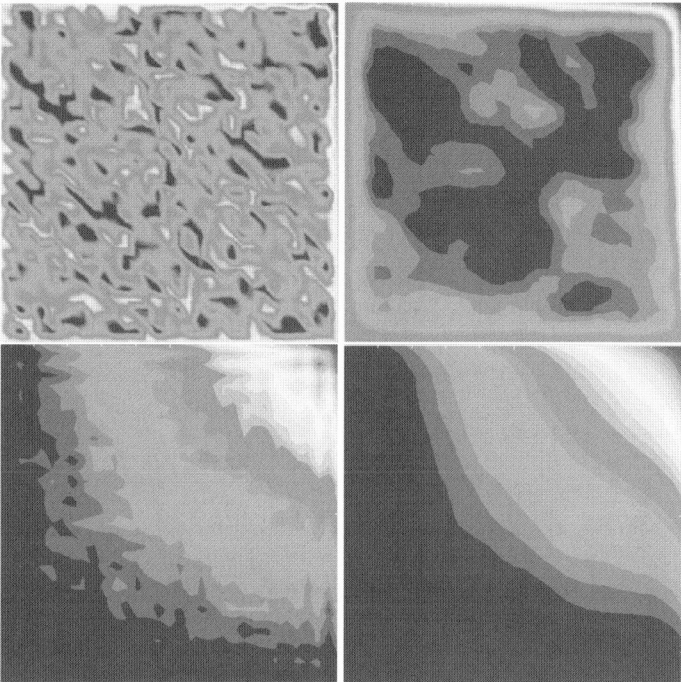

Fig. 4.5. The upper left picture shows the initial vector. It is a random vector that should contain "all possible" errors. In the upper right picture the solution after one symmetric Gauss–Seidel sweep is displayed. It is clear that the random high-frequency behaviour in the initial solution has been effectively removed. The picture down to the left shows the solution after the coarse grid correction. The smooth components of the solution have improved dramatically. In the last picture the solution after the post smoothing is displayed. The solution is now very close to the actual solution. (For the color version, see Figure A.9 on page 291).

simpler domains Ω_1 and Ω_2. Later, in particular from the 1980s and onwards, researchers discovered that the technique was a powerful algorithm for the efficient solving of linear systems that arise from the discretizations of various PDEs, particularly on parallel computers. For a more detailed description of domain decomposition methods, see [22] and [128].

We begin this section by reviewing the results from Schwarz's early paper [124], because it is a very good illustration of the method. Schwarz was interested in finding the solution to the following Poisson problem[5]

$$-\Delta u = f, \quad \text{in } \Omega,$$
$$u = g, \quad \text{on } \partial\Omega,$$

where the domain $\Omega = \Omega_1 \cup \Omega_2$ is depicted in Figure 4.6.

---

[5] Schwarz actually studied the Laplace equation, i.e., the Poisson problem with a homogeneous right-hand side, $f = 0$. However, the algorithm works for any f.

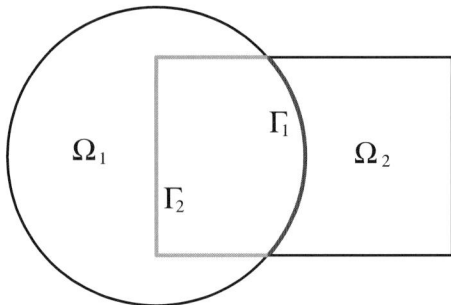

Fig. 4.6. Schwarz problem domain.

The solutions of the Poisson problem on the simpler domains Ω_1 and Ω_2 were known. Schwarz's powerful idea was to perform an iteration that repeatedly reuses the solutions in Ω_1 and Ω_2 to get the proper boundary conditions on Γ_1 and Γ_2, and thereby the solution in the compound domain Ω. The algorithm can be summarized as follows. First, the solution in Ω_1 is computed as the solution of the following problem:

$$-\Delta u_1^n = f, \quad \text{in } \Omega_1,$$
$$u_1^n = g, \quad \text{on } \partial\Omega_1 \backslash \Gamma_1,$$
$$u_1^n = u_2^{n-1}, \quad \text{on } \Gamma_1.$$

The unknown boundary condition on Γ_1 is based on the previous solution in the domain Ω_2 or an initial guess. Once u_1^n is computed, it is used on the unknown boundary Γ_2 such that u_2^n can be computed by

$$-\Delta u_2^n = f, \quad \text{in } \Omega_2,$$
$$u_2^n = g, \quad \text{on } \partial\Omega_2 \backslash \Gamma_2,$$
$$u_2^n = u_1^n, \quad \text{on } \Gamma_2.$$

After the new solution, u_2^n, is computed, the next iteration starts with the computation of u_1^{n+1} with new boundary conditions based on u_2^n and the iteration continues. Schwarz was able to prove that this iteration converges to the actual solution in Ω.

We will now generalize the alternating Schwarz algorithm to handle many subdomains and make it suitable for parallel computers. However, we will not go into the parallelization here. This is a large field of its own and is covered in Chapter 6. Instead, we will focus on the mathematical properties of the algorithm. Let Ω_h be a triangulation of Ω and furthermore let $\Omega^1, \ldots, \Omega^p$ be an overlapping subdivision of Ω_h, where p will typically equal the number of processors on the parallel computer. The subdomains are usually constructed as follows. Let $\hat{\Omega}^1, \ldots, \hat{\Omega}^p$ be a partition of non-overlapping domains such that $\Omega = \hat{\Omega}^1 \cup \ldots \cup \hat{\Omega}^p$ and $\hat{\Omega}^i \cap \hat{\Omega}^j = 0$ for $i \neq j$. Each of the subdomains $\hat{\Omega}^i$ is then extended with a distance βH such that the subdomains become overlapping.

A convenient assumption is that the subdomains are nested in the sense that $\Omega^i \subset \Omega_h$. This assumption is not necessary, but simplifies the exposition. Inherited from the nestedness of the grids there exists a restriction matrix $R_i : \Omega_h \to \Omega^i$, such that

$$R_i(x_j) = 1, \quad \text{if } x_j \in \Omega^i, \tag{4.101}$$

$$R_i(x_j) = 0, \quad \text{elsewhere.} \tag{4.102}$$

Here, x_j are the vertices (or nodal points) in the grid Ω_h. The corresponding interpolation operator is simply the transposed of the restriction matrix, R_i^T. Finally, the subdomain matrices are defined by

$$A_i = R_i A R_i^T.$$

With the above definitions of A_i, R_i and R_i^T we can now state the additive Schwarz algorithm.

The Additive Schwarz Algorithm

for $i = 1, \ldots, p$:
$$u^{n+1} = u^n - \tau R_i^T A_i^{-1} R_i (Au^n - b)$$

The additive Schwarz algorithm can be expressed as a preconditioner, B_a, for the Richardson iteration (4.27), as follows:

$$B_a = \sum_{i=1}^{p} R_i^T A_i^{-1} R_i. \tag{4.103}$$

Notice also that the previously computed u^n is used in each step. This means that the algorithm is parallel by nature. Furthermore, if A is symmetric then so is B_a. It may be necessary to adjust τ to ensure convergence.

The multiplicative Schwarz algorithm is very similar to the additive Schwarz algorithm. The only difference is that the most recently computed values of u are always used. Further, the multiplicative Schwarz is convergent with $\tau = 1$ and therefore the τ parameter is usually avoided.

The Multiplicative Schwarz Algorithm

for $i = 1, \ldots, p$:
$$u^{n+\frac{i}{p}} = u^{n+\frac{i-1}{p}} - R_i^T A_i^{-1} R_i (Au^{n+\frac{i-1}{p}} - b)$$

The multiplicative Schwarz can also be represented in terms of a preconditioned Richardson iteration. We have the identity

$$I - B_m A = (I - R_p^T A_p^{-1} R_p) \cdots (I - R_1^T A_1^{-1} R_1), \tag{4.104}$$

where B_m is the preconditioner in the Richardson iteration (4.27).

The multiplicative Schwarz algorithm is generally more efficient than the additive Schwarz algorithm on a scalar computer. However, the multiplicative Schwarz is a sequential algorithm and is not suitable for parallel computers. In its present form, the algorithm is not symmetric. However, a symmetric version can be made by one standard multiplicative Schwarz iteration followed by an additional iteration with the domains numbered backwards. These algorithms are generalizations of the block Jacobi (additive Schwarz) and block Gauss–Seidel (multiplicative Schwarz) to overlapping blocks.

There is one major problem with the algorithms as stated above which is that their efficiency depends on the number of subdomains p. In fact, the efficiency deteriorates quite rapidly as p increases. A cheap, but efficient solution to the problem of the p dependency is to introduce a coarse grid Ω_H, with characteristic grid size H, where H is much larger than h. This grid can be very coarse (in contrast to multigrid methods). We refer to the matrix on the coarse grid as A_H and the restriction operator is R_H. Notice that the restriction operator is not on the form (4.101)–(4.102), but is instead similar to the restriction operator used for multigrid methods.

The above definition of the Schwarz algorithms can easily employ coarse grid correction. Instead of the previous numbering where $\Omega^1, \ldots, \Omega^p$ is an overlapping partition of Ω_h; we number the domains as $\Omega^2, \ldots, \Omega^{p'}$, where $p' = p + 1$. In addition, the coarse grid is used. Let $\Omega^1 = \Omega_H$. With this numbering of the domains and the corresponding matrices A_i and R_i, the above algorithms extend directly to the case with a coarse grid, given that p is replaced with p'.

The following results are known for the additive and multiplicative Schwarz preconditioners. For the additive Schwarz preconditioner with a coarse grid correction, B_a, the condition number of $B_a A$ is independent of h. In fact,

$$K(B_a A) \leq C(1 + \beta^{-1}).$$

Notice that this does not mean that the additive Schwarz algorithm with coarse grid correction is necessarily convergent. However, τ can be chosen such that the convergence rate,

$$\rho_A = \|I - \tau B_a A\|_A < 1,$$

is independent of h and H. The choice of τ is dealt with in more detail later, in Section 4.6.

On the other hand, the multiplicative Schwarz method with coarse grid correction is convergent and the convergence rate, $\rho_A < 1$, is independent of h and H, where

$$\rho_A = \|I - B_m A\|_A.$$

The proof can be found in, e.g, [22] and [128]. Without a coarse grid correction the convergence rate typically deteriorates as $\mathcal{O}(H^{-2})$.

There are also nonoverlapping domain decomposition algorithms where the underlying domains are not overlapping. Moreover, the domains may not even be motivated geometrically. We will not go into these algorithms here, instead we refer to, e.g., [22] and [128].

4.6 Preconditioning Revisited

4.6.1 Idea

We have now introduced the classical iterations, the multigrid method, domain decomposition and the Conjugate Gradient method as separate methods. It is now time to combine them to derive a framework suitable for handling the equations of the Bidomain model. The gluing concept is *preconditioning*.

As before, we want to solve the linear system

$$Au = b. \tag{4.105}$$

We saw earlier that the efficiency of both the Richardson and the Conjugate Gradient method depends on the condition number of A, which was typically of order h^{-2}. The idea of preconditioning is simply to multiply (4.105) with a matrix or linear operator B to obtain an equivalent system,

$$BAu = Bb, \tag{4.106}$$

where B is usually called the *preconditioner*. The system (4.106) has the same solution as (4.105) provided that B has full rank[6]. The hope is then that (4.106) is easier to solve than (4.105). The preconditioner B can be any matrix or linear operator that in some sense resembles A^{-1}. For instance, it may be based on an approximate factorization of A, or it may be a sweep with the Jacobi or the multigrid method. What is crucial is that B should be a good approximation of A^{-1} and also inexpensive with respect to storage and evaluation.

Earlier, we considered order-optimal solution algorithms, and we will now relate these algorithms to *order-optimal preconditioners*. First, we need the concept of spectral equivalence, which describes what a "good" approximation of A^{-1} is.

4.6.2 Spectral Equivalence

Notice that A and B are families of matrices with respect to the triangulations Ω_h of Ω, where h approaches zero in the limit. It is, therefore, common to let A and B have the subscript h, A_h and B_h.

The two linear operators or matrices, A_h^{-1} and B_h, which are assumed to be symmetric and positive definite[7], are spectrally equivalent (independent of h) if there exist constants c_1 and c_2, independent of h such that

$$c_1(A_h^{-1}v, v) \le (B_h v, v) \le c_2(A_h^{-1}v, v), \quad \forall v. \tag{4.107}$$

Alternatively, we may express this property as

$$c_1(A_h v, v) \le (A_h B_h A_h v, v) \le c_2(A_h v, v), \quad \forall v. \tag{4.108}$$

---

[6] A matrix has full rank if it is invertible.

[7] If A_h and B_h are symmetric and positive definite then so are A_h^{-1} and B_h^{-1}.

It is known that if A_h^{-1} and B_h are spectrally equivalent, the condition number of $B_h A_h$ is bounded by the constants c_1 and c_2,

$$K(B_h A_h) \leq \frac{c_2}{c_1}. \tag{4.109}$$

It is common to denote spectral equivalence of A_h^{-1} and B_h by $A_h^{-1} \sim B_h$.

With the definition of spectral equivalence we can now state what a well-designed preconditioner should look like:

- B_h should be spectrally equivalent to A_h^{-1},
- the evaluation of B_h on a vector v, $B_h v$, should cost $\mathcal{O}(N)$ operations,
- the storage of B_h should be similar to the storage of A_h, $\mathcal{O}(N)$ floating point numbers.

In what follows we will see that an order-optimal preconditioner leads to an order-optimal solution algorithm, in the case of both the Richardson and the Conjugate Gradient methods.

4.6.3   The Richardson Iteration Re-Revisited

We can now derive what the spectral equivalence of B and A^{-1} means in the context of the preconditioned Richardson iteration. We remember from (4.28) that the error at the n iteration, e^n, could be stated in terms of the error at the previous iteration, e^{n-1}:

$$e^n = (I - \tau B A)e^{n-1}. \tag{4.110}$$

The convergence rate in the A-norm is

$$\rho_A = \|I - \tau B A\|_A, \tag{4.111}$$

and the error reduction can be estimated as

$$\|e^n\|_A \leq \rho_A \|e^{n-1}\|_A.$$

Because BA is symmetric with respect to the A inner product, ρ_A can be stated in terms of the eigenvalues of BA, μ_i,

$$\rho_A = \|I - \tau B A\|_A = \sup_{\mu_i} |1 - \tau \mu_i|.$$

Hence, ρ_A is a linear polynomial in μ_i and its maximum is obtained at the extreme values of μ_i,

$$|1 - \tau \mu_0| \quad \text{or} \quad |1 - \tau \mu_N|,$$

where μ_0 and μ_N are the smallest and largest eigenvalues, respectively. We choose τ as the minimizer of $|1 - \tau \mu_i|$. The minimum is obtained when

$$1 - \tau \mu_0 = \tau \mu_N - 1.$$

which makes

$$\tau = \frac{2}{\mu_0 + \mu_N}$$

the optimal choice. With this choice of τ, ρ_A is

$$\rho_A = 1 - \tau\mu_0 = 1 - \frac{2\mu_0}{\mu_0 + \mu_N} = \frac{\mu_N - \mu_0}{\mu_N + \mu_0} = \frac{K-1}{K+1}, \qquad (4.112)$$

and we have the corresponding error estimate,

$$\|e^n\|_A \le \left(\frac{K-1}{K+1}\right)^n \|e^0\|_A.$$

Hence, when A and B^{-1} are spectrally equivalent, the condition number K is independent of h, and therefore the convergence rate, ρ_A, is independent of h (with a reasonable choice of τ).

4.6.4   Preconditioned Conjugate Gradient Method

In this section, we will extend the Conjugate Gradient method with a preconditioner. This is nearly always done in practice, at least when CG is used to compute the numerical solution of PDE problems. First, observe that BA is not necessarily symmetric, even if A and B are. Symmetry is required to ensure convergence of the Conjugate Gradient method, as we saw in Section 4.3. However, the algorithm can be stated in any inner product. The inner product of choice when using a preconditioner is $(\cdot, \cdot)_{B^{-1}}$ defined as

$$(u, v)_{B^{-1}} = (B^{-1}u, v).$$

The preconditioner B is positive definite and symmetric, which means that B^{-1} also is positive definite and symmetric. Consequently, also B^{-1} defines an inner product. Moreover, BA is obviously symmetric in the B^{-1} inner product, because

$$(BAu, v)_{B^{-1}} = (Au, v) = (u, Av) = (u, BAv)_{B^{-1}}.$$

Still, we do not want to form either B^{-1} or BA. In fact, when using either multigrid or domain decomposition as preconditioner, the only available action is evaluation on a vector, Bv. Fortunately, the Conjugate Gradient method applied to BA in the B^{-1} inner product can be stated similarly to the Conjugate Gradient algorithm on page 119. The only difference is the use of an additional vector and the evaluation of B.

The Preconditioned Conjugate Gradient Algorithm
Let $A \in \mathbb{R}^{N,N}$, $B : \mathbb{R}^N \rightarrow \mathbb{R}^N$, $f \in \mathbb{R}^N$, $u^0 \in \mathbb{R}^N$, and $0 < \varepsilon < 1$ be given. The matrix A and the preconditioner B are supposed to be symmetric and positive definite.
$u = u^0$
$r = f - Au$
$s = Br$
$p = s$
$\rho_0 = (s, r)$
$k = 0$
While $\rho_k / \rho_0 > \varepsilon$ **do**

$$
\begin{array}{lll}
z & = Ap & (4.113) \\
t & = Bz & (4.114) \\
\gamma & = (p, z) & (4.115) \\
\alpha & = \rho_k / \gamma & (4.116) \\
u & = u + \alpha p & (4.117) \\
r & = r - \alpha z & (4.118) \\
s & = s - \alpha t & (4.119) \\
\rho_{k+1} & = (s, r) & (4.120) \\
\beta & = \rho_{k+1} / \rho_k & (4.121) \\
p & = s + \beta p & (4.122) \\
k & = k + 1 & (4.123)
\end{array}
$$

end

4.6.5 Convergence Analysis Revisited

The convergence analysis in Section 4.3 extends directly to the case with a preconditioner, but the condition number is $K = K(BA)$ instead of $K(A)$:

$$
\|e^n\|_A \leq \left(\frac{\sqrt{K} - 1}{\sqrt{K} + 1} \right)^n \|e^0\|_A.
$$

In the case where $B \sim A^{-1}$, $K \leq \frac{c_2}{c_1}$ by (4.109) and is bounded independently of h. This is always an improvement over the preconditioned Richardson iteration,

$$
\frac{\sqrt{K} - 1}{\sqrt{K} + 1} = \frac{K - 1}{K + 2\sqrt{K} + 1} < \frac{K - 1}{K + 1}.
$$

However, if K is small, the improvement is not large. This is the case for the Poisson equation, because the multigrid methods or domain decomposition methods are highly efficient. In fact, in the case of multigrid preconditioning, a convergence rate of about $\frac{1}{10}$ should be expected. Using the relation (4.112) the condition number can be estimated to $K \approx 1.2$, which makes $\sqrt{K} \approx 1.1$. Moreover, each iteration with the Conjugate Gradient method is slightly heavier than an Richardson iteration and there is, therefore, not much to gain by accelerating an efficient multigrid method with the Conjugate Gradient method. However, if robustness is an issue, the preconditioned Conjugate Gradient method should be considered. This will be exemplified in the next section.

4.6.6 Variable Coefficients

In this section we will briefly consider the case of elliptic equations with a variable coefficient. The reason for this study is that the electrical conductivity in the heart and body vary spatially. In fact, the ratio between the largest and the smallest conductivities may be almost as large as 100 (between blood and bone) and the conductivity is discontinuous. Here, we will consider how such variations affect the multigrid and domain decomposition performance. The model problem is

$$-\nabla \cdot (M\nabla u) = f, \quad \text{in } \Omega, \qquad (4.124)$$
$$u = 0, \quad \text{on } \partial\Omega, \qquad (4.125)$$

where $M(x)$ in general is a matrix that describes the electrical conductivity of the media. Let the corresponding linear system be

$$Au = b.$$

Both multigrid and domain decomposition have problems removing the error associated with the jumps in M. In fact, even divergence may occur, depending on the jump and the geometry of M, see also [135].

By contrast, the preconditioned Conjugate Gradient method always converges, given that the matrix and the preconditioner are positive definite and symmetric. However, the efficiency of the Conjugate Gradient method depends on the condition number. We know from Sections 4.4 and 4.5 that efficient multigrid and domain decomposition algorithms can be made for discretizations of the Poisson problem

$$-\Delta u = f, \text{ in } \Omega, \qquad (4.126)$$
$$u = 0, \text{ on } \partial\Omega. \qquad (4.127)$$

The corresponding linear system is denoted by

$$Cu = c.$$

In this case, we know that it is possible to construct a preconditioner, B, based on, e.g., multigrid or domain decomposition techniques, that is spectrally equivalent to the inverse of C,

$$c_0(Bv, v) \le (C^{-1}v, v) \le c_1(Bv, v), \quad \forall v,$$

where c_0 and c_1 are independent of h. Furthermore, C and A are spectrally equivalent, because

$$m_0 \int_\Omega \nabla u \nabla v \, dx \leq \int_\Omega M \nabla u \nabla v \, dx \leq m_1 \int_\Omega \nabla u \nabla v \, dx,$$

where

$$m_0 = \min(M), \quad m_1 = \max(M).$$

Because $A \sim C \sim B^{-1}$, we end up with the following estimate on the condition number:

$$K(BA) = \frac{m_1}{m_0} \frac{c_1}{c_0}. \tag{4.128}$$

Hence, it appears that an order-optimal preconditioner for the Poisson problem yields an order-optimal preconditioner also in the case with a variable coefficient. However, the condition number depends on the coefficient M.

Finally, we remark that although we in the above analysis assumed that the multigrid or the domain decomposition algorithms employed matrices based on discretizations of the Poisson equation, this is not necessarily best in practice. Instead, one should base the multigrid or the domain decomposition solvers on matrices containing the variable coefficient on all levels. In this case, one does not have an equally solid theoretical framework as for the Poisson equation, but the preformance of the preconditioners, in particular in combination with the Conjugate Gradient method, are usually quite good.

4.6.7 Numerical Experiments

In this example we will consider three different solution algorithms; 1) multigrid as a stand-alone solver, with matrices based on the problem with variable coefficient (4.124), 2) multigrid as a preconditioner for the Conjugate Gradient method with matrices based on (4.124), and 3) multigrid as a preconditioner with matrices based on the Laplacian (4.126). Let the model problem be

$$-\nabla \cdot (M \nabla u) = f, \quad \text{in } \Omega,$$
$$u = 0, \quad \text{on } \partial\Omega,$$

where Ω is the unit square and the conductivity M is given as

$$M(x, y) = \begin{cases} m, & \text{if } 0.45 \leq x \leq 0.55 \text{ and } 0.1 \leq y \leq 0.9, \\ n, & \text{else,} \end{cases} \tag{4.129}$$

and f is given as

$$f(x, y) = \begin{cases} 1 & \text{if } \sqrt{(x - 0.3)^2 + (y - 0.3)^2} \leq 0.1, \\ 0 & \text{else.} \end{cases} \tag{4.130}$$

The conductivity M and the source function f are shown in Figure 4.7.

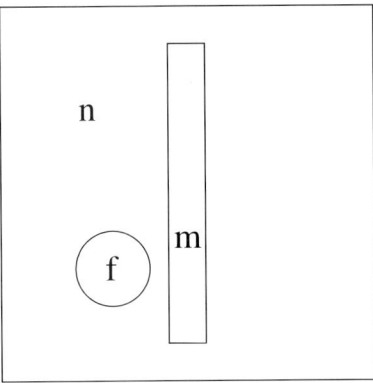

Fig. 4.7. The jump in the conductivity M.

The grid hierarchies are made as successive refinements of a 2×2 partition of the unit square with bilinear elements. The initial guess is a highly oscillating random function and we use $\frac{\|r_k\|}{\|r_0\|} \leq 10^{-8}$ as the stopping criterion. We let $m = 1, 1/10, 1/100, 1/1000$, while $n = 1$. As seen in Table 4.6, multigrid even diverges, if the jump in the conductivity is as large as 1000. However, when the jump is less than or equal to 100, which is the case in our body, all methods converge[8]. The multigrid preconditioner based on the matrices with variable coefficient is performing better than the other methods.

Table 4.6. The number of iterations with respect to M, h and the different multigrid solvers. The '-' means that the solver diverges.

$h \backslash M$	MG-Variable Coeff.				CG/MG-Laplacian				CG/MG-Variable Coeff.			
	1	10	100	1000	1	10	100	1000	1	10	100	1000
2^{-3}	6	7	7	7	5	13	14	14	5	5	5	5
2^{-4}	6	13	17	18	5	18	27	30	5	6	7	7
2^{-5}	7	10	16	-	5	24	44	54	5	7	9	15
2^{-6}	7	12	15	-	5	23	41	53	5	8	10	16
2^{-7}	7	12	14	-	5	27	54	81	5	8	9	14
2^{-8}	7	16	20	-	5	27	55	75	5	9	10	15

---

[8] One should notice that if the size of the elements vary, then this variation induces extra variation in the numerical conductivity.

4.7   The Monodomain Model

Recall the equation for the Monodomain model introduced in Chapter 2;

$$u_t - \nabla \cdot (M\nabla u) = f(u), \quad \text{in } \Omega, t > 0,$$
$$u(0) = u_0, \quad \text{in } \Omega, t = 0,$$
$$u(x) = 0, \quad \text{on } \partial\Omega,$$

where u is the transmembrane potential, M is a scaled conductivity, and $f(u)$ is a nonlinear function. One way to solve the equation is to split it in two parts:

$$u_t = \nabla \cdot (M\nabla u), \quad \text{in } \Omega, t > 0, \tag{4.131}$$

and

$$u_t = f(u), \quad \text{in } \Omega, t > 0, \tag{4.132}$$

and assign suitable boundary and initial conditions to these equations.

The main reason for performing the splitting is that efficient solution algorithms exist for both (4.131) and (4.132). Solution algorithms for (4.132) are described in Chapter 5. Here, we will focus on (4.131) and consider both multigrid and domain decomposition.

Using either implicit Euler or Crank–Nicholson for the time discretization and either FDM or FEM in space, we need to solve the following linear system at each time step:

$$(I + \Delta t A)u^k = u^{k-1}, \tag{4.133}$$

where A is similar-to the positive Laplacian. Notice that this linear system corresponds to the discretization of a reaction-diffusion equation:

$$u - \epsilon \Delta u = f, \tag{4.134}$$

where $\epsilon \ll 1$. One of the characteristics of this equation is that boundary layers may be present as ϵ approaches zero. This causes the solution u to be less regular than for standard elliptic equations, which again causes a deterioration of the approximation on coarse grids. The quality of the coarse-grid correction is important for both multigrid and domain decomposition, but is expected to decrease in the case of (4.134). This is not the case for the system (4.133), due to the particular right-hand side, which is the solution at the previous time step, u^{k-1}.

In the following sections we will describe multigrid and domain decomposition preconditioners that are independent of both h and Δt, in the sense that

$$c_0(Bu, u) \le ((I - \Delta t A)^{-1}u, u) \le c_1(Bu, u), \quad \forall u,$$

where c_0 and c_1 are independent of h and Δt.

Finally, we remark that the difference between two time steps decreases as $\Delta t \to 0$,

$$\|u^k - u^{k-1}\| \le C\Delta t.$$

Therefore, the solution from the previous time step, $k-1$, provides a very good start vector at the next time step, k. With this start vector, the error reduction required by the linear solvers is only $\sim\Delta t$. Hence, when using iterative methods, the linear system at each time step becomes less expensive to solve as Δt decreases.

4.7.1 Multigrid

The multigrid methods considered in Section 4.4 extend to the problem (4.133) c.f. [7], [107] and [134] and are independent of both Δt and h. This is described below.

As earlier, we have a sequence of nested quasi-uniform grids, $\Omega_0 \subset \Omega_1 \subset \ldots \subset \Omega_L$. For each grid the corresponding linear system is

$$(I - \Delta t A_J)u_J^k = u_J^{k-1}, \quad J = 0, \ldots, L. \tag{4.135}$$

The smoothers, restriction and interpolation operators are defined as in Section 4.4.

In Section 4.4 we saw that multigrid is an efficient algorithm for solving the following equation:

$$cu - \nabla \cdot (a\nabla u) = f, \tag{4.136}$$

where c and a are bounded below and above. However, we did not mention the case where $a \to 0$. This is the case here. We begin by considering the situation when $a = 0$. (In this case the solution has already been found at the previous time step and there is no need to solve a linear system. Look aside from this point.) The solution of (4.136) reduces to solving a linear system where the matrix is a mass matrix (with boundary conditions). This can be done optimally with standard classical iterations, because the mass matrix has a condition number independent of h and is diagonally dominant.

Hence, in the limit case of $a = 0$, the coarse grid correction is not needed. This is the general case with this equation. The performance of the smoothers improves as Δt decreases. In fact, it has been shown in [107] that the improved smoothing balance the loss in the approximation on the coarse grid, such that multigrid is an order-optimal method independently of Δt, even for the more difficult case of the reaction-diffusion equation. In Table 4.7 we see that the performance improves as Δt decreases for (4.133).

4.7.2 Numerical Experiments

Let the model problem be

$$u - \epsilon\Delta u = 1, \quad \text{in } \Omega,$$
$$u = 0, \quad \text{on } \partial\Omega.$$

The linear system is solved with multigrid as a stand-alone solver. The grid hierarchies are constructed as successive refinements of a 2×2 partition of the unit square with bilinear elements. The initial guess is a highly oscillating random function. We

Table 4.7. The number of iterations as a function of ϵ and h.

$h\backslash\epsilon$	1	2^{-1}	2^{-2}	2^{-3}	2^{-4}	2^{-5}	2^{-6}	2^{-7}
2^{-3}	4	4	3	3	4	3	2	2
2^{-4}	4	4	4	4	4	4	3	2
2^{-5}	5	5	5	4	5	5	4	3
2^{-6}	5	5	5	5	5	5	4	4
2^{-7}	4	4	4	5	4	4	4	4
2^{-8}	4	4	4	4	4	4	4	4
2^{-9}	4	4	4	4	4	4	4	4
2^{-10}	4	4	4	4	4	4	4	4

remark that although an oscillating random function will not be used in practice because the solution on the previous time step is a very good start vector, it is a good stress test for the multigrid method. We use $\frac{\|r_k\|}{\|r_0\|} \leq 10^{-6}$ as the stopping criterion. Multigrid seems to be robust with respect to both h and ϵ; in fact, the performance of multigrid seems to improve slightly as ϵ decreases.

4.7.3 Domain Decomposition

The domain decomposition methods considered in Section 4.5 extend to the problem (4.133) (c.f. [18], [19] and [22]) and are independent of both Δt and h. This is described in what follows.

As earlier in Section 4.5, the domain is divided into p overlapping subdomains, $\Omega_h = \Omega_h^1 \cup \ldots \cup \Omega_h^p$, with an additional coarse grid Ω_H. The restriction and interpolation operators, R_i and R_i^T, are derived from the geometrical relation between the subdomain Ω_h^i and Ω_h, with no reference to the matrix to be solved. Therefore, these operators can also be used in this application. This also applies to the restriction and interpolation operators for the coarse grid. Therefore, the subdomain matrix is constructed as

$$(I - \Delta t A)_i = R_i (I - \Delta t A) R_i^T.$$

With these subdomain matrices, the Schwarz algorithms in Section 4.5 can be used directly.

In the previous section, where we considered the multigrid methods, we noticed that the smoothers improved as $\Delta t \to 0$. In addition, for domain decomposition methods, the error appears to be become more local as Δt decreases. In fact, if $\Delta t \leq CH^2$, the coarse grid solver is not necessary; see [18] and [19].

4.8 The Bidomain Model

In the previous sections we saw that it is possible to construct order-optimal preconditioners for both of the matrices A and $I - \Delta t A$, where A was similar to a discrete Laplacian. However, the Bidomain model contains a system of PDEs in which each component is similar to either A or $I - \Delta t A$. It is not easy to extend the theoretical framework developed for A and $I - \Delta t A$ to this model problem directly, but it is possible. However, that is not what we will do here. Instead, we will show that these components can be reused in the powerful concept of (block) preconditioning, to give an order-optimal preconditioner.

Block preconditioners for the Bidomain model are described in, e.g., [96], [111], [130]. It can be proven, see [95], that both the block Jacobi preconditioner and the symmetric block Gauss–Seidel preconditioners are order-optimal. We briefly review the theory for the block Jacobi preconditioner here. Numerical experiments, using a combination of multigrid and domain decomposition on a parallel computer, are presented in Section 6.8.4 in Chapter 6. Block preconditioners for other systems of PDEs have been considered in, e.g., [5], [35], [36], [75], [97], [122], and [138].

To clarify the description we restate the equations that were derived in Chapter 2,

$$v_t = \nabla \cdot (M_i \nabla v) + \nabla \cdot (M_i \nabla u),$$
$$0 = \nabla \cdot (M_i \nabla v) + \nabla \cdot ((M_i + M_e) \nabla u).$$

In Chapter 3, these equations were discretized by the finite element method in space and a Crank–Nicholson scheme in time, such that we arrive at the following system of algebraic equations:

$$I v^n + \frac{\Delta t}{2} A_i v^n + \frac{\Delta t}{2} A_i u^n = I v^{n-1} - \frac{\Delta t}{2} A_i v^{n-1} - \frac{\Delta t}{2} A_i u^{n-1},$$
$$\frac{\Delta t}{2} A_i v^n + \frac{\Delta t}{2} A_{i+e} u^n = -\frac{\Delta t}{2} A_i v^{n-1} - \frac{\Delta t}{2} A_{i+e} u^{n-1}.$$

Recall that the matrices A_i and A_{i+e} are discretizations of the above differential operators, i.e.,

$$A_i = (\nabla \cdot M_i \nabla)_h,$$
$$A_{i+e} = (\nabla \cdot M_{i+e} \nabla)_h,$$

and I is the mass matrix in the finite element method. Later we will also need the discrete Laplacian,

$$A_0 = \Delta_h.$$

The right-hand sides are not of particular importance when considering preconditioners. In fact, we construct the preconditioners such that they handle any right-hand side. That is, letting b^{n-1} and c^{n-1} to be the respective right-hand sides,

$$\begin{pmatrix} b^{n-1} \\ c^{n-1} \end{pmatrix} = \begin{pmatrix} I v^{n-1} - \frac{\Delta t}{2} A_i v^{n-1} - \frac{\Delta t}{2} A_i u^{n-1} \\ -\frac{\Delta t}{2} A_i v^{n-1} - \frac{\Delta t}{2} A_{i+e} u^{n-1} \end{pmatrix},$$

then we can write (4.137) as

$$\begin{pmatrix} I + \frac{\Delta t}{2} A_i & \frac{\Delta t}{2} A_i \\ \frac{\Delta t}{2} A_i & \frac{\Delta t}{2} A_{i+e} \end{pmatrix} \begin{pmatrix} v^n \\ u^n \end{pmatrix} = \begin{pmatrix} b^{n-1} \\ c^{n-1} \end{pmatrix}. \tag{4.137}$$

The matrix in (4.137) is symmetric and positive definite. This makes the Conjugate Gradient method an appropriate iterative solver, when combined with a suitable preconditioner. In what follows we will describe a block preconditioner, where the blocks are constructed by preconditioners for A_i and $I + \frac{\Delta t}{2} A_{i+e}$. These were described earlier in this chapter.

We will now describe an order-optimal preconditioner for the Bidomain model. The matrix in (4.137) is

$$T = \begin{pmatrix} I + \frac{\Delta t}{2} A_i & \frac{\Delta t}{2} A_i \\ \frac{\Delta t}{2} A_i & \frac{\Delta t}{2} A_{i+e} \end{pmatrix}. \tag{4.138}$$

In what follows we will show that T is spectrally equivalent to the inverse of the block Jacobi preconditioner. Let

$$S = \begin{pmatrix} I + \frac{\Delta t}{2} A_0 & 0 \\ 0 & \frac{\Delta t}{2} A_0 \end{pmatrix}. \tag{4.139}$$

The exact Jacobi preconditioner will then be S^{-1}, but, of course, inverting S is too costly. However, in Sections 4.4, 4.5 and 4.7, order-optimal preconditioners for A_0 and $I + \frac{\Delta t}{2} A_0$ were described. Therefore, we know that it is possible to construct an order-optimal preconditioner for S. This is, indeed, the main motivation behind constructing the preconditioner on this block diagonal form. Let this preconditioner be R, defined as

$$R = \begin{pmatrix} \widehat{(I + \frac{\Delta t}{2} A_0)}^{-1} & 0 \\ 0 & \widehat{(\frac{\Delta t}{2} A_0)}^{-1} \end{pmatrix}, \tag{4.140}$$

where the notation $\widehat{(\cdot)}^{-1}$ denotes an approximate inverse. Spectral equivalence is associative, in the sense that if A, B, and C are three matrices and $A \sim B$ and $B \sim C$, then $A \sim C$. We have that $R^{-1} \sim S$. Therefore, it remains to show that $S \sim T$, and this would imply that $R^{-1} \sim T$. Then, we can deduce that the condition number of the preconditioned system is bounded[9],

$$K(RT) \leq c. \tag{4.141}$$

where c is independent of h and Δt.

Hence, the concept of block preconditioning allow us to build an order-optimal preconditioner by reusing standard algorithms based on multigrid and domain decomposition for linear scalar PDEs that are either elliptic or parabolic. These methods have been studied extensively in the literature.

The last piece of the puzzle is to prove that S and T are spectrally equivalent, $S \sim T$, c.f. Section 4.6.2. This is achieved in the rest of this section.

Assumptions We assume that there exist constants c_0 and c_1, independent of h and Δt, such that

$$c_0(A_0 v, v) \leq (A_\alpha v, v) \leq c_1(A_0 v, v), \tag{4.142}$$

for $\alpha = i, e$.

---

[9] Notice that we assume that S, T and R^{-1} are spectrally equivalent independent of h and Δt, c.f. Section 4.6.2.

The assumption simply states that the matrix generated by the Laplace operator is spectrally equivalent to the matrices generated by the variable coefficient problems.

Notice that, since

$$(A_{i+e}u, u) = (A_i u, u) + (A_e u, u),$$

we have from (4.142) that

$$2c_0(A_0 v, v) \le (A_{i+e}u, u) \le 2c_1(A_0 v, v). \tag{4.143}$$

The proof is split in two parts, the upper and the lower bound. We start with the upper bound.

Upper bound

Let

$$W = \begin{pmatrix} v \\ u \end{pmatrix}.$$

We start by proving that

$$(TW, W) \le c(SW, W), \quad \forall W,$$

for a suitable choice of c (which is independent of h and Δt). First, we compute a more direct expression for (TW, W),

$$
\begin{aligned}
(TW, W) &= \left(\begin{pmatrix} I + \frac{\Delta t}{2} A_i & \frac{\Delta t}{2} A_i \\ \frac{\Delta t}{2} A_i & \frac{\Delta t}{2} A_{i+e} \end{pmatrix} \begin{pmatrix} v \\ u \end{pmatrix}, \begin{pmatrix} v \\ u \end{pmatrix} \right) \\
&= \left(\left((I + \tfrac{\Delta t}{2} A_i)v + \tfrac{\Delta t}{2} A_i u, \tfrac{\Delta t}{2} A_i v + \tfrac{\Delta t}{2} A_{i+e} u \right), \begin{pmatrix} v \\ u \end{pmatrix} \right) \\
&= ((I + \frac{\Delta t}{2} A_i)v, v) + \Delta t (A_i v, u) + \frac{\Delta t}{2} (A_{i+e} u, u).
\end{aligned}
$$

The corresponding expression for the preconditioner is

$$
\begin{aligned}
(SW, W) &= \left(\begin{pmatrix} I + \frac{\Delta t}{2} A_0 & 0 \\ 0 & \frac{\Delta t}{2} A_0 \end{pmatrix} \begin{pmatrix} v \\ u \end{pmatrix}, \begin{pmatrix} v \\ u \end{pmatrix} \right) \\
&= \left(\left((I + \tfrac{\Delta t}{2} A_0)v, \tfrac{\Delta t}{2} A_0 u \right) \begin{pmatrix} v \\ u \end{pmatrix} \right) \\
&= ((I + \frac{\Delta t}{2} A_0)v, v) + \frac{\Delta t}{2} (A_0 u, u).
\end{aligned}
$$

By using (4.142) and (4.143), we get

$$
(TW, W) \leq \left(\left(I + \frac{\Delta t}{2} A_i \right) v, v \right) + \frac{\Delta t}{2} (A_{i+e} u, u) + \Delta t (A_i v, u)
$$

$$
+ \frac{\Delta t}{2} (A_i (v - u), (v - u))
$$

$$
= \left(\left(I + \frac{\Delta t}{2} A_i \right) v, v \right) + \frac{\Delta t}{2} (A_{i+e} u, u) + \Delta t (A_i v, u)
$$

$$
+ \frac{\Delta t}{2} (A_i v, v) + \frac{\Delta t}{2} (A_i u, u) - \Delta t (A_i v, u)
$$

$$
= ((I + \Delta t A_i) v, v) + \Delta t (A_i u, u) + \frac{\Delta t}{2} (A_e u, u)
$$

$$
\leq c_1 ((I + \Delta t A_0) v, v) + c_1 \Delta t (A_0 u, u) + c_1 \frac{\Delta t}{2} (A_0 u, u)
$$

$$
\leq 3 c_1 \left\{ ((I + \frac{\Delta t}{2} A_0) v, v) + \frac{\Delta t}{2} (A_0 u, u) \right\}.
$$

Hence

$$
(TW, W) \leq 3 c_1 (SW, W). \tag{4.144}
$$

Lower bound

$$
(TW, W) = \left(\left(I + \frac{\Delta t}{2} A_i \right) v, v \right) + \Delta t (A_i v, u) + \frac{\Delta t}{2} (A_{i+e} u, u)
$$

$$
= (Iv, v) + \frac{\Delta t}{2} (A_i (v + u), v + u) + \frac{\Delta t}{2} (A_e u, u)
$$

$$
\geq c_0 (v, v) + \frac{\Delta t c_0}{2} (A_0 (v + u), v + u) + \frac{\Delta t c_0}{2} (A_0 u, u)
$$

$$
\geq c_0 (v, v) + \frac{\Delta t c_0}{2} (A_0 (v + u), v + u) + \frac{\Delta t c_0}{2} (A_0 u, u)
$$

$$
- \frac{\Delta t c_0}{2} \left(A_0 \left(\varepsilon u + \frac{1}{\varepsilon} v \right), \varepsilon u + \frac{1}{\varepsilon} v \right),
$$

where $\varepsilon > 0$ is to be determined. Here

$$
\left(A_0 \left(\varepsilon u + \frac{1}{\varepsilon} v \right), \varepsilon u + \frac{1}{\varepsilon} v \right) = \varepsilon^2 (A_0 u, u) + \frac{1}{\varepsilon^2} (A_0 v, v) + 2 (A_0 v, u),
$$

so

$$
(TW, W) \geq c_0 \left\{ (v, v) + \frac{\Delta t}{2} (A_0 v, v) + \frac{\Delta t}{2} (A_0 u, u) \right.
$$

$$
+ \Delta t (A_0 v, u) + \frac{\Delta t}{2} (A_0 u, u) - \frac{\Delta t}{2} \varepsilon^2 (A_0 u, u)
$$

$$
\left. - \frac{\Delta t}{2 \varepsilon^2} (A_0 v, v) - \Delta t (A_0 v, u) \right\}
$$

$$
= c_0 \left\{ (v, v) + \frac{\Delta t}{2} (1 - \frac{1}{\varepsilon^2}) (A_0 v, v) + \frac{\Delta t}{2} (2 - \varepsilon^2) (A_0 u, u) \right\}.
$$

By picking $\varepsilon^2 = 3/2$, we have $1 - \frac{1}{\varepsilon^2} = 1/3$ and $2 - \varepsilon^2 = 1/2$, and then

$$(TW, W) \geq c_0 \left\{ (v, v) + \frac{\Delta t}{6}(A_0 v, v) + \frac{\Delta t}{4}(A_0 u, u) \right\}$$

$$\geq \frac{c_0}{3} \left\{ \left(\left(I + \frac{\Delta t}{2} A_0 \right) v, v \right) + \frac{\Delta t}{2}(A_0 u, u) \right\}$$

$$= \frac{c_0}{3}(SW, W). \tag{4.145}$$

Summarizing the results from (4.144) and (4.145), we have

$$\frac{c_0}{3}(SW, W) \leq (TW, W) \leq 3c_1(SW, W).$$

Hence, T and S are spectrally equivalent. This implies that the condition number of $S^{-1}T$ is bounded. In fact,

$$\kappa(S^{-1}T) = 9\frac{c_1}{c_0}.$$

An interesting consequence of this result is that we can obtain a relation between the computational cost of the bidomain model and the monodomain model; see [133].

Chapter 5

Solving Systems of ODEs

The operator splitting algorithms introduced in Chapter 3 reduced the solution of the bidomain equations to solving linear PDE systems and nonlinear systems of ODEs. Techniques for discretizing the PDE system were presented in Chapter 3, while techniques for solving the resulting linear systems were discussed in Chapter 4. What remains to have a complete computational method for the bidomain model is to find an efficient method for solving the nonlinear ODE systems. Note that the spatial discretization of the bidomain equations results in one ODE system for each node in the finite element grid. Realistic simulations may require several millions of nodes, and it is therefore of the utmost importance to solve the ODE systems efficiently.

Solvers for systems of ODEs is a large research field, and there exists a huge amount of available literature on theoretical and computational aspects of the various solvers. In this chapter, we provide only a brief introduction to the most widely used classes of solvers, and comment on their applicability for solving the cell model ODEs. For a thorough presentation of ODE solvers, the reader is referred to, e.g., [58,59].

5.1 Simple ODE Solvers

In Chapter 3 we introduced the so-called θ-rule for time discretization of PDEs. The θ-rule is a general formulation from which well known methods such as the forward and backward Euler methods method are derived as special cases. In this chapter, we will provide a short description of these simple discretization techniques for ordinary differential equations. Although none of these techniques are among the most suitable for solving the cell model ODE systems, they serve as a foundation for describing the more advanced methods.

5.1.1 The Euler Methods

Consider a general initial value problem of the form

$$\frac{dy}{dt} = f(t, y), \quad t > t_0, \tag{5.1}$$

$$y(t_0) = y_0, \tag{5.2}$$

where y_0 is a given initial value. The simplest algorithm for numerical solution of this problem is the forward Euler method. This method can be motivated in a number of different ways. In Chapter 3 it was derived as a special case of the θ-rule, in which the time derivative was simply replaced by a finite difference approximation, and a weighted average was used to approximate the right hand side. Here, we apply a slightly different approach. Assuming that the value of y is known at $t = t_n$, the solution at $t = t_{n+1}$ can be found by integrating (5.1). We have

$$\int_{t_n}^{t_{n+1}} \frac{dy}{dt} dt = \int_{t_n}^{t_{n+1}} f(t,y)dt,$$

which gives

$$y(t_{n+1}) = y(t_n) + \int_{t_n}^{t_{n+1}} f(t,y)dt. \tag{5.3}$$

In general, the integral on the right hand side cannot be computed directly and needs to be approximated. A simple approximation is obtained by inserting the known values t_n, y_n into the function $f(t,y)$. The integrand then becomes constant, and the integral is approximated by

$$\int_{t_n}^{t_{n+1}} f(t,y)dt \approx \Delta t f(t_n, y_n),$$

where $\Delta t = t_{n+1} - t_n$. Inserting this into (5.3) gives

$$y(t_{n+1}) = y(t_n) + \Delta t f(t_n, y_n), \tag{5.4}$$

which is the forward Euler method.

The forward Euler method can also be derived by performing a Taylor series expansion of the solution around $t = t_n$. Truncating the series after the first two terms then gives (5.4). Using this approach, it is also very easy to see that the local error of the forward Euler method is proportional to the square of the time step Δt. This leads to the well known fact that the forward Euler method is first-order accurate, i.e., the accumulated error after $n \sim \Delta t^{-1}$ steps is proportional to Δt.

Starting from (5.3), different approximations of the right hand side integral will yield numerical methods with different properties. For instance, the left end-point approximation used above may be replaced by

$$\int_{t_n}^{t_{n+1}} f(t,y)dt \approx \Delta t f(t_{n+1}, y_{n+1}),$$

which approximates the integral using the values at the right end-point. This gives the backward Euler method

$$y(t_{n+1}) = y(t_n) + \Delta t f(t_{n+1}, y_{n+1}), \tag{5.5}$$

where we see that the unknown value y_{n+1} occurs as an argument to the function $f(t,y)$. For this method it is therefore not possible to derive an explicit formula for $y(t_{n+1})$. Instead, we are required to solve a system of generally nonlinear algebraic equations.

The backward Euler method can also be derived from a Taylor series expansion around the point t_{n+1}. With this approach, it is easy to see that this method is also first-order accurate.

More accurate ODE solvers are obtained by using more accurate approximations of the right hand side integral in (5.3). For instance, approximating the integral with the trapezoidal rule, i.e.

$$\int_{t_n}^{t_{n+1}} f(t, y)dt \approx \Delta t \frac{f(t_n, y_n) + f(t_{n+1}, y_{n+1})}{2},$$

gives the second-order accurate Crank-Nicolson scheme.

As described in Chapter 3, the two Euler methods and the Crank-Nicolson method are all special cases of a general scheme known as the θ-rule. In the present setting, the θ-rule is derived by approximating the integral as a weighted average of the two endpoint values:

$$\int_{t_n}^{t_{n+1}} f(t, y)dt \approx \Delta t[(1 - \theta)f(t_n, y_n) + \theta f(t_{n+1}, y_{n+1})],$$

for $\theta \in [0, 1]$.

5.1.2 Stability Analysis for the Euler Methods

The order of accuracy is not the only important property of a numerical method. In many cases, the stability of the method is at least as important. In particular, this is the case for the equations that describe the cellular reactions, because the multiple time scales of these problems tend to make them *stiff*. The concept of stiffness, although very often described in terms of multiple time scales, may be difficult to define precisely. For homogeneous, linear equations, stiffness can be related to the eigenvalues of the problem. Consider, for instance, the Dahlquist test equation:

$$y' = \lambda y, \tag{5.6}$$
$$y(0) = 1, \tag{5.7}$$

where λ may be a complex number. As described in [6], this equation is stiff for an interval $[0, b]$ if the real part of λ satisfies

$$b\mathbb{R}(\lambda) \ll -1.$$

Note that we assume
$$\mathbb{R}(\lambda) \leq 0,$$

since this is required for the equation to be stable. For the general non-linear problem (5.1)–(5.2), the stiffness can be related to the eigenvalues λ_i of the local Jacobian matrix J, defined by

$$J_{ij} = \frac{\partial f_i(t, y)}{\partial y_j}.$$

The problem is stiff for an interval $[0, b]$ if

$$b \min_i \mathbb{R}(\lambda_i) \ll -1.$$

More pragmatic definitions of stiffness have been given by, for instance, Curtiss and Hirschfelder [27] and Ascher and Petzold [6], who define an equation to be stiff if the time step needed to maintain stability of an explicit method is much smaller than the time step dictated by the accuracy requirements. This indicates that a problem's being stiff is not simply a function of the equation itself, but also of the interval of integration and of the chosen accuracy requirement. A detailed discussion of stiff ODE systems is given in, e.g., [6,59].

The stability of ODE solvers can be studied by investigating their behaviour when applied to the test problem (5.6)–(5.7). For instance, applying one time step of the forward Euler method to (5.6)–(5.7), gives the approximate solution

$$y(\Delta t) = 1 + \lambda \Delta t = R(\lambda \Delta t),$$

where we have defined

$$R(z) = 1 + z. \tag{5.8}$$

The function $R(z)$ is called the stability function for the forward Euler method. The stability function can be defined in the same way for other methods, by applying one time step of the method to the test problem (5.6)–(5.7). For instance, for the backward Euler method we get

$$y(\Delta t) = \frac{1}{1 - \lambda \Delta t},$$

so the stability function for this method is

$$R(z) = \frac{1}{1 - z}.$$

By similar calculations it is easy to see that the stability function for the θ-rule is

$$R(z) = \frac{1 + (1 - \theta)z}{1 - \theta z},$$

which for $\theta = 0$ and $\theta = 1$ corresponds to the stability functions of the forward and backward Euler methods, as expected.

As the name suggests, the stability function gives an indication of the stability of the method. As noted above, the analytical solution of (5.6)–(5.7) is stable for $\mathbb{R}(\lambda) \leq 0$. Ideally, we want a numerical method to retain this property, i.e., we want the numerical solution to be stable at least for $\mathbb{R}(\lambda) \leq 0$. Clearly, the numerical solution of (5.1)–(5.2) will be bounded if we have

$$|R(z)| \leq 1, \tag{5.9}$$

for $z = \lambda \Delta t$. Since we allow λ to be complex, z will be a complex number. The region in the complex plane for which (5.9) holds is called the region of absolute

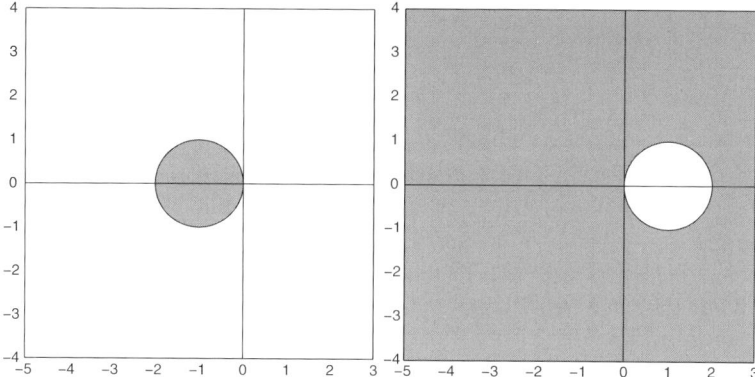

Fig. 5.1. The shaded area is the stability domain for the forward Euler method (left) and the backward Euler method (right).

stability, or simply the stability domain, for the method. More precisely, the stability domain S for an ODE solver with stability function $R(z)$ is defined as

$$S = \{z \in C; |R(z)| \leq 1\}. \qquad (5.10)$$

The stability domain for the forward Euler method is a circle with centre in $\mathbb{R}(z) = -1, \Im(z) = 0$, and radius 1, as shown in the left panel of Figure 5.1. The right panel of Figure 5.1 shows the stability domain for the backward Euler method. The method is stable for all values of $z = \lambda \Delta t$ lying *outside* the circle with centre $(1, 0)$ and radius 1. The backward Euler method is hence very stable.

An attractive property of a numerical method is that for the test equation (5.6)–(5.7), it reproduces the essential stability property of the analytical solution, i.e. it is stable for all step sizes when $\mathbb{R}(\lambda) \leq 0$. Methods having this stability property are referred to as A-stable methods [28]. In other words, an A-stable method is a method which is stable on the entire negative half-plane, i.e. its stability domain satisfies

$$S \supset C^- = \{z; \mathbb{R}(z) \leq 0\}.$$

From the stability domains plotted in Figure 5.1 we see that the backward Euler method is A-stable, while the forward Euler method is not. The θ-rule is A-stable for $\theta \in [1/2, 1]$.

Although A-stability is a desirable property of a numerical method for stiff systems, it is not always sufficient. For instance, although for an A-stable method we know that as long as all eigenvalues λ_i have a nonpositive real part, oscillations in the solution will not grow. However, the definition of A-stability only required $|R(z)| \leq 1$. This may present a problem in the very stiff case, where $\lambda \ll -1$. Then, the analytical solution of (5.6)–(5.7) decays very rapidly, implying $y(t_n + \Delta t) \ll y(t_n)$ for a small time step Δt. The definition of A-stability only ensures $y(t_n + \Delta t) \leq y(t_n)$, and does therefore not rule out situations where

$y(t_n + \Delta t) \approx y_n$. This potential problem has motivated the concept of L-stability. A method is said to be L-stable if it is A-stable and, in addition,

$$\lim_{z \to -\infty} R(z) = 0. \tag{5.11}$$

See, e.g., [59] for more details. L-stability is a highly attractive feature for very stiff problems, since reasonably accurate solutions can be obtained without resolving the sharpest transition layers. An inspection of the stability functions listed above reveals that the backward Euler method satisfies (5.11) and is hence L-stable. The forward Euler method is, of course, not L-stable because (i) it is not A-stable and (ii) its stability function fails to satisfy (5.11). The θ method is L-stable for $\theta \in (1/2, 1]$. For $\theta = 1/2$, the only choice of θ that yields second-order accuracy, this method is hence A-stable but not L-stable. For a thorough discussion of stability properties for stiff problems we refer the reader to [59] and [6].

5.2   Higher-Order Methods

We saw in the previous section that the two Euler methods for solving ODEs had very different properties in terms of stability, but they are both only first-order accurate. The Crank-Nicolson/trapezoidal rule obtained by the θ-rule for $\theta = 1/2$ increased the accuracy to order 2. We recall that the operator splitting algorithms introduced in Chapter 3 were only second-order accurate, so it may seem sufficient to solve the resulting ODE systems with second-order accuracy. However, as noted above, the Crank-Nicolson scheme is A-stable but not L-stable. L-stability is an attractive feature when solving stiff systems, and this leads us to consider more advanced ODE solvers.

5.2.1   Multistep Methods

Several techniques exist for improving the accuracy of the ODE solvers described above. We have already introduced the idea of improving the accuracy by using more accurate approximations of the integral in (5.3). The two Euler methods were based on using a constant approximation of the right hand side function f in the interval from t_n to t_{n+1}, while a linear approximation based on the trapezoidal rule yielded a second-order accurate numerical method. A natural extension of these ideas is to interpolate the solution using polynomials of higher order.

We now assume that we know not only the solution at the previous step, y_n, but also the preceding solutions $y_{n-k+1}, \ldots, y_{n-1}$ for some integer k. The integrand in (5.3) can then be replaced by a polynomial $p(t)$, which interpolates the known points $(t_i, f_i), i = n - k + 1, \ldots, n$. This leads to a family of methods known as explicit Adams methods. The accuracy of the methods is limited by the fact that the integration interval lies outside the interpolation interval of the polynomial. The accuracy may be increased by replacing $p(t)$ with a polynomial that also interpolates the unknown point (t_{n+1}, f_{n+1}). This gives the implicit Adams methods.

The Adams methods belong to a class of methods called multistep methods, because the solution y_{n+1} is computed based not only on y_n, but also on the solution at previous steps. A general multistep method can be written as

$$\alpha_0 y_{n+1} + \alpha_1 y_n + \ldots + \alpha_k y_{n-k+1} = \Delta t (\beta_0 f_{n+1} + \ldots + \beta_k f_{n-k+1}), \quad (5.12)$$

where k denotes the number of steps in the method, and $\alpha_i, \beta_i, i = 1, \ldots, k$ are given, method-specific coefficients. For simplicity we use the notation $f_i = f(t_i, y_i)$.

The Adams methods are widely used for solving non-stiff ODEs, but another class of multistep methods, known as backward differentiation formulae (BDF), are more popular for stiff systems. The idea behind these methods is different from the Adams methods, in that they are not derived by approximating the integral in (5.3). Instead, the solution is approximated by a polynomial $q(t)$ that interpolates the points $(t_i, y_i), i = n-k+1, \ldots, n+1$. The polynomial $q(t)$ is often represented in terms of backward differences, which gives rise to the name of the methods. As above, the values $y_i, i = n-k+1, \ldots, n$ are assumed to be known. The unknown y_{n+1} is computed so that the polynomial $q(t)$ satisfies the differential equation in at least one point, i.e.,

$$\frac{dq(t_{n+1-r})}{dt} = f(t_{n+1-r}, y_{n+1-r}),$$

where r is normally either zero or one. Setting $r = 1$ gives explicit formulas for y_{n+1}. As above k denotes the number of interpolation points in $q(t)$; hence the number of steps in the method. For instance, choosing $r = 1$ and $k = 1$ gives the explicit Euler method. For $r = 0$ the BDF method is implicit, and if the step size is constant it may be written as

$$\alpha_0 y_{n+1} + \alpha_1 y_n + \ldots + \alpha_k y_{n-k+1} = \Delta t \beta_0 f_{n+1}. \quad (5.13)$$

Here the coefficients $\alpha_0, \ldots, \alpha_k$ are constants. Implicit BDF methods are of order k and have good stability properties, and hence are very suitable for solving stiff ODE systems. For $k = 1$ the method is equal to the backward Euler method, for which the good stability properties were demonstrated above. The stability of the method decreases as k is increased, and for $k \geq 7$ the methods are unstable; see, e.g., [59] for details.

The multistep methods are based on knowing the solution at the time points $t_{n-k+1}, \ldots, t_n$. Obviously, this causes problems when starting a computation, because we normally only have the initial value y_0 available. Different methods exist to overcome this problem. One is to use another method of sufficient order to compute the first $k - 1$ steps, and then start the multistep method. Another possibility is to start the method with $k = 1$, and increase the order as the necessary information becomes available. The low order used for the first steps must be compensated by using smaller time steps in this region. The problem of starting, and restarting, a multistep method is handled efficiently and automatically by modern ODE software.

Normally, storing the values of y for k time points, as required by a k-step method, does not cause any problems. For very big systems of ODEs the required memory may become significant, but compared to other methods of high order the memory requirement is not particularly large. Recall, however, that the spatial discretization of the bidomain model leads to one ODE system for each node in the finite element grid. Realistic computations require up to several millions of grid nodes, and each ODE system may consist of 30 or more ODEs. In this context, the memory requirement of the multistep methods becomes very significant. This potential storage problem has led us to focus mainly on other methods as subproblem solvers in the operator splitting algorithm presented in Chapter 3. For simulations of a single cell, BDF methods have proven to be highly efficient, but for full-scale simulations of the heart it may be more suitable to employ methods that only require storing the solution at one time step.

5.2.2 Runge-Kutta Methods

It is possible to increase the accuracy of the numerical integration in (5.3) without interpolating the solution over a large number of points. A second-order method was obtained above by using a trapezoidal rule to approximate the integral, and methods of higher order can be constructed by using even more accurate techniques for numerical integration. One possibility is to compute a number of intermediate values of $f(t, y)$ in the interval $[t_n, t_{n+1}]$, and approximate the integral as a weighted sum of these values. This leads to the large family of methods known as Runge-Kutta (RK) methods, which includes both implicit and explicit methods that vary significantly with respect to their accuracy and stability. Denoting the intermediate approximations of $f(t, y)$ by k_i, the approximation of the right hand side integral in (5.3) becomes

$$\int_{t_n}^{t_{n+1}} f(t, y)dt \approx \Delta t \sum_{i=1}^{s} b_i k_i,$$

for given weights b_i and a given number of stages s. The values k_i are normally referred to as the stage derivatives of the method. Obviously, a higher number of stages enables a better approximation of the integral and therefore higher accuracy of the method. A general RK method with s stages can be written as

$$k_i = f\left(t_n + c_i \Delta t, y_n + \Delta t \sum_{j=1}^{s} a_{ij} k_j\right), \text{ for } i = 1, \ldots, s \qquad (5.14)$$

$$y_{n+1} = y_0 + \Delta t \sum_{i=1}^{s} b_i k_i. \qquad (5.15)$$

Here c_i, b_i, a_{ij}, for $i, j, = 1, \ldots, s$ are method-specific, given coefficients. All RK methods can be written in this form, and a method is uniquely determined by the number of stages s and the values of the coefficients. In the ODE literature one often sees these coefficients specified in the form of a *Butcher tableau*. The Butcher tableau of a general RK method is written as

$$
\begin{array}{c|ccc}
c_i & a_{11} & \cdots & a_{1s} \\
\vdots & \vdots & & \vdots \\
c_s & a_{s1} & \cdots & a_{ss} \\
\hline
& b_1 & \cdots & b_s
\end{array}
$$

and offers a short notation for the method.

An explicit RK (ERK) method is one in which $a_{ij} = 0$ for $j \geq i$. From the formulas presented above, we see that the expression for each stage derivative k_i then only includes previously computed stage derivatives. We therefore have an explicit formula for each k_i, and there is no need for solving nonlinear equations. ERK methods are simple to implement, and the amount of computational work for each time step is fairly small. The methods can also be constructed to have a high order of accuracy, and are therefore very popular methods for solving systems of ODEs. Numerous explicit RK methods have been derived, see e.g. [58], with large variations in number of stages and order of accuracy. For instance, the forward Euler method is a one-stage ERK method with first-order accuracy. A two-step method with second-order accuracy is given by the Butcher tableau

$$
\begin{array}{c|cc}
0 & 0 & 0 \\
1 & 1 & 0 \\
\hline
& 1/2 & 1/2
\end{array}
$$

which inserted into (5.14)–(5.15) gives

$$
\begin{aligned}
k_1 &= f(t_n, y_n), \\
k_2 &= f(t_n + \Delta t, y_n + \Delta t k_1), \\
y_{n+1} &= y_0 + \frac{\Delta t}{2}(k_1 + k_2).
\end{aligned}
$$

This method can be seen as an explicit version of the trapezoidal rule introduced above, and is often referred to as the improved Euler method. An example of a widely used ERK method is one derived by Dormand and Prince (see, e.g., [30,58]), with six stages and order five. Its coefficients are

$$
\begin{array}{c|cccccc}
0 & & & & & & \\
\frac{1}{5} & \frac{1}{5} & & & & & \\
\frac{3}{10} & \frac{3}{40} & \frac{9}{40} & & & & \\
\frac{4}{5} & \frac{44}{45} & -\frac{56}{15} & \frac{32}{9} & & & \\
\frac{8}{9} & \frac{19372}{6561} & -\frac{25360}{2187} & \frac{64448}{6561} & -\frac{212}{729} & & \\
1 & \frac{9017}{3168} & -\frac{355}{33} & \frac{46732}{5247} & \frac{49}{176} & -\frac{5103}{18656} & \\
\hline
& \frac{35}{384} & 0 & \frac{500}{1113} & \frac{125}{192} & -\frac{2187}{6784} & \frac{11}{84}
\end{array}
$$

where obvious zeros have been omitted.

From the fact that the ERK methods are indeed explicit, and the brief stability analysis we presented above for the forward Euler method, it might be supposed that these methods are not very stable. A quick inspection of the stability properties of the methods confirms this supposition. The stability function for the improved Euler method defined above is

$$R(z) = 1 + z + \frac{z^2}{2},$$

while the stability function of the Dormand-Prince method is

$$R(z) = 1 + z + \frac{z^2}{2} + \frac{z^3}{6} + \frac{z^4}{24} + \frac{z^5}{120} + \frac{z^6}{600}.$$

In fact, it can be shown (see, e.g., [59]), that the stability function for an ERK method with s stages is a polynomial with degree at most s. Since no polynomial can satisfy the condition (5.10) for all z having a negative real part, we may conclude that no A-stable explicit Runge-Kutta method exists. The stability domains of the improved Euler method and the Dormand-Prince method are shown in Figure 5.2. We see that the methods are slightly more stable than the forward Euler method, but it is easily seen that they are not A-stable.

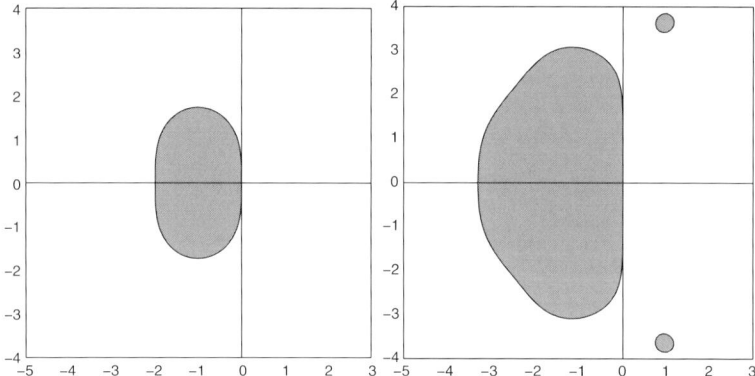

Fig. 5.2. The left plot shows the stability domain for the improved Euler method, and the right plot is the stability domain of the 5th order explicit Dormand-Prince method.

Implicit RK (IRK) methods are RK methods where at least one of the coefficients $a_{ij}, j \geq i$ is non-zero. From (5.14)–(5.15), we see that it will then no longer be possible to compute all k_i with explicit formulae. At least one stage derivative must be computed by solving a generally nonlinear algebraic equation. If the method is applied to a system of ODEs, we need to solve a system of nonlinear equations.

Many special cases of IRK methods exist, differing in a sense in their degree of implicitness. Methods where all coefficients $a_{ij}, i, j = 1, \ldots, s$ are non-zero are known as fully implicit methods. At the other end of the scale, there are methods for which $a_{ij} = 0$ for $j > i$, i.e. if a_{ij} is viewed as a matrix, all entries above the diagonal are zero. These methods are referred to as diagonally implicit (DIRK) methods or semi-explicit methods. Special cases of these methods exist, for instance if all the diagonal entries a_{ii} are equal, i.e. $a_{ii} = \lambda, i = 1, \dots, s$. These methods are called singly diagonally implicit (SDIRK) methods, and offer some computational advantages that will be discussed below. Furthermore, a special version of the SDIRK methods is obtained if the coefficient a_{11} is zero. In this case, the first stage of the method is explicit, while the subsequent steps are all implicit. The SDIRK methods generally require fewer computations for each time step than the fully implicit methods, a difference that will be discussed in greater detail in Section 5.3 below. However, the accuracy and stability properties are also not as good as for the fully implicit methods. The stability function of all implicit RK methods can be written on the form

$$R(z) = \frac{P(z)}{Q(z)}, \tag{5.16}$$

where P and Q are polynomials of degree at most s for a fully implicit method, and at most $s - 1$ for a DIRK method, see e.g. [59].

It is easily seen from (5.16) that the IRK methods may give better stability properties than explicit methods. Consider, for instance, the backward Euler method, which is the simplest possible IRK method. As described above this method has a stability function on the form (5.16) with $P(z)$ of degree zero and $Q(z)$ of degree one, and the method was found to be both A-stable and L-stable. A popular group of fully implicit RK methods are derived using Radau quadrature formulae, see e.g. [17]. A three-stage method of order five is given by

$$k_1 = f(t_n + c_1 \Delta t, y_n + \Delta t (a_{11} k_1 + a_{12} k_2 + a_{13} k_3)), \tag{5.17}$$
$$k_2 = f(t_n + c_2 \Delta t, y_n + \Delta t (a_{21} k_1 + a_{22} k_2 + a_{23} k_3)), \tag{5.18}$$
$$k_3 = f(t_n + c_3 \Delta t, y_n + \Delta t (a_{31} k_1 + a_{32} k_2 + a_{33} k_3)), \tag{5.19}$$
$$y_{n+1} = y_n + \Delta t (b_1 k_1 + b_2 k_2 + b_3 k_3), \tag{5.20}$$

with coefficients specified in Appendix C. The stability function of this method is given by

$$R(z) = \frac{1 + 2z/5 + z^2/20}{1 - 3z/5 + 3z^2/20 - z^3/60},$$

and the region of z-values that give $R(z) \le 1$ is plotted in the left panel of Figure 5.3. More methods of the Radau family, as well as other fully implicit RK methods, are discussed in [59]. An example of a third-order SDIRK method (see [78]), with an explicit first step is

$$
\begin{array}{c|cccc}
0 & & & & \\
c_2 & a_{21} & \gamma & & \\
c_3 & a_{31} & a_{32} & \gamma & \\
c_4 & a_{41} & a_{42} & a_{43} & \gamma \\
\hline
 & b_1 & b_2 & b_3 & b_4
\end{array}
$$

where the values of the coefficients can be found in Appendix C. Inserted into the RK formula, this gives

$$k_1 = f(t_n, y_n) \tag{5.21}$$
$$k_2 = f(t_n + c_2 \Delta t, y_n + \Delta t(a_{21}k_1 + \gamma k_2)), \tag{5.22}$$
$$k_3 = f(t_n + c_3 \Delta t, y_n + \Delta t(a_{31}k_1 + a_{32}k_2 + \gamma k_3)), \tag{5.23}$$
$$k_4 = f(t_n + c_4 \Delta t, y_n + \Delta t(a_{41}k_1 + a_{42}k_2 + a_{43}k_3 + \gamma k_4)), \tag{5.24}$$
$$y_{n+1} = y_n + \Delta t(b_1 k_1 + b_2 k_2 + b_3 k_3 + b_4 k_4)). \tag{5.25}$$

The stability domain of the method is shown in the right panel of Figure 5.3. We see that both of the implicit RK methods are very stable, although they are both slightly less stable than the backward Euler method. It is easy to see that the methods are A-stable, and it can be shown that their stability functions also satisfy (5.11), so both methods are L-stable. These methods therefore seem quite suitable for solving stiff problems. Their performance when applied to the cell model equations will be investigated below.

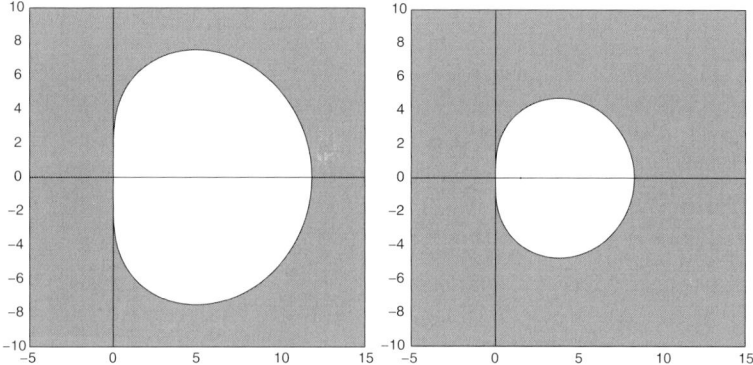

Fig. 5.3. Stability domains for the fifth-order Radau method (left) and the third-order SDIRK method (right).

5.3 Solving Nonlinear Equations

As noted above, all the implicit methods require some degree of equation solving for each time step. This is the main drawback of implicit methods, adding both to the computational cost and the complexity of implementing the methods. Solving the nonlinear equation systems efficiently is crucial for the overall performance of the ODE solver, and so choosing the right method will be important. The standard technique for solving nonlinear algebraic systems of equations is Newton's method, but many variations of this method have been derived and applied in ODE solvers.

5.3.1   Newton's Method

As an example, consider again the simplest implicit method available; the backward Euler method. This method requires solving algebraic systems of the form (5.5) for each time step. This equation system can be written in the form

$$y_{n+1} - \Delta t f(y_{n+1}) - y_n = 0. \tag{5.26}$$

Since f is generally nonlinear, this equation must be solved with an iterative technique, typically some form of Newton's method. For a general (scalar) equation of the form

$$g(u) = 0,$$

Newton's method involves making an appropriate initial guess u^0, and performing iterations on the form

$$u^{l+1} = u^l - \frac{g(u^l)}{g'(u^l)} \tag{5.27}$$

until a given convergence criterion is satisfied. Each iteration represents a linearization of the nonlinear equation around the current solution u^l. If the initial guess u^0 is sufficiently close to the exact solution, Newton's method yields a quadratic convergence, i.e.,

$$\|u^{l+1} - u\| \le C\|u^l - u\|^2, \tag{5.28}$$

for some constant C. We want to terminate the iterations when u^l is sufficiently close to the exact solution u. However, since u is not known, the convergence criteria are normally based on the residual $g(u^l)$ or the difference $u^{l+1} - u^l$. Commonly used variants include

$$\|u^{l+1} - u^l\| \le \epsilon_u,$$

and

$$\|g(u^{l+1})\| \le \epsilon_r,$$

where ϵ_u and ϵ_r are specified tolerances.

When the backward Euler method is applied to a system of ODEs, (5.26) will be a system of nonlinear equations. In this case, (5.27) cannot be applied directly, and the Newton iteration takes a slightly different form. Given a first guess y_{n+1}^0, the solution y_{n+1} is found through iterations involving two steps:

1. Solve the linear system

$$(I - \Delta t J(y^l))\delta y^{l+1} = -(y_{n+1}^l - \Delta t f(y_{n+1}^l) - y_n), \tag{5.29}$$

   where $J(y^l)$ is the Jacobian of f evaluated at y^l, and I is the identity matrix.

2. Set

$$y_{n+1}^{l+1} = y_{n+1}^l + \delta y^{l+1}. \tag{5.30}$$

For large systems of ODEs it may become quite expensive to compute the Jacobian $J(y^l)$ and solve the linear system (5.29) for each iteration, and these tasks may easily dominate the computation load of solving an ODE system. Various techniques have been proposed to reduce this cost, resulting in different forms of modified Newton methods. A simple approach commonly used in ODE solvers is to keep the Jacobian constant for several iterations, and even for several consecutive time steps. This reduces the convergence of the method, but this is compensated by the fact that each iteration is considerably less expensive. The most obvious cost reduction comes from the reduced number of Jacobian evaluations, which are quite expensive, but keeping the Jacobian constant also offers benefits when solving the linear system in (5.29). For ODE systems of moderate size, this system is normally solved by LU factorization followed by forward-backward substitution. If J is kept constant for several iterations, there is no need to perform a new LU factorization for each iteration. The factorized matrix can hence be reused for several iterations, reducing the work of each iteration to a simple forward-backward substitution. If Δt is kept constant, the LU factorization of $(I - \Delta t J)$ may also be reused for several time steps.

5.3.2 Newton's Method for Higher-Order Solvers

We have seen that the backward Euler method requires solving one nonlinear system of size n for each time step. The more advanced implicit methods require solving similar systems, but there are important differences between the methods. Recall the formulation of the implicit BDF methods, given in (5.13). This nonlinear equation system can be written as

$$\alpha_k y_{n+1} - \Delta t \beta_0 f(y_{n+1}) + \alpha_{k-1} y_n + \ldots + \alpha_0 y_{n-k+1} = 0.$$

We see that for any number of steps k this system is very similar to the equation system that arises from the backward Euler method. This is also reflected in the linear system that must be solved for each Newton iteration, which is given by

$$(\alpha_k I - \Delta t \beta_0 J(y_{n+1}^l)) \delta y^{l+1} = -(\alpha_k y_{n+1}^l - \Delta t f(y_{n+1}^l) + \alpha_{k-1} y_n + \ldots + \alpha_0 y_{n-k+1}).$$

Note that the evaluation of the right hand side involves only one evaluation of the function f. This is an attractive feature of the BDF methods; high order is obtained by solving nonlinear systems similar to those arising from the backward Euler, of size n and where each Newton iteration involves only one evaluation of the right hand side function f.

Recall that for a fully implicit RK method all the coefficients a_{ij} in (5.14) are non-zero. This implies that each stage derivative k_i depends implicitly on all other stage derivatives. All the stage derivatives must hence be computed simultaneously, and for an s-stage method applied to a system of n ODEs, this involves solving a system of sn nonlinear algebraic equations. For large systems this task may become computationally demanding, and constitutes the main drawback of implicit RK methods compared to multistep methods such as the BDF.

Consider, for instance, the Radau method defined by (5.17)–(5.20). Eqs. (5.17)–(5.19) is a coupled system of nonlinear equations that must be solved to determine the stage derivatives k_1, k_2, and k_3. Contrary to the backward Euler and the BDF methods, we are hence required to solve a system of $3n$ nonlinear equations. The linear system arising from applying Newton iterations to (5.17)–(5.19) is given by

$$\begin{pmatrix} I - a_{11}\Delta t J & a_{12}\Delta t J & a_{13}\Delta t J \\ a_{21}\Delta t J & I - a_{22}\Delta t J & a_{23}\Delta t J \\ a_{31}\Delta t J & a_{32}\Delta t J & I - a_{33}\Delta t J \end{pmatrix} \begin{pmatrix} \delta k_1^{l+1} \\ \delta k_2^{l+1} \\ \delta k_3^{l+1} \end{pmatrix} = \begin{pmatrix} F_1 \\ F_2 \\ F_3 \end{pmatrix}, \qquad (5.31)$$

with

$$F_i = -k_i^l + f\left(t_n + c_i\Delta t, y_n + \Delta t \sum_{j=1}^{n} a_{ij}k_j^l\right). \qquad (5.32)$$

In addition to the larger linear systems to be solved, we see from (5.31)–(5.32) that each iteration also requires three evaluations of the right hand side function f. This is an additional drawback compared to the BDF methods, which only required one function evaluation for each iteration. Fortunately, fully implicit RK methods require only a few stages to obtain high order, and their stability properties make them well suited for stiff computations.

For ODE systems of moderate size, it is possible to solve the linear system in (5.31) with a direct method, even though it is larger than that of the backward Euler and BDF methods. The considerations described above regarding the cost of computing and factorizing the Jacobian will also apply in this case. A simplified Newton algorithm, where the factorized Jacobian is reused for several iterations, is normally the most efficient technique for the fully implicit RK methods as well.

Earlier, we introduced DIRK and SDIRK methods as an alternative to fully implicit methods. The main reason for applying (and developing) these methods is to reduce the nonlinear equation solving required for fully implicit RK methods. With a diagonally implicit method the stage derivatives $k_1, \ldots, k_s$ may be determined one by one, although each of them must be computed by solving a system of nonlinear equations. The size of the equation systems that must be solved is the same as for the backward Euler and BDF methods, but several such systems must be solved for each time step. Consider, for instance, the four-stage SDIRK method described above. Here the first stage is explicit, so k_1 can be computed directly. The nonlinear systems required to determine k_2, k_3 and k_4 are given by (5.22)–(5.24). When solving these systems with Newton's method they lead to linear systems of the form

$$(I - \gamma\Delta t J)\delta k_i^l = -k_i^l + f\left(t_n + c_i\Delta t, y_n + \Delta t \sum_{j=1}^{i-1} a_{ij}k_i + \Delta t\gamma k_i\right),$$

for $i = 2, 3, 4$. We see that the matrix of this linear system is independent of whether we solve for k_2, k_3 or k_4. This is an important feature of the SDIRK methods. For a general DIRK method, with γ replaced by diagonal coefficients a_{ii} that are different for each stage, we see that the matrix of the linear system will also be different for each stage. This will limit the benefits of using modified Newton techniques, since the changed matrix would require a new LU factorization at each stage. For an SDIRK method the factorized matrix may be reused for all stages in a time step, to be recomputed only if convergence is slow. In theory, the possibility of choosing the parameters a_{ii} independently allows the design of methods with better accuracy and stability properties, compared to restricting the methods to $a_{ii} = \gamma$. However, in practice this additional freedom does not purchase much, and the SDIRK methods are usually a better alternative.

5.4 Automatic Time Step Control

In practical computations, one is interested in obtaining a certain accuracy with the minimum amount of computational work. For a given method, this implies that we want to use the largest possible value of the time step Δt. The step size will obviously depend on the demanded accuracy, as well as the order of the applied method, but it will also be influenced by the characteristics of the solution. In quiescent areas large steps can be taken without introducing a large error, while if there are fast variations in the solution the time step has to be small. Obviously, the solution of a nonlinear ODE system may show fast variations in some regions, and be very smooth in others. This is illustrated well by the cell model ODEs, see e.g. Figure 5.4, where the solution changes extremely quickly in the upstroke phase and is nearly stationary in the plateau and resting phases. This motivates methods that can adjust the time step to the properties of the solution. Commonly referred to as adaptive methods or methods with automatic time step control, this is an important part of successful ODE software.

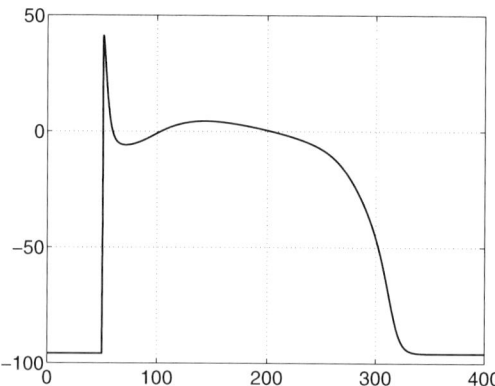

Fig. 5.4. The action potential produced by the cell model by Winslow et al., clearly showing very fast variations in the upstroke and smoother behaviour in other intervals.

For many problems, automatic step size control can be based on certain charac-teristics of the solution. For instance, in the case of the cell model ODEs, it is easy to construct an algorithm that checks for large variations in v, corresponding to the upstroke, and chooses a smaller time step in this region. This approach was used by Rush and Larsen [121] in their clever algorithm for solving the cell model ODEs. It may work well for many problems, but its applicability is not very general, and the criteria for choosing the time step must be chosen carefully based on known properties of the problem to be solved. A much more general approach to adaptive time steps is obtained by basing the step size on estimates of the error introduced in each step. The length of each time step can then be chosen such that this local error falls below a certain threshold. While this approach offers algorithms for adaptive time steps that are capable of handling any system of differential equations, a ques-tion arises as to how the error estimate should be computed. Since the true solution is not known, the error can, of course, not be computed directly, and must instead be based on numerical estimates. A common technique is to approximate the error using two numerical solutions of different accuracy, and this will be the basis of all the approaches discussed in this section.

A simple approach to estimating the local error, which works for any numerical method, is to use step doubling; see, e.g., [6]. Assume that for a given initial value $y(t_0) = y_0$ we perform two steps of length Δt with a method of order p. The error after the first step is given by

$$e_1 = \|y_1 - y(t_0 + \Delta t)\| = \|\psi(t_n, y(t_n))\|\Delta t^{p+1} + O(\Delta t^{p+2}),$$

where ψ is called the principal error function; see, e.g., [6,58] for further details. The error e_2 after two steps consists of the error carried over from the first step as well as the error introduced in the second step, see [58]. We here assume (see [6]), that the error after two steps is twice as large as that after one step. We have

$$e_2 = \|y(t_0 + 2\Delta t) - y_2\| = 2\|\psi(t_n, y(t_n))\|\Delta t^{p+1} + O(\Delta t^{p+2}). \quad (5.33)$$

Similarly, we can estimate the error after one step of length $2\Delta t$ as

$$\tilde{e}_2 = \|y(t_0 + 2\Delta t) - \tilde{y}_2\| = \|\psi(t_n, y(t_n))\|(2\Delta t)^{p+1} + O(\Delta t^{p+2}). \quad (5.34)$$

From (5.33) and (5.34), we get

$$\|y_2 - \tilde{y}_2\| = (2^p - 1)2\|\psi(t_n, y(t_n))\|h^{p+1},$$

and this can be used to eliminate the unknown ψ from e_2, to give

$$e_2 = \frac{\|y_2 - \tilde{y}_2\|}{2^p - 1}.$$

The estimated local error can be compared to an error tolerance *tol* specified for the given application. If we have

$$e_2 \leq tol \quad (5.35)$$

the step is accepted and the step size for the next step is computed from

$$h_{new} = h \left(\frac{tol}{e_2} \right)^{1/(p+1)} . \tag{5.36}$$

If the estimated error does not satisfy (5.35), the step is rejected and is recomputed with a smaller step. The new step may also, in this case, be based on a formula such as (5.36). In practice, (5.36) is normally modified slightly, by introducing a safety factor to keep the steps slightly below the optimal size, and limiting the amount by which the step size may vary from one step to the next. See, e.g., [58,6] for further details.

The step-doubling procedure offers a good estimate of the local error, but it is also quite expensive. Consequently, automatic step size control in modern ODE software is normally based on other estimates of the local error. For multistep methods, a number of factors interplay to make the issue of adaptive time steps fairly complicated. First, the methods described above were based on interpolating values $y_{n+1}, \ldots, y_{n+1-k}$ (BDF) or $f_{n+1}, \ldots, f_{n+1-k}$ (Adams) with a polynomial. For a constant step size this leads to a constant set of coefficients α_i, β_i in (5.12), whereas for a variable time step the coefficients must be recomputed for each step. Second, recall that the multistep methods also have variable order. It is therefore possible to vary the order in addition to the step size to satisfy a given error tolerance. The best combination of method order and step size will depend on the (local) behaviour of the solution. In spite of these difficulties, many successful ODE solvers are based on multistep methods. The ability to adjust the order as well as the time step enables very flexible solvers, and various techniques have been developed to handle the problem related to recomputing the coefficients. Multistep methods also offer a fairly simple way to estimate the local errors. Since the main focus of the present text is on Runge-Kutta methods, we refer the reader to, e.g., [6] and [58,59] for a description of automatic step size control for multistep methods.

Implicit and explicit Runge-Kutta methods offer some advantages over the multistep methods in terms of automatic step size control. Since they are one-step methods, there is no need to recompute coefficients or perform additional interpolation when the step size is changed. However, while multistep methods offer a direct estimate of the local truncation error and thereby the local error (see [6]), this is not readily available for the Runge-Kutta methods. We therefore return to the idea introduced above; to compute two numerical solutions with different accuracy, and use the difference to estimate the local error. One possibility was the step-doubling technique introduced above, but a more practical alternative is to use two numerical solutions computed with the same step size, but with a different order of accuracy. For a numerical method of order p, we can use a numerical solution computed with a method of order $p + 1$ to estimate the local error. Assume that y_1 is a numerical solution computed with a method of order p, and $\tilde{y}_1$ is a solution obtained with a method of order $p + 1$. The difference $\|y_1 - \tilde{y}_1\|$ then provides an estimate of the local error of y_1. Using this estimate of the local error, the time step can be controlled by formulae such as (5.36).

The remaining issue is how we should compute the two different solutions y_1 and $\tilde{y}_1$. If we are to apply two entirely different methods to obtain these solutions, the adaptive time step algorithm becomes very expensive. Fortunately, the error estimate can often be obtained by so-called embedded methods. For a given RK method, an embedded method is a method that uses the same stage computations as the original method, but obtains a different order of accuracy. Since most of the work in RK methods is in the stage computations, error estimates based on embedded methods are very cheap to evaluate. Introducing an embedded error estimator involves augmenting (5.14)–(5.15) with an additional line, yielding

$$k_i = f\left(t_n + c_i \Delta t, y_n + \Delta t \sum_{j=1}^{s} a_{ij} k_j\right) \quad \text{for } i = 1, \ldots, s \tag{5.37}$$

$$y_{n+1} = y_0 + \Delta t \sum_{i=1}^{s} b_i k_i, \tag{5.38}$$

$$\hat{y}_{n+1} = y_0 + \Delta t \sum_{i=1}^{s} \hat{b}_i k_i. \tag{5.39}$$

Although the idea is to reuse the same stage computations to compute both $\hat{y}_{n+1}$ and y_{n+1}, it is not uncommon to introduce one additional stage in the method to obtain the error estimate. For instance, the Dormand-Prince method described above can be equipped with an error estimator by introducing one additional stage, which yields a seven-stage method with coefficients:

0							
$\frac{1}{5}$	$\frac{1}{5}$						
$\frac{3}{10}$	$\frac{3}{40}$	$\frac{9}{40}$					
$\frac{4}{5}$	$\frac{44}{45}$	$-\frac{56}{15}$	$\frac{32}{9}$				
$\frac{8}{9}$	$\frac{19372}{6561}$	$-\frac{25360}{2187}$	$\frac{64448}{6561}$	$-\frac{212}{729}$			
1	$\frac{9017}{3168}$	$-\frac{355}{33}$	$\frac{46732}{5247}$	$\frac{49}{176}$	$-\frac{5103}{18656}$		
1	$\frac{35}{84}$	0	$\frac{500}{1113}$	$\frac{125}{192}$	$-\frac{2187}{6784}$	$\frac{11}{84}$	
y_n	$\frac{35}{384}$	0	$\frac{500}{1113}$	$\frac{125}{192}$	$-\frac{2187}{6784}$	$\frac{11}{84}$	0
$\hat{y}_n$	$\frac{5179}{57600}$	0	$\frac{7571}{16695}$	$\frac{393}{640}$	$-\frac{92097}{339200}$	$\frac{187}{2100}$	$\frac{1}{40}$

We see that the seventh step is only used only for the error estimator $\hat{y}_n$. A particular feature of this method, and of many other explicit RK methods, is that the highest-order method is used to advance the solution, while the lower-order method is used only for error estimation. A Runge-Kutta method with an embedded error estimator is often referred to as an RK pair of order $n(m)$, where n is the order of the main method and m the order of the error estimator.

Implicit RK methods can also be equipped with embedded methods. Consider, for instance, the SDIRK method given by (5.21)–(5.25). This can be extended to an

SDIRK pair of order 3(2) of the form

$$
\begin{array}{c|cccc}
0 & & & & \\
c_2 & a_{21} & \gamma & & \\
c_3 & a_{31} & a_{32} & \gamma & \\
c_4 & a_{41} & a_{42} & a_{43} & \gamma \\
\hline
 & b_1 & b_2 & b_3 & b_4 \\
\hline
 & \hat{b}_1 & \hat{b}_2 & \hat{b}_3 & \hat{b}_4
\end{array}
$$

where again the coefficients can be found in Appendix C. This method uses the same technique as the Dormand-Prince pair; the highest-order method is used to advance the solution while the lowest-order method is used only for error estimation. Similar embedded methods can be added to other implicit methods, such as the Radau method presented above; see [59] for details.

Although the idea of error estimation and time step control that is based on embedded methods is very simple, there are numerous matters to be considered regarding its practical use. While (5.36) offers a very simple formula for the next step size, improved formulae have been derived that may give better results, in particular for stiff problems. Also particular to stiff problems is the fact that the stability properties of both the main method and the embedded error estimator must be good, in order to ensure reliable error estimates. The reader is referred to [6] and [58,59] for a detailed discussion of automatic time step control for ODEs.

5.5   The Cell Model Equations

The main purpose of this chapter is to introduce ODE solvers suitable for solving the ODE systems that result from the operator splitting approach described in 3. Quite a few models are listed in Chapter 2, although they represent only a small selection of the wide range of existing models for cellular activity. The models are dramatically different in terms of complexity and physiological realism, and they also possess very different characteristics in terms of stiffness and stability properties. We will here demonstrate the properties of some of the methods described above, by applying them to a small selection of the cell model ODE systems.

5.5.1   Explicit versus Implicit Methods

The simplest model we considered in Chapter 2 was the two-variable FitzHugh-Nagumo model, given by (2.45)–(2.46). To simplify the discussion, we repeat the equations here:

$$
\frac{dv}{dt} = \frac{c_1}{v_{amp}^2}(v - v_{rest})(v - v_{th})(v_{peak} - v)
$$

$$
- \frac{c_2}{v_{amp}}(v - v_{rest})w + I_{app}, \tag{5.40}
$$

$$
\frac{dw}{dt} = b(v - v_{rest} - c_3 w). \tag{5.41}
$$

The values of the model parameters are listed in Chapter 2. Although this is a very simple cell model, the ODE system is nonlinear, and so the stability properties are somewhat more complicated to analyze than for linear equations. As described above, the stiffness of a nonlinear ODE system is related to the eigenvalues of the Jacobian of the right hand side function f.

For a simple system such as (5.40)–(5.41), it is possible to compute analytical expressions for the components of the Jacobian. Recall the definition $J_{ij} = \partial f_i(y)/\partial y_j$, where f is the vector-valued right hand side function and y is the vector of unknowns. However, for the more complex models described in Chapter 2 it becomes difficult to compute the Jacobian matrix analytically. Consider, for instance, the model by Winslow et al. [145], where the right hand side function f has more than 30 components, involving highly complicated expressions. One option for computing the Jacobian and its eigenvalues is to use automatic differentiation tools present in software such as Maple and Mathematica [94,100], but even then the resulting expressions may become quite difficult to handle. Another option is to estimate the eigenvalues numerically. This can be accomplished by simply running a normal simulation with a given model, and storing the values of all variables at all time steps. The Jacobian at each time step can then be evaluated by finite differences, and its eigenvalues can be computed by using, for instance, Matlab [101] tools.

Table 5.1. The largest negative eigenvalues occurring in some popular cell model systems, along with the time step required for stability of the forward Euler method. The unit for the time step is milliseconds.

Model	λ_{min}	Δt_{max}
Fitzhugh-Nagumo	-0.146	13.74
Beeler-Reuter	-82.15	0.0243
Luo-Rudy phase 2	-365	0.0055
Winslow et al.	-19167	$1.04 \cdot 10^{-4}$

Table 5.1 shows numerically estimated eigenvalues for a few popular cell model systems. We only show the real part of the largest negative eigenvalue that occurs during a typical simulation, since this gives a reasonable indication of the stiffness of each model[1]. We see that all the models are considerably stiffer than the Fitzhugh-Nagumo model, but the model by Winslow et al. stands out as particularly stiff. Recall from Section 5.1.2 that to obtain stable solutions the time step must be chosen such that the complex number $\lambda\Delta t$ lies inside the stability region of the applied method. The rightmost column shows the largest time steps that satisfy this criterion

---

[1] The stiffness of a system is related to the existence of fast and slow time scales, and therefore to the ratio between the smallest and largest eigenvalues. However, all the models considered here contain some eigenvalues quite close to zero, and the largest negative eigenvalue therefore gives a reasonable picture of the stiffness.

for the forward Euler method. We see that, for instance, for the model by Winslow et al., the largest possible time step predicted by this analysis is $1.04 \cdot 10^{-4}$ ms. Numerical experiments with the forward Euler method applied to this model reveal that a time step up to $\Delta t = 1.07 \cdot 10^{-4}$ gives a satisfactory solution, while for $\Delta t = 1.08 \cdot 10^{-4}$ the solution becomes unstable. The results are similar for the other cell models, which indicates that the time step estimates predicted from the eigenvalues are quite sharp.

It is easily seen that the forward Euler method is not very efficient for a stiff system such as the model by Winslow et al. Even though each time step is very cheap, the extremely small values required give rise to unacceptable computation times. Applying more advanced explicit solvers does not help, since they offer a very small increase in stability at the cost of significantly increased computational effort per step. Consider, for instance, the Dormand-Prince method described above, with the stability domain plotted in Figure 5.2. For $\lambda = -19167$, we must choose $\Delta t \approx 1.56 \cdot 10^{-4}$, which is a very small improvement considering that each step of this method is approximately six times as expensive as for the forward Euler method.

A nonlinear ODE system can be stiff in some regions and nonstiff in others. The time steps presented in Table 5.1 are dictated by the interval where the models are stiffest, and larger steps may be allowed in other regions. The performance of explicit methods may therefore be enhanced by using adaptive time stepping technique, but in practice the performance increase is limited. Although the stiffness of the models varies through a typical simulation interval, the more complex models tend to be quite stiff through all of the action potential. Adaptive time stepping algorithms used with explicit solvers therefore typically demand very small time steps, leading to quite inefficient computations.

Figure 5.5 illustrates the different stability properties of explicit and implicit RK solvers. The top row shows the action potential of the modified FitzHugh-Nagumo model described above, computed with the explicit Dormand-Prince method of order 5(4) and the SDIRK method of order 3(2) given by (5.21)–(5.25). Both solvers use automatic step size control with the same relative error tolerance ($tol = 10^{-4}$), and all computations are performed using the ODE solver library Godess; see [108]. The explicit solver uses just 35 steps, marked by crosses in the figure, to compute a solution within the required local error tolerance. The SDIRK method requires 98 steps to give a solution with the same error tolerance. This result is not surprising since we have seen that the FitzHugh-Nagumo equations are not stiff, implying that the time steps are chosen based on accuracy requirements rather than stability. The Dormand-Prince method has a higher order of accuracy than the SDIRK method, and therefore requires fewer steps. One time step of the Dormand-Prince method, consisting of seven explicit stages, is also significantly cheaper than that of the SDIRK method, which consists of one explicit and three implicit stages. Hence, for the nonstiff FitzHugh-Nagumo equations, explicit solvers are clearly the best choice.

The bottom row of Figure 5.5 shows the completely different results obtained for the cell model by Winslow et al. Here, the explicit solver requires 54162 time steps,

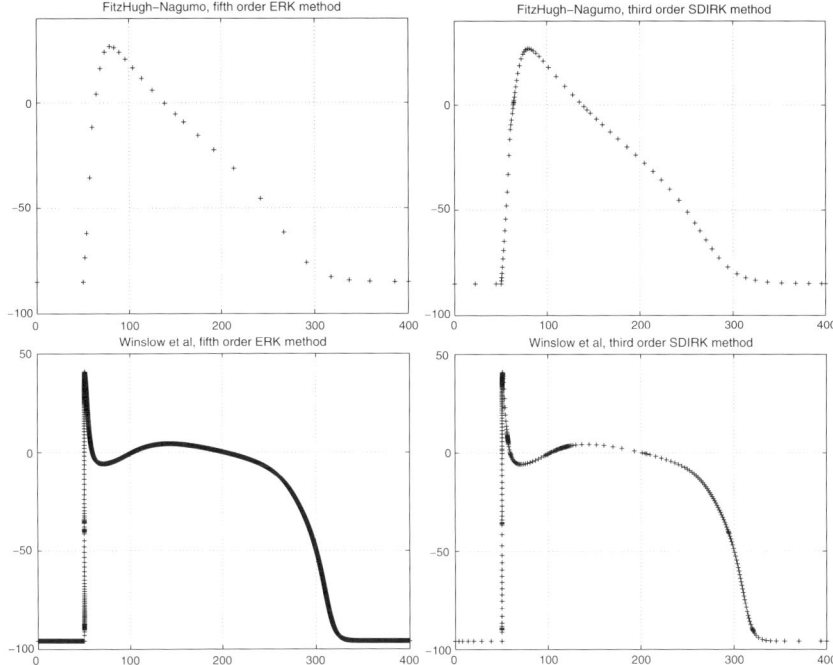

Fig. 5.5. Illustration of time steps used by explicit and implicit methods, for two different cell model systems. Top row: The FitzHugh-Nagumo equations solved with the 5(4) Dormand-Prince pair (left) and an SDIRK method of order 3(2) (right). Bottom: The model by Winslow et al. solved with the Dormand-Prince method (left) and the SDIRK method (right).

leading to a very inefficient computation. The implicit solver gives more reasonable step sizes, 432 steps in total, dictated by accuracy requirements rather than stability. Although each time step of the SDIRK method is more expensive than for the explicit solver, the CPU time for solving the system was reduced by approximately 97 % when switching to the SDIRK scheme.

5.5.2 Simulation Results

If one is interested in modelling only a single cell, efficiency of the numerical method is not a very important issue. Although some methods will lead to impractical computation times when applied to complex cell models like the model by Winslow et al., more advanced explicit methods with adaptive time step control typically give acceptable results. After all, although the CPU time can easily be reduced by a factor of 20–30 by applying a more appropriate stiff solver, this is of limited value if the computation can be performed by an explicit solver in less than 30 seconds. In many cases, this increase in speed is not considered to be worth the increased complexity of implementing an implicit solver.

The situation is entirely different when simulating the electrical activity in the complete heart, or in large volumes of heart tissue. As described in Chapter 2 the operator splitting algorithm reduced the nonlinear PDE system to systems of linear PDEs and nonlinear ODEs. Furthermore, the spatial discretization resulted in one ODE system per node in the computational grid. The number of nodes can easily exceed ten million, and for the more complex cell model systems it is easy to see that this represents a substantial computational effort. Choosing an efficient solver then becomes very important for reducing the CPU time of the complete computation. The relative importance of solving the ODEs efficiently will also be influenced by the complexity of the applied PDE model. The scalar PDE in the monodomain model is considerably less demanding to solve than the system of two PDEs that occurs in the bidomain model, while the effort for solving the ODEs will be virtually the same regardless of which PDE model is used. Using the monodomain model in combination with one of the more realistic cell models can lead to computations that are completely dominated by time spent for solving ODEs.

A number of things must be considered when the ODE systems are solved as part of an operator splitting algorithm. First, recall that the accuracy of the operator splitting was limited to order two. It is therefore not very useful to apply ODE solvers with high order of accuracy, because the global accuracy will still be limited by the splitting error. Second, the time step of the ODE solver will be limited by the time step of the operator splitting algorithm. A typical time step used for the splitting algorithm and the PDE solver is 0.125 ms, which means that a simulation interval of 400 ms, as used in Figure 5.5, will require at least 3200 time steps. This will reduce the efficiency of the implicit methods, which normally require much fewer steps than this, but will have virtually no effect on the performance of explicit solvers. A more detailed discussion of issues related to ODE solvers used for complete heart simulations can be found in [131].

In order to illustrate the performance of the different ODE solvers we have performed a number of 2D experiments with different choices of cell models. The geometry and fibre orientation used for the computations is the same as in Chapter 3.4.2, depicted in Figure 3.10. The experiments were performed using a grid with 15607 nodes in the heart, and we did not include a surrounding body, since this is of limited interest when comparing ODE solvers. The experiments were performed on Itanium 1.3 GHz processors with 4 GBytes of memory.

Simulations were performed using both the monodomain model and the bidomain model, combined with the cell models of FitzHugh-Nagumo [40], Beeler-Reuter [9] and Winslow et al. [145]. The results obtained using an explicit solver for the ODE systems are shown in Table 5.2. We see that the CPU time required for solving the ODE systems is dramatically different for the three models. Solving the ODE systems is an almost insignificant task for the FitzHugh-Nagumo model, but completely dominates the CPU time for the model by Winslow et al. We see that the time required to solve the PDEs is not much affected by the choice of cell model, but the CPU time required for the ODEs is different depending on whether the model is monodomain or bidomain. This is somewhat surprising, since the ODE systems to be solved are independent of the applied PDE model, but a likely explanation is

Table 5.2. CPU time spent on the two main parts of the solution process, using a fifth-order ERK method for the ODEs, and a finite element method combined with multigrid block preconditioning for the PDEs. Results are shown for both the monodomain and the bidomain model.

	Bidomain model		Monodomain model	
Cell model	PDE	ODE	PDE	ODE
FitzHugh-Nagumo	2002	35	427	34
Beeler-Reuter	2060	877	430	826
Winslow et al.	2052	32187	425	43651

Table 5.3. CPU time spent on the two main parts of the solution process, using a third-order SDIRK method for solving the ODE systems. Results are shown for both the monodomain model and the bidomain model.

	Bidomain model		Monodomain model	
Cell model	PDE	ODE	PDE	ODE
FitzHugh-Nagumo	2000	202	427	206
Beeler-Reuter	2079	1346	430	1330
Winslow et al	2053	4344	426	4944

that the two models give different propagation speeds. This will affect the number of ODE systems that are in the different phases of the action potential, and since the solvers choose the time step adaptively this will affect the total computation time.

Table 5.3 shows similar CPU results, obtained using the SDIRK 3(2) method described above to solve the ODEs. We see that the results are roughly what we would expect from the single cell experiments. The SDIRK method gives an increase in CPU time both for the FitzHugh-Nagumo model and the Beeler-Reuter model, and a dramatic performance improvement for the model by Winslow et al. For this model the time required to solve the ODEs is reduced by 85-90 %, resulting in a 80% reduction in total CPU time for the bidomain model, and an almost 90 % reduction for the monodomain model. Using appropriate stiff ODE solvers is hence almost a necessity for using this advanced model in full-scale heart simulations, and clearly worth the additional complexity of implementing the implicit solver. However, our experiments also show that even for a fairly realistic heart cell model such as the Beeler-Reuter model, an explicit ODE solver is clearly the best choice. Hence, to achieve the greatest possible efficiency, one must base the choice of ODE solver on the properties of the applied cell model.

Chapter 6

Large-Scale Electrocardiac Simulations

In the preceding chapters, we have discussed various numerical techniques for solving the different parts of our mathematical model problem. Now it is time to turn our attention to simulating the complete mathematical model. First, we will explain the diverse computational tasks that constitute an electrocardiac simulator. Then, we will estimate the computational resources needed to carry out high-resolution simulations. It will be shown that *parallel computing* is an essential technique for large-scale electrocardiac simulations. Thereafter, this chapter will focus on the general principles for parallelizing a sequential simulator, as well as on major software components needed to achieve this end. These software components will be described in a general setting, without reference to specific programming languages. On the other hand, a simple pseudo language will be used to explain certain details when necessary. Some performance measurements of electrocardiac simulations will also be presented.

6.1 The Electrocardiac Simulator

Let us first recall the complete mathematical model from Chapter 2. In the heart domain H, the Bidomain equations are valid and consist of a parabolic PDE and an elliptic PDE. The transmembrane potential v and the extracellular potential u_e are the primary unknowns. For treating the ionic current $I_{\text{ion}}(s, v, t)$, which is involved in the parabolic PDE, we adopt a system of ODEs that evolves v together with a vector of membrane state variables s. In the torso domain T, a single elliptic PDE models the distribution of the electrical potential u_T within T. This PDE, which models what is called the forward problem, is coupled with the Bidomain equations through suitable boundary conditions on ∂H. For a detailed description of the complete mathematical model, we refer the reader to Chapter 2.

Throughout this chapter, the term *electrocardiac simulation* refers to the numerical solution of the complete mathematical model from $t = 0$ until a given stopping time, where the initial conditions are given in terms of prescribed values for v and s at $t = 0$. Among the three PDEs involved in the complete mathematical model, i.e., (2.66)–(2.68), the parabolic equation (2.66) requires special care, due to both the time dependency and its nonlinearity. Using an operator-splitting technique (see Chapter 3), we split the parabolic equation into two parts:

$$\frac{\partial v}{\partial t} = -I_{\text{ion}}(s, v, t) \tag{6.1}$$

$$\text{and} \qquad \frac{\partial v}{\partial t} = \nabla \cdot (M_i \nabla v) + \nabla \cdot (M_i \nabla u_e). \tag{6.2}$$

The first part of the splitting will become one equation of the ODE system modelling v and s, whereas the second part will be discretized together with the other two elliptic PDEs.

6.1.1 The Numerical Strategy

The overall numerical strategy is based on dividing the time domain into discrete time levels:

$$0 = t_0 < t_1 < t_2 \cdots$$

During each time step, $t_{l-1} < t \le t_l$, the already computed solutions v^{l-1}, u_e^{l-1}, u_T^{l-1}, and s^{l-1} are used as the starting values for computing v^l, u_e^l, u_T^l, and s^l. Based on a flexible θ-rule, where $0 \le \theta \le 1$, we can dissect the computational work per time step into three substeps:

1. Solve the ODE system for $t \in (t_{l-1}, t_{l-1} + \theta(t_l - t_{l-1})]$. This needs to be done at every computational point in H, using v^{l-1} and s^{l-1} as the initial values. As pointed out in Chapter 5, implicit ODE-solving algorithms are usually needed for treating this stiff ODE system. This might involve an adaptive time stepping strategy with variable inner time steps. The computed results from this substep are an intermediate transmembrane potential solution $\tilde{v}^{l-1}$ and an updated $\tilde{s}^{l-1}$ vector.

2. Solve a system of three discretized PDEs for $t \in (t_{l-1}, t_l]$, where (6.2) replaces the original parabolic PDE (2.66), and $\tilde{v}^{l-1}$, u_e^{l-1}, and u_T^{l-1} are used as the initial values. The computational work of this substep amounts to solving a global system of linear equations, which comprises all the degrees of freedom of the electrical potential fields v, u_e, and u_T. As stated in Chapter 3, finite differences and the θ-rule are applied in the temporal discretization, and finite elements are used in the spatial discretization. Combining the unknowns of u_e and u_T into one vector u, we can derive a linear system that has a 2×2-blocked structure; see Chapter 4. The resulting 2×2 block linear system can be solved by preconditioned CG iterations. The computed results from this substep are u_e^l and u_T^l, plus a new intermediate transmembrane potential solution $\hat{v}^l$.

3. Solve the ODE system in the same way as in the first substep, for $t \in (t_{l-1} + \theta(t_l - t_{l-1}), t_l]$, using the initial values $\hat{v}^l$ and $\tilde{s}^{l-1}$. The computed results become v^l and s^l.

6.1.2 Software Components of the Electrocardiac Simulator

In order to run electrocardiac simulations based on the above numerical strategy, we must implement the numerics as a computer program, hereafter called the electrocardiac simulator. The following software components constitute such a simulator:

- *The preprocessing component* is responsible for choosing the simulation-related parameters, such as θ, the size of the time step, and the stopping criterion for

Table 6.1. An example of measuring the CPU usage (in seconds) by the main computational tasks during one time step of a small-scale electrocardiac simulation.

Task	Solving ODEs	Discretizing PDEs	Solving 2×2 system
CPU	99.31	67.04	134.37

the CG iterations. This information can be gathered by prompting a user for answers through a menu interface. The preprocessing component is also responsible for constructing the computational grids for H and T. The other tasks of the preprocessing component include building up the internal data structure, preparing the conductivity fields, and enforcing the initial conditions.

- *The ODE component* is responsible for solving the ODE system, which is based on a chosen ionic current model and an implicit ODE-solving algorithm. Each time this component is invoked, it traverses the heart domain and solves the ODE system for each point in the heart grid.

- *The PDE component* is responsible for solving the PDEs. The work of this component involves carrying out finite element discretizations and running preconditioned CG iterations for the resulting 2×2 block linear system.

- *The time-stepping component* is responsible for the entire time integration process. It invokes the ODE and PDE components during each time step to compute v^l, u_e^l, u_T^l, and s^l, using v^{l-1}, u_e^{l-1}, u_T^{l-1}, and s^{l-1} as the initial values. In addition, it is responsible for storing the computed results as data files at chosen time levels.

- *The postprocessing component* is responsible for summarizing the computational results and releasing the memory allocated for the internal data structure.

Table 6.1 indicates the relative time consumption by the main computational tasks during one time step of a typical simulation. The measurements are obtained on a single Pentium III 1 GHz processor, where the electrocardiac simulation uses a 3D heart grid with $82,768$ points and a 3D body grid (comprising H and T) with $219,398$ points. This particular example uses the Winslow model for the ionic current I_{ion} and an implicit ODE solver, see Chapter 5. The 2×2 block system is solved by 20 preconditioned CG iterations.

Conventionally, the electrocardiac simulator is implemented for single-processor computers. In the following section, we will argue that such a sequential implementation has obvious limits regarding the size of feasible simulations. For real-life electrocardiac simulations, the sequential electrocardiac simulator must be parallelized. This topic is discussed in Sections 6.3–6.9.

6.2 Requirements for Large-Scale Simulations

It should not be difficult to agree on the following observation: computing the electrical activity in the heart and torso is a highly complicated task. This is due to rapidly changing cardiac processes, complex geometry, anisotropic muscle tissue properties, etc. Such a simulation task requires very high resolutions in both time and space.

According to [55,16], a temporal resolution of 0.1 ms and a spatial resolution of 0.2 mm seem to be desirable for highly accurate simulations. For the temporal discretization, this means that the simulation of a single heart beat should be divided into about 10^4 time steps, where the computational cost per time step depends on the number of degrees of freedom involved. In the following, we will show roughly what the spatial resolution of 0.2 mm means for the electrocardiac simulator.

6.2.1 The Memory Requirement

To achieve the desired spatial resolution of 0.2 mm, it can be estimated that the 3D domain of the heart should contain around 5×10^7 grid points. In addition, a considerable amount of grid points should also be placed in the torso domain, although the spatial resolution there does not need to be so high as that in the heart. For simplicity, let us assume that 10^7 grid points are needed in the torso. Then, for the electrocardiac simulator that solves the Bidomain equations coupled with the forward problem in the torso, we can roughly estimate the minimum amount of data needed as follows:

- Consider the ODE system that is applicable in the heart domain H:

$$\frac{\mathrm{d}v}{\mathrm{d}t} = -I_{\mathrm{ion}}(s, v, t),$$
$$\frac{\mathrm{d}s}{\mathrm{d}t} = F(s, v, t).$$

If the number of state variables is 30, e.g., as in the Winslow model, then there are in total $30 \times 5 \times 10^7 = 1.5 \times 10^9$ degrees of freedom. This is because an ODE system needs to be solved at each grid point in the heart, and it is necessary to store all these s solutions as the initial values for subsequent ODE solves.

- For the 2×2 block linear system

$$\begin{bmatrix} I + \theta \Delta t A_v & \theta \Delta t \hat{A}_v \\ \theta \Delta t \hat{A}_v^T & \theta \Delta t \hat{A}_u \end{bmatrix} \begin{bmatrix} v \\ u \end{bmatrix} = \begin{bmatrix} b \\ 0 \end{bmatrix}, \tag{6.3}$$

which arises from discretizing the coupled PDEs, see Chapter 3, the vector v is of length 5×10^7, and the vector u is of length 6×10^7. The right-hand side vectors are of the same lengths as v and u. Therefore, v, u, and the right-hand side vectors require in total

$$5 \times 10^7 + 6 \times 10^7 + 5 \times 10^7 + 6 \times 10^7 = 2.2 \times 10^8$$

floating point numbers. Suppose that there are on average 20 nonzero entries per row in the four sparse matrices that constitute the 2×2 block matrix in (6.3). Then, the entire 2×2 block matrix has in total

$$\left(5 \times 10^7 + 5 \times 10^7 + 5 \times 10^7 + 6 \times 10^7\right) \times 20 = 4.2 \times 10^9$$

nonzero entries, each represented as a floating point number. We remark that the off-diagonal blocks $\hat{A}_v$ and $\hat{A}_v^T$ have essentially the same number of nonzero entries as A_v.

Therefore, for setting up the ODE and PDE systems alone, it needs as many as

$$(1.5 + 0.22 + 4.2) \times 10^9 = 5.92 \times 10^9$$

floating point numbers. In the standard case that each floating point number in double precision occupies 8 bytes in the computer memory, these floating point numbers amount to more than 45 GB! We can of course utilize the fact that the 2×2 block matrix is symmetric and reduce its storage by half. Nevertheless, there are many other memory consuming quantities needed by the electrocardiac simulator, such as

- some internal vectors needed by a Krylov solver,
- some additional matrices and vectors associated with the linear systems for the coarser grid levels in a multigrid preconditioner,
- some additional matrices working as the restriction and interpolation operators between grid levels inside the multigrid preconditioner,
- the x-, y-, and z-coordinates of all the grid points,
- each grid point in the heart grid has to store its associated conductivity tensors M_i and M_e (note that each symmetric 3×3 tensor needs to store six values),
- each grid point in the torso grid has to store its associated conductivity tensor M_T,
- the element-to-point mapping information (represented as integer arrays) for all the finite element grids; e.g., a 3D grid with E tetrahedral elements (four points per tetrahedron) needs a mapping array of length $4E$,
- the locations (represented as integer arrays) of the nonzero entries in all the involved sparse matrices, etc.

Consequently, the actual amount of memory needed by a sufficiently accurate electrocardiac simulation, which satisfies the desired spatial resolution of $0.2\,\text{mm}$, is likely to be several times the estimate of $45\,\text{GB}$. Clearly, there exists no single-processor computer that has this amount of memory. In fact, even for a grid having one tenth of the desired spatial resolution, e.g., with 5×10^6 degrees of freedom for v, the memory requirement is still too high for a single-processor computer.

6.2.2 Realistic Estimates for Memory and Time Usage

By extrapolating a memory or time usage model built on the basis of measuring small-scale simulations, we can come up with realistic estimates for large-scale simulations. Let us look at Table 6.2 that shows some actual measurements of memory and CPU usage. These measurements are obtained by running the first time step of a 3D electrocardiac simulation on several small grids. We have used the Winslow

Table 6.2. Some measurements of memory and CPU usage associated with running the first time step of a 3D electrocardiac simulation. The measurements are obtained on a single R14000 600 MHz processor.

Grid levels	N	Memory usage	CPU usage	CPU/N
1	39,589	55 MB	83.77 s	2.12×10^{-3} s
2	302,166	425 MB	491.63 s	1.63×10^{-3} s
3	1,552,283	2254 MB	3261.64 s	2.10×10^{-3} s

model for the ionic current I_{ion}, whereas CG iterations are used for solving the 2×2 block linear system. The blockwise multigrid preconditioner (see Chapter 5) is used for the CG iterations, and convergence is claimed when the L_2-norm of the residual is reduced by a factor of 10^4. The symbol N denotes the total number of unknowns in v and u. Calculations show that the measured memory requirement in Table 6.2 satisfies the following model:

$$\mathrm{Memory}(N) = 1.22 \times 10^{-3} \cdot N^{1.012} \ \mathrm{MB}. \tag{6.4}$$

Therefore, for the desired high spatial resolution with 5×10^7 heart grid points plus 10^7 torso grid points, i.e., $N = 5 \times 10^7 + 6 \times 10^7 = 1.1 \times 10^8$, the memory requirement can be estimated to be about $160 \, \mathrm{GB}$. Note that this measurement-based estimate is almost four times the above theoretical estimate of the minimum memory requirement ($45 \, \mathrm{GB}$).

Another observation from Table 6.2 seems to suggest that the CPU usage per unknown does not increase with N. This nice property is due to the use of the order-optimal blockwise multigrid preconditioner. We also remark that the first time step of an electrocardiac simulation is more time consuming than the subsequent time steps. This is due to the work on determining the sparsity pattern of A_{v} and $\hat{A}_{\mathrm{u}}$, diverse data storage allocations, and an especially costly finite element discretization and assembly of the 2×2 block system. Let us for the sake of argument assume that, using software optimization and a faster processor, we can improve the average CPU usage model per time step from Table 6.2 by a factor of more than twenty, i.e.,

$$\mathrm{CPU}(N) = 1.0 \times 10^{-4} \cdot N \ \ \mathrm{seconds}. \tag{6.5}$$

Even so, the value of $N = 1.1 \times 10^8$ will mean that an average CPU usage per time step is 1.1×10^4 seconds, i.e., more than three hours. If five thousand time steps are needed for simulating a full cardiac cycle, the required total CPU usage on a single-processor computer will be 636 days, almost two years! This is clearly a formidable computational task for a single-processor computer.

We can conclude that for large-scale electrocardiac simulations, the huge demand of memory calls for the use of multiple memory units, whereas the huge amount of computational work also suggests employing many processors. Therefore, we have to resort to a "special" computer that has multiple processors and memory units. The simulation program may thus be fitted inside the collective memory of such a computer, and the computation time can be shortened due to the involvement of multiple processors. This motivates the adoption of parallel computing, which will be explained in Sections 6.3–6.5. Of course, readers who are already familiar with the subject may proceed directly to Section 6.6.

6.3 Introduction to Parallel Computing

Parallel computing roughly refers to involving multiple processors in solving a global computational task, where some kind of collaboration exists between the processors. The first motivation for adopting parallel computing is the wish of shortening the computation time. The second motivation, especially of relevance for memory-intensive computations such as solving PDEs, is the need of larger memory capacity than that of a single-processor computer.

6.3.1 Hardware and Programming Models

With respect to hardware, parallel computing relies on computing systems that in some way integrate multiple processors and have a total memory capacity equivalent to the sum of many single-processor memory units. There are many variants of such parallel computing systems. At one end of the spectrum, there exist so-called *shared-memory* parallel computers, on which there is a single global memory space shared equally by all the processors. We remark that the single global memory space on a shared-memory parallel computer is enabled either directly by hardware technology or by a combination of hardware and software technologies.

Besides such shared-memory systems, most of the other parallel computers are based on the concept of *distributed memory*. In a distributed-memory parallel computer, each processor has its local memory space that can not be accessed directly by other processors. Collaboration between the processors implies some kind of inter-processor communication of information, on some kind of network. Distributed-memory systems can be further categorized with regard to the involved communication network. Two examples are expensive massively parallel processing systems that use a tightly coupled high-speed network, and clusters of PCs that often use a cheap local area network. Assuming the same strength of the processors, we note that the better the performance of the communication network, the better the possibility of achieving good overall performance of parallel computing.

Different types of hardware naturally give rise to different models of parallel programming. For distributed-memory systems, the standard programming model is *message passing*, where processors "talk to each other" by sending and receiving short or long messages. Here, the messages simply refer to arrays of values, such as characters, integers, floating point numbers, etc. Communication in this programming model can be of two types. The first type of communication is so-called *one-to-one*, where one processor sends a message to another receiving processor. On the other hand, in *collective* communication, a group of processors participate collectively in exchanging messages with each other inside the group. Typical collective communication operations include one-to-all operations such as "broadcast", all-to-one operations such as "gather", and all-to-all operations such as "all-reduce" (see, e.g., [110]). For example, finding the sum of P numbers that are distributed on P processors can be done by an all-reduce add operation. The obtained result is then available on all the processors. See Section 6.4.2 for a concrete example.

Message-passing programs consist of standard sequential statements that are interleaved with additional communication commands. During a parallel execution, each processor runs a copy of the parallel program independently. The communication commands ensure necessary information exchange and synchronization between the processors. We mention here that the *message passing interface* (MPI) library is the de-facto standard of message passing commands, see [53,41,110]

As counterpart to the message-passing programming model, the *shared-memory programming model* is achievable on shared-memory systems. In this programming model, programs have mainly an appearance of sequential execution, which spawns multiple "execution threads" whenever entering a code region typically involving computation over long loops, and returns to the single "execution thread" immediately after exit. Necessary information exchange and synchronization between the processors are normally hidden in some special system calls, such as those in the OpenMP standard (see [23,109]). However, this parallel programming model will not be adopted in this chapter due to the following reasons:

- Distributed-memory systems are the dominating hardware platforms.

- Message-passing programs can execute on both distributed-memory and shared-memory systems. But shared-memory programs rely completely on shared-memory platforms.

- Although the shared-memory programming model is easy to use, it is often not flexible enough to treat complicated situations. On the other hand, message-passing programming is a versatile tool at the cost of some additional programming effort.

- Performance of message-passing programs is normally better than that of shared-memory programs.

Therefore, we will only consider the message-passing programming model in the remainder of this chapter, and the sequential electrocardiac simulator will be parallelized in this style.

6.3.2   Division of Work and the Resulting Overhead

We recall that the motivation for adopting parallel computing is the wish to reduce computation time and memory requirement per processor. This is in principle achieved by employing multiple processors and dividing the global computational work and data among them. In the message-passing programming model, the way of dividing the work and data must be decided and explicitly implemented by the programmer.

However, continuing the discussion of work division, we need to mention the non-parallelizable tasks. There is always a portion of a sequential program that is not parallelizable, meaning that for certain tasks it is not possible to reduce the computation time by use of multiple processors. Such non-parallelizable tasks typically contain either too few operations, or operations that are subject to an execution sequence dependency, thus destroying the parallelism. Two examples of non-parallelizable tasks are adding two scalar numbers and reading command-line user inputs. During parallelization, non-parallelizable tasks have to be either handled by one "master" processor or replicated on all the processors. Quite often, the percentage of non-parallelizable work within a large-scale sequential program is very small and often diminishes when the size of the global computational task increases.

Having mentioned the non-parallelizable tasks, let us now move on to the discussion of work division. When we handle the parallelizable portion of a global computational task, it is important to use a "fair" division scheme for all the processors. It is also important to note that the division scheme should be enforced in the very beginning of a parallel program and kept, if possible, throughout the execution. This approach avoids unnecessary work associated with re-division and shuffling local data between the processors. Obviously, we would like to divide a parallelizable task into P equal-sized parts, so that the processors can finish their assigned work simultaneously, without having to spend much time on waiting for each other. This issue is often referred to as *load balancing*. However, it is not enough to only have an equal division of the work and data, we should also try to minimize the overhead due to parallelization. Some common cases of parallelization-caused overhead are discussed below.

Overhead due to Communication. Even an idealized procedure for sending a message from processor A to processor B involves several substeps. First, processor A writes the message into a local array. Then, the system might have to copy the message into a system message buffer. Finally, the message is transferred over the network and copied into a waiting array on processor B. A simple cost model for one-to-one transfer of a message of length L can be formulated as:

$$\tau_C(L) = \tau_0 + \xi L, \tag{6.6}$$

where τ_0 is the so-called *latency*, which represents the system startup time for communication. The coefficient $1/\xi$ is often referred to as the *bandwidth*, which indicates the rate at which messages can be exchanged between two processors. We remark that the unit of message length is typically one floating point number in double precision. An all-to-all communication operation is typically decomposed into $\log_2 P$ sub-stages, where each sub-stage involves a two-way one-to-one message exchange. Thereby, the cost of an all-to-all communication is typically $\log_2 P$ times the cost of (6.6).

Overhead due to Synchronization. Very often, it is impossible to achieve an entirely equal division of work among the processors, possibly due to some unstructured nature of the computations. Also, if the local computation has a dynamic work load, perfect load balance is not achievable. So it may sometimes happen that the processors are executing their copy of a parallel program at their own pace. However, in many cases of numerical computation, such as iterations of a Krylov subspace solver, we can not allow some processors to progress too far ahead of the others. Synchronization is thus needed, which can be achieved either implicitly or explicitly through a special system call. An example of implicit synchronization is to combine it with a collective communication. In general, the worse the situation of uneven load balance, the larger the synchronization overhead.

Overhead due to Duplicated Computation. Ideally, the work of a global computation should be divided equally and also *disjointly* among the processors. It is however not always possible to achieve a disjoint division. As an example, let us consider the division of a finite element grid. While it is possible to divide the elements disjointly, the grid points lying on the element boundaries between the neighbors have to be replicated. This often gives rise to a situation where the sum of all the local computations exceeds the amount of the original global computation. We remark that the overlapping domain decomposition methods (see Section 4.5) requires even more duplicated computation, because the elements from neighboring subdomains should have a certain number of overlapping layers. Moreover, overhead of memory usage is often associated with duplicated computations.

Considering the different types of overhead that may be introduced by parallelization, we arrive at the following principles for dividing the computational work of a sequential program:

- A chosen division of the work and data should be enforced in the beginning of a parallel program and remain unchanged, if possible, throughout the whole parallel execution. In this way, each processor can concentrate on its assigned local data and computation, without having to shuffle data with the neighbors.

- Preferably, we should avoid situations where a global data structure is first built on one master processor and then divided and distributed to the other processors. Each processor should be responsible for building up its own local data structure.

- The "areas" of the local computations should be approximately the same. This is for reducing the overhead due to synchronization.

- The "shapes" of the local computations should be of such a form that the number of "neighbors" each processor has is small and the size of upcoming communication messages is also small. This is for reducing the overhead due to communication and possibly also duplicated local computation.

6.3.3 Speedup and Parallel Efficiency

To measure the quality of a parallelized program, we introduce the important concept of *speedup*, which is defined as

$$S(P) = \frac{T(1)}{T(P)}. \tag{6.7}$$

Here, P is the number of processors involved, and $T(1)$ denotes the time usage of the original sequential program. Moreover, $T(P)$ denotes the time usage of the parallelized program using P processors. In case of different time usages by the P processors, which actually is quite usual, $T(P)$ refers to the time usage by the "slowest" processor.

As we have seen in Section 6.3.2, parallelization normally introduces different types of overhead. Therefore, the following relation is generally valid:

$$T(P) \geq \frac{T(1)}{P}.$$

Consequently, we normally have

$$S(P) = \frac{T(1)}{T(P)} \leq \frac{T(1)}{\frac{T(1)}{P}} = P. \tag{6.8}$$

Another equivalent metric for the quality of a parallelization is *parallel efficiency*, which is defined as

$$\eta(P) = \frac{S(P)}{P}. \tag{6.9}$$

Following (6.8), we can arrive at

$$\eta(P) \leq \frac{P}{P} = 1.$$

This means that the best result of parallelization is that we achieve 100% parallel efficiency. However, due to the possible existence of overhead in a parallelized program, we should normally be satisfied with an efficiency close to 100%. In general, the efficiency is a decreasing function with respect to P.

Remarks. There is one factor that actually has the potential of improving the parallel efficiency when more processors are used. The possibility of local data being able to fit entirely inside the cache[1] of a processor increases when the size of the local computation decrease, i.e., when P increases. We may find cases where the sequential computation is too large to fit in a cache, while using a certain number of processors can considerably improve the local cache performance, therefore reducing $T(P)$ noticeably. Nevertheless, we are mainly concerned with large-scale parallel computations that have too large local computations to be located entirely inside a cache. So the positive cache effect on the parallel efficiency is not considered in this chapter.

---

[1] A cache can be regarded as a special memory area located between a processor and its main memory, see, e.g., [47]. The memory size of a cache is relatively small but its access speed is much higher than that of the main memory. If the data of a small computational task can be programmed to fit entirely inside the cache, exceptionally good performance may arise.

For the parallel solution of a global computational task of a fixed size, increasing the number of processors P normally results in a reduced parallel efficiency $\eta(P)$. This is because that the amount of different overhead normally does not scale down proportionally with $1/P$. Therefore, the parallel efficiency $\eta(P)$ normally decreases with increasing P. On the other hand, for a fixed number of processors P, if the size of the global computation increases, the amount of local computation increases proportionally. The percentage of overhead in $T(P)$ normally decreases in this case. Consequently, the parallel efficiency $\eta(P)$ normally increases together with the size of the global computation.

6.4   Two Simple Examples

Now let us look at two simple examples of parallelization. The purpose is to let the reader familiarize with the division of work, speedup analysis, and a pseudo programming language.

6.4.1   Adding Two Vectors

We first consider a global computational task of adding two long vectors $u = (u_1, u_2, \ldots, u_N)$ and $v = (v_1, v_2, \ldots, v_N)$, both of length N. The result of the addition is to be stored in another vector of the same length: $w = (w_1, w_2, \ldots, w_N)$. It is obvious that adding two vectors can be done in the following way, entry by entry:

$$w_i = u_i + v_i \quad \forall i \in \mathcal{I}, \tag{6.10}$$

where $\mathcal{I}$ is an index set containing $1, 2, \ldots, N$.

Note that the above N adding operations of two scalar values need not be executed in an increasing order of the index i. Actually, any random order of traversing $\mathcal{I}$ produces the same numerical result. That is, the N scalar adding operations can be executed completely independently of each other. This immediately gives rise to parallelism.

The first step of parallelization is to divide the global computational work among P processors. For simplicity, let us divide the work disjointly. So processor number p is responsible for N_p adding operations, where

$$N = \sum_{p=1}^{P} N_p.$$

There are, however, many different ways of achieving this division of work and data. One approach is a cyclic distribution of the entries, such that entries number $1, P+1, 2P+1, \ldots$ are given to processor number 1, and entries number $2, P+2, 2P+2, \ldots$ are given to processor number 2, and so on. Another approach is a blockwise distribution where entries number $1, 2, 3, \ldots, N_1$ are given to processor number 1, and entries number $N_1 + 1, N_1 + 2, N_1 + 3, \ldots, N_1 + N_2$ are given to processor number 2, and so on.

Here, it is not important to know exactly which scheme is used to divide the global work. This is because adding two vectors is a frequently encountered operation in numerical computations, so the division of work and data is typically "inherited" from a preceding computational task. For the moment, it suffices with an assumption that there exists a set of local indices $\mathcal{I}_p = \{1, 2, \ldots, N_p\}$, $1 \leq p \leq P$. The P values of N_p should be either identical or have very small differences. This is for achieving a good load balance. In fact, the result of a division of data can be expressed by the following mapping:

$$i \rightarrow (p, j), \quad i \in \mathcal{I}, \ 1 \leq p \leq P, \text{ and } j \in \mathcal{I}_p. \tag{6.11}$$

We remark that a disjoint division of data implies a unique mapping from i to (p, j). Very often, the mapping (6.11) is not explicitly used in every part of a parallel program. On each processor, a copy of the parallel (message-passing) program works primarily with its assigned local data. Each processor distinguishes itself from other processors by its unique rank p. In this example, the local data for each processor are three local vectors u_p, v_p, and w_p. If we assume that the three local vectors are ready, typically "inherited" from a preceding computational task in the same parallel program, a pseudo code segment for a parallel addition of two vectors can be as follows:

Parallel_Add(u_p, v_p, w_p)
Input: u_p and v_p on processor number p
Output: w_p on processor number p
    for all $j \in \mathcal{I}_p$
        $w_p(j) := u_p(j) + v_p(j)$
    end for

Speedup Analysis. For this extremely simple parallelization example, there is no need to use communication between the processors. We can derive the following model for $T(P)$:

$$T(P) = \max_p N_p \tau_A, \tag{6.12}$$

where τ_A loosely denotes the cost of one floating point arithmetic operation. In fact, τ_A also includes the cost of reading data from the memory and writing them back. This is because the data structure for almost any PDE simulator is in form of long (and multi-dimensional) arrays, so memory access tightly accompanies the arithmetic operations.

Similarly, the time usage by a sequential addition is

$$T(1) = N \tau_A. \tag{6.13}$$

Therefore, the speedup $S(P)$ becomes

$$S(P) = \frac{T(1)}{T(P)} = \frac{N \tau_A}{\max_p N_p \tau_A} = \frac{N}{\max_p N_p}. \tag{6.14}$$

In case of a perfect load balance and a disjoint division of data, i.e., $\max_p N_p = N/P$, we can see that (6.14) is reduced to $S(P) = P$. In other words, the following three factors together give rise to a perfect speedup result: the absence of communication overhead, a disjoint division of work, and identical amount of local work per processor.

Non-Disjoint Division of Data. Although the mapping (6.11) does not appear in the pseudo code of **Parallel_Add**, it is implied by the "inherited" division of work and data, i.e., $\mathcal{I}_p$. Very often, it suffices with such a "distributed view" of a parallel program, without having to know how exactly a global index i can be mapped to (p, j) in every part of the parallel program. We have so far assumed a disjoint division of data, which means that the mapping (6.11) is unique and $N = \sum N_p$. However, in many large-scale parallel programs, the division of work and data is non-disjoint (see Section 6.3.2). That is, there is a small amount of overlap between the local vectors on different processors and thus $N < \sum N_p$. A natural question is whether a non-disjoint division affects the implementation of **Parallel_Add**. The answer is that **Parallel_Add** remains the same for a non-disjoint division, even though the mapping (6.11) is no longer unique and the size of $\mathcal{I}_p$ becomes larger. We require, of course, that if an entry of u or v is distributed to more than one processor, the value of the entry should be replicated on all the involved processors. Regarding the parallel efficiency, the local work load is increased due to some duplicated computation, so the speedup $S(P)$ becomes less perfect.

One solution to the above problem of duplicated computation is to use an additional disjoint division $\mathcal{I}_p^d \subset \mathcal{I}_p$ and thus let processor number p only compute a portion of w_p, $j \in \mathcal{I}_p^d$. Then, the processors exchange results with each other to fill the rest of w_p. Note that we can not simply skip the computation for these entries of w_p, because they are typically needed by other parts of a parallel program. This solution, however, is typically too expensive due to the new communication overhead.

6.4.2 Inner Product

As another example of parallelization, we consider the inner product of two vectors: $u = (u_1, u_2, \ldots, u_N)$ and $v = (v_1, v_2, \ldots, v_N)$. That is, we want to compute

$$c = (u, v) \equiv \sum_{i=1}^{N} u_i \cdot v_i. \tag{6.15}$$

To reveal parallelism for this case, we give an equivalent reformulation of (6.15), i.e., starting with $c = 0$ and accumulating the value of c by

$$c = c + u_i \cdot v_i \quad \forall i \in \mathcal{I}, \tag{6.16}$$

where $\mathcal{I} = \{1, 2, \ldots, N\}$. The purpose with (6.16) is to show that the contribution from $u_i \cdot v_i$ can be accumulated by traversing $\mathcal{I}$ in any order.

Following the previous example, we assume again that there exists a division of work and data, in form of $\mathcal{I}_p$ on each processor. The basic idea for doing a parallel inner product is to first let each processor compute a portion of c and then use communication and some follow-up additions to obtain the correct value of c. Assuming that the two local vectors u_p and v_p are available, arising from an inherited disjoint division of data, we can write the following pseudo code segment for the computation of a parallel inner product:

Parallel_Inner(u_p, v_p, c)
Input: u_p and v_p on processor number p
Output: c on every processor
    $c_p := 0$
    for all $j \in \mathcal{I}_p$
        $c_p := c_p + u_p(j) \cdot v_p(j)$
    end for
    $c := c_p$
    for $1 \leq i \leq P$ and $i \neq p$
        Send(c_p,i)
        Receive(c_i,i)
        $c := c + c_i$
    end for

In the above code segment, we have used Send(c_p,i) to denote a communication operation of sending the value c_p to processor number i. Similarly, Receive(c_i,i) denotes a communication operation of receiving the value c_i from processor number i.

Speedup Analysis. The main difference between the present example and the previous one is that communication is now involved. Each processor has to exchange $P - 1$ small messages, each containing only one scalar value, with $P - 1$ other processors. Therefore, the parallel time usage model $T(P)$ is as follows:

$$T(P) = \left(\max_p (2N_p - 1)\tau_A \right) + (P - 1)\left(\tau_C(1) + \tau_A \right), \qquad (6.17)$$

where we have used the cost model for one-to-one communication (6.6). In addition, we have assumed that one multiplication operation takes the same amount of time (τ_A) as one addition. In case of a perfect load balance $\max_p N_p = N/P$ and also

Table 6.3. The wall-clock time measurements of Parallel_Inner on a Linux cluster.

	$N = 10^6$	$N = 10^7$
$P = 1$	4.605365×10^{-2} s	4.548842×10^{-1} s
$P = 2$	2.499276×10^{-2} s	2.287801×10^{-1} s
$P = 4$	1.505762×10^{-2} s	1.161581×10^{-1} s
$P = 8$	8.052347×10^{-3} s	5.977540×10^{-2} s
$P = 16$	6.118934×10^{-3} s	3.206090×10^{-2} s

$N \gg P$, the speedup can be found as

$$S(P) = \frac{T(1)}{T(P)} = \frac{(2N-1)\tau_A}{\left(\max_p(2N_p-1)\tau_A\right) + (P-1)\left(\tau_C(1) + \tau_A\right)}$$

$$= \frac{(2N-1)\tau_A}{\left(\frac{2N}{P} - 1\right)\tau_A + (P-1)\left(\tau_C(1) + \tau_A\right)}$$

$$\approx \frac{P}{1 + \frac{P(P-1)}{2N}\left(\frac{\tau_C(1)}{\tau_A} + 1\right)} < P. \qquad (6.18)$$

It can be observed from (6.18) that the speedup result is not perfect due to communication overhead and a very small amount of duplicated computation, which is associated with the follow-up additions $c = \sum c_p$. Two factors decide the actual quality of the speedup: the size of the global computational task N and the communication speed τ_C. More specifically, the following two observations are in general true for any parallel program:

– The smaller the ratio $\frac{P}{N}$, the better the speedup result. This means that a larger global work load is favorable for speedup.

– The larger the ratio $\frac{\tau_C(1)}{\tau_A}$, the worse the speedup result. This means that a relatively slow communication speed is not favorable for speedup.

Let us consider Parallel_Inner(u_p, v_p, c) for two cases: $N = 10^6$ and $N = 10^7$. The chosen parallel computing system is a Linux cluster consisting of Pentium III 1 GHz processors, which are connected through a fast Ethernet network. For the Pentium processors, the value of τ_A is measured as about 2.3×10^{-8} s, which is mainly determined by the memory speed. For the Linux cluster, which uses MPI for communication, the cost of exchanging one message containing one double-precision number, i.e., $\tau_C(1)$, is measured as about 1.4×10^{-4} s. The wall-clock timing results for Parallel_Inner are displayed in Table 6.3, and the associated parallel efficiency values are showed in Figure 6.1. We can observe from Figure 6.1 that a larger value of N is clearly favorable for achieving better speedup and parallel efficiency.

Collective Communication. We can see that the communication part in Parallel_Inner becomes quite costly if P is large. This is because handling $P - 1$ different messages on each processor is not a simple task. To improve the performance

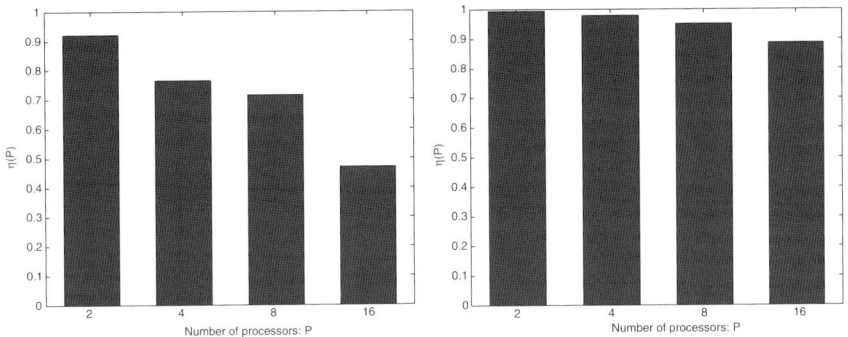

Fig. 6.1. The parallel efficiency of Parallel_Inner on a Linux cluster for two cases: $N = 10^6$ (left) and $N = 10^7$ (right).

of such a frequently encountered operation, we should communicate by means of a global reduction operation, which combines the message exchanges and the follow-up additions in Parallel_Inner into one special command: AllReduce_Add(c_p, c). This all-to-all communication command typically involves $\log_2 P$ sub-stages, thus having an improved cost model, e.g., $\log_2 P(\tau_C(1) + \tau_A)$. Using AllReduce_Add, we can write an improved parallel inner product as follows:

Parallel_Inner2(u_p, v_p, c)
Input: u_p, v_p on processor number p
Output: c on every processor
    $c_p := 0$
    for all $j \in \mathcal{I}_p$
        $c_p := c_p + u_p(j) \cdot v_p(j)$
    end for
    AllReduce_Add(c_p, c)

Non-Disjoint Division of Data. In case of a non-disjoint division of u and v, neither Parallel_Inner nor Parallel_Inner2 will compute the correct result c. This is because some of the entries in u_p and v_p are also duplicated on other processors. One solution to this problem is to find another local index set $\mathcal{I}_p^d \subset \mathcal{I}_p$ on processor number p. The requirement is that $\mathcal{I}_p^d$, $1 \leq p \leq P$, should form a disjoint division of the global data u and v. Then, we just need to replace $\mathcal{I}_p$ with $\mathcal{I}_p^d$ in Parallel_Inner2.

6.5   Domain-Based Parallelization

In the preceding section, we have looked at parallel versions of a vector addition and an inner product. Both operations are frequently encountered in the computational task of solving PDEs. So far it has been said that the division of the global data and work is "inherited" from other parts of a parallel program. In this section, we will show how the division of data and work can be achieved.

We recall that a PDE problem normally has a spatial domain Ω in which the PDE is valid. For parallelizing a PDE simulator, the division of the global work and data is most naturally realized by partitioning the global solution domain Ω into P subdomains. Then, each subdomain Ω_p, $1 \leq p \leq P$, becomes the "territory" of one processor, so that each processor can construct its own local data restricted to Ω_p. The global data are represented by the set of subdomain data collectively. There is normally no need to construct and store the global data physically, they exist only in a "logical" sense. The division of the global data is thus enforced, and the division of the global work arises accordingly, such that each processor only does computations using its local data. These local computations are done concurrently in parallel, interleaved with inter-processor communication when necessary. The collective effect of these coordinated local computations becomes the same as using one processor for the entire global work. We mention that this domain-based parallelization approach is quite different from using the shared-memory model, where the global data are physically constructed and the work load is divided typically at the loop/array level.

6.5.1   Division of Data for FDM

Normally, the finite difference method (FDM) uses a uniform or structured lattice grid covering the global solution domain Ω. Solutions are sought on each of the grid points and computations typically involve FDM stencils that couple each grid point with a few neighboring points.

The division of the global work and data in the FDM is achievable by partitioning the grid points evenly between the processors. Using cutting lines or planes, see Figure 6.2, we produce subdomains also of a rectangular shape. So the original sequential FDM program can in principle be re-applied for local computations on the subdomains. Additional inter-processor communication commands must be inserted at the appropriate locations. We note that the cutting lines/planes are located between the grid points, so that all the grid points are divided disjointly between the subdomains. The computational task on each subdomain is to find the solution on the local grid points. We will use the term *internal boundary points* to refer to the grid points in a subdomain that lie immediately beside the cutting lines/planes. We also remark that a FDM lattice grid can be partitioned using cutting lines of lower spatial dimensions. For example, a 3D lattice grid can also be partitioned one-dimensionally or two-dimensionally. A lower dimensional partitioning results in fewer neighbors, but larger volume of communication.

The above disjoint partitioning of the grid points and the consequent disjoint division of the global computational work are straightforward. However, we can observe from Figure 6.2 that computations associated with the internal boundary points require values on some grid points lying in the neighboring subdomains. Such values should therefore be provided by the neighboring subdomains, i.e., each subdomain needs to send the values on its internal boundary points to its neighbors and also receive the required values. This exchange of values is carried out by several one-to-one communication operations. Let us consider subdomain number one

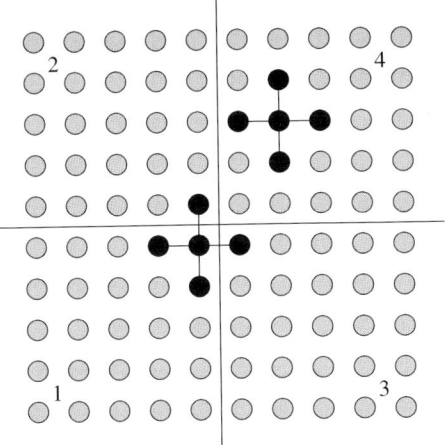

Fig. 6.2. Partitioning of the points of a FDM lattice grid and data dependency of a five-point FDM stencil.

in Figure 6.2 for example. Every time the subdomains have updated values on their local points, subdomain number one has to send the new values on its right-side internal boundary nodes, as an array, to subdomain number three. In return, subdomain number one should receive an array, containing the new values on the left-side internal boundary nodes in subdomain number three. A similar two-way one-to-one data exchange has to be done between subdomains number one and number two.

To handle the above type of communication, one may extend the local data structure of each subdomain to include one layer of so-called *ghost boundary points*. Of course, if a FDM stencil covers more than one layer of the surrounding points, more layers of ghost boundary points should be included in a subdomain. This is done in order to obtain a convenient storage of the values associated with the ghost boundary points. It may also prevent special implementations of the local computations for the internal boundary points. A sample partitioning of a FDM lattice grid is shown in Figure 6.3. We stress that the responsibility of finding solutions on the ghost boundary points belongs to the neighboring subdomains. The ghost boundary points only participate in local computations associated with the internal boundary points. The subdomain computational points arise from a disjoint division of the global computational work.

For explicit finite difference schemes, a parallelized program typically involves the same type of loops as in the original sequential program. The only differences are that the loops are now restricted to the subdomain computational points, and communication must be added for updating the ghost boundary points, as shown above. Section 6.6 presents such an example.

For implicit finite difference schemes, which involve the solution of a global linear system in form of $Ax = b$, each subdomain builds its local linear system

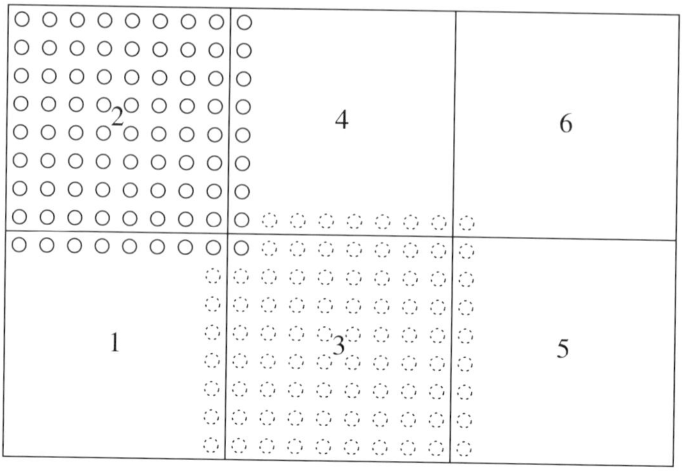

Fig. 6.3. An example of partitioning a global FDM lattice grid into subdomain lattice grids, each with one layer of ghost boundary points.

associated with the division of the global data:

$$A_p x_p = b_p. \qquad (6.19)$$

This local discretization arises from traversing the computational points in the subdomain, while applying the FDM stencil. The number of equations in (6.19) is determined by the number of subdomain computational points, i.e., *not* including the ghost boundary points. The same is the length of the right-hand side vector b_p, whereas the vector of unknowns x_p includes the ghost boundary points. This is because the FDM stencil relates the ghost boundary points with the internal boundary points. So the subdomain system matrix A_p has more columns than rows. Of course, the local system (6.19) can not be solved independently. We have to solve the P subdomain local systems collectively to find the global solution, which is always distributed as x_p, $1 \le p \le P$. A parallel iterative procedure is most appropriate for such a distributed linear system. Section 6.7 demonstrates how the kernel linear algebra operations can be parallelized using such distributed matrices and vectors.

6.5.2 Division of Data for FEM

The FEM typically uses unstructured computational grids, where elements are the basic computational units. Following the approach of domain-based parallelization, we need to partition the global finite element grid into P subdomain grids, so that the elements are divided between the subdomains. This can be a complicated process when the global finite element grid is unstructured. As discussed in Section 6.3.2, we want the number of elements per subdomain to be approximately the same, and the "shape" of the subdomains to be such that the parallelization-induced communication overhead is limited. We will not discuss further the topic of partitioning

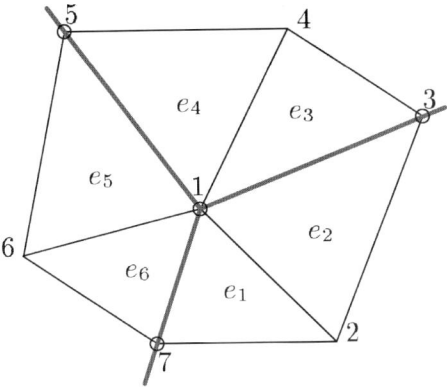

Fig. 6.4. A non-overlapping partitioning of a finite element grid.

unstructured finite element grids, but rather refer the reader to, e.g., [39,70,140]. Throughout the following text, it will be assumed that any finite element grid can be partitioned into a set of subdomain grids, with sufficiently good load balance.

When a set of subdomain grids is ready, parallelization of the finite element discretization and assembly process is straightforward. No communication is needed at this stage. Each processor simply restricts its operations to the subdomain grid. The result is that the global linear system is distributed as a set of local linear systems arising from the subdomains. It is important to note that the global linear system needs only to exist *logically*, not physically as a whole. During the solution of the distributed global linear system, the work of each processor consists of local linear algebra operations and message exchanges with its neighbors. The details will be presented in Section 6.7.

Non-Overlapping Subdomains. Unlike the cutting lines or planes that are used to divide the points in a FDM lattice grid, the cutting curves or surfaces for partitioning a global finite element grid can not "cut through" the elements. The cutting curves/surfaces must coincide with some of the element edges or sides. These element edges or sides constitute the internal boundaries between the neighboring subdomains. Therefore, although the elements can be divided disjointly between the subdomains, i.e., there is no overlap in terms of the elements, neighboring subdomains always share some grid points that lie on the internal boundaries. Figure 6.4 depicts such a situation, where elements e_1 and e_2 belong to one subgrid, e_3 and e_4 belong to another subgrid, whereas e_5 and e_6 belong to a third subgrid. Grid points that are marked with a circle in Figure 6.4 lie on the internal boundaries, and are therefore shared between neighboring subdomains. Consequently, when the global vectors are divided using this partitioning, some entries of each subdomain vector have to be replicated among the neighbors. The details on how to divide the data and computation associated with a global matrix or vector are given in Section 6.7.

Overlapping Subdomains. Very often, the mathematical design of a numerical component, such as a parallel preconditioner, requires that the neighboring subdomains should have some amount of overlap between each other. That is, there should be at least one layer of overlapping elements between two neighboring subdomains. This typically arises from two non-overlapping subdomains by "pushing" the internal boundary of one subdomain towards the interior of its neighbor. An equivalent requirement for overlapping subdomains is that every internal boundary point in one subdomain must have at least one of its duplicates as an interior point in a neighboring subdomain. In a sense, the internal boundary points in overlapping subdomains have the same role as the ghost boundary points in a parallelized FDM program. This is because the responsibility of finding the solution on an internal boundary point is now given to a neighboring subdomain, which has the duplicated point lying in its interior, see Section 6.7. Normally, an overlapping partitioning arises from enlarging a set of non-overlapping subdomains. Consequently, the number of duplicated grid points will be higher than that in the non-overlapping case. This results in an increased amount of local computations, thus a decrease in the speed-up. We leave the details on parallelizing linear algebra operations associated with an overlapping partitioning to Section 6.7.

6.5.3 Summary of Principles

In the following, we summarize the principles of the domain-based parallelization approach to be used in the present chapter:

- The general hardware model of distributed memory is assumed.
- The division of work and data is done through grid partitioning, which in fact gives a domain division.
- A subdomain works only with its local matrices and vectors. It has no direct access to other subdomains' local data.
- A processor is assigned with one subdomain. There is no master-slave relation between the processors, which execute the same simulation code. However, the progression of code may differ between the processors at a given time instant. Synchronization is enforced when necessary, either explicitly by a specific command or implicitly as a result of some collective communication operations.
- Communication between neighboring subdomains is either in form of one-to-one exchange of arrays of values, or in form of collective communication operations involving very short messages, such as AllReduce_Add.

6.6 Explicit FDM in Parallel

In order to illustrate the parallelization procedure for a sequential FDM program, let us consider the following simple parabolic PDE:

$$\frac{\partial v}{\partial t} = \nabla^2 v \quad \text{in } \Omega \times (0, T], \tag{6.20}$$

$$\frac{\partial v}{\partial n} = 0 \quad \text{on } \partial\Omega, \tag{6.21}$$

$$v(x, 0) = v^0(x) \quad x \in \Omega, \tag{6.22}$$

where n in (6.21) is the outward unit normal vector on $\partial\Omega$. We remark that (6.20) can be considered as a simplified version of the Monodomain equation (2.35), after we have chosen $M_i \equiv 1$ and ignored the ionic current.

6.6.1 Discretization by Finite Differences

For simplicity, let us assume that Ω is the unit box $(x, y, z) \in [0, 1]^3$. In addition, let us use the Forward Euler scheme for the temporal discretization and a seven-point FDM stencil in the spatial discretization. Suppose we use a uniform $N \times N \times N$ spatial grid, i.e.,

$$\Delta x = \Delta y = \Delta z = \frac{1}{N - 1} = h$$

and each grid point is identified by a triplet of subscripts (i, j, k), such that

$$x_i = (i - 1)\Delta x, \quad y_j = (j - 1)\Delta y \quad z_k = (k - 1)\Delta z, \quad 1 \leq i, j, k \leq N.$$

Treating the Boundary Conditions. Since the boundary conditions (6.21) are of the Neumann type, it is convenient to introduce a layer of ghost boundary points around the computational points. That is, we introduce

$$x_{0,j,k}, \quad x_{N+1,j,k}, \quad 1 \leq j, k \leq N,$$
$$y_{i,0,k}, \quad y_{i,N+1,k}, \quad 1 \leq i, k \leq N,$$
$$z_{i,j,0}, \quad z_{i,j,N+1}, \quad 1 \leq i, j \leq N.$$

Using the above ghost boundary points, we can discretize the homogeneous Neumann boundary condition on the left side $x = 0$ by

$$\left.\frac{\partial v}{\partial x}\right|_{x=0} \approx \frac{v_{2,j,k} - v_{0,j,k}}{2\Delta x} = 0 \quad \Rightarrow \quad v_{0,j,k} = v_{2,j,k}, \quad 1 \leq j, k \leq N.$$

Using a superscript l to denote the time level, such that $t^l = l\Delta t$, we can write the numerical strategy for solving (6.20)–(6.22) as follows:

1. Enforce the initial condition (6.22) at $t = 0$.

$$v_{i,j,k}^0 = v^0(x_i, y_j, z_k), \quad 1 \le i, j, k \le N. \tag{6.23}$$

2. Enforce the boundary conditions (6.21) at $t = 0$.

$$\begin{aligned}
v_{0,j,k}^0 &= v_{2,j,k}^0, & 1 \le j, k \le N, \\
v_{N+1,j,k}^0 &= v_{N-1,j,k}^0, & 1 \le j, k \le N, \\
v_{i,0,k}^0 &= v_{i,2,k}^0, & 1 \le i, k \le N, \\
v_{i,N+1,k}^0 &= v_{i,N-1,k}^0, & 1 \le i, k \le N, \\
v_{i,j,0}^0 &= v_{i,j,2}^0, & 1 \le i, j \le N, \\
v_{i,j,N+1}^0 &= v_{i,j,N-1}^0, & 1 \le i, j \le N.
\end{aligned}$$

3. Repeat time stepping until $t = T$ is reached. The work at each time step consists of updating values on the computational points, i.e.,

$$\frac{u_{i,j,k}^{l+1} - u_{i,j,k}^l}{\Delta t} = \tag{6.24}$$

$$\frac{u_{i,j,k-1}^l + u_{i,j-1,k}^l + u_{i-1,j,k}^l - 6u_{i,j,k}^l + u_{i+1,j,k}^l + u_{i,j+1,k}^l + u_{i,j,k+1}^l}{h^2}$$

for $1 \le i, j, k \le N$, and enforcing the boundary conditions afterwards:

$$\begin{aligned}
v_{0,j,k}^{l+1} &= v_{2,j,k}^{l+1}, & 1 \le j, k \le N, \\
v_{N+1,j,k}^{l+1} &= v_{N-1,j,k}^{l+1}, & 1 \le j, k \le N, \\
v_{i,0,k}^{l+1} &= v_{i,2,k}^{l+1}, & 1 \le i, k \le N, \\
v_{i,N+1,k}^{l+1} &= v_{i,N-1,k}^{l+1}, & 1 \le i, k \le N, \\
v_{i,j,0}^{l+1} &= v_{i,j,2}^{l+1}, & 1 \le i, j \le N, \\
v_{i,j,N+1}^{l+1} &= v_{i,j,N-1}^{l+1}, & 1 \le i, j \le N.
\end{aligned} \tag{6.25}$$

6.6.2 A Sequential Program

A straightforward implementation of the above numerical strategy should use two arrays $\bar{v}$ and v, which are of the dimension $(N+2) \times (N+2) \times (N+2)$. The lower bound of the array indices is 0 and the upper bound is $N+1$. The array $\bar{v}$ contains values on the computational points plus the ghost boundary points at time level l, whereas v stores the values for time level $l+1$. In addition, three one-dimensional arrays x, y, and z are used to store the coordinates of the computational points.

Before showing the main program Sequential_Heat, let us first introduce two assistant functions: Enforce_BC and Update, which can be used to enforce the boundary conditions (6.21) and update the computational points following (6.24), respectively.

Enforce_BC(v, N_1, N_2, N_3)
Input: v, N_1, N_2, N_3
Output: v
    for all $(j, k) \in [1, N_2] \times [1, N_3]$
        $v(0, j, k) := v(2, j, k)$
        $v(N_1 + 1, j, k) := v(N_1 - 1, j, k)$
    end for
    for all $(i, k) \in [1, N_1] \times [1, N_3]$
        $v(i, 0, k) := v(i, 2, k)$
        $v(i, N_2 + 1, k) := v(i, N_2 - 1, k)$
    end for
    for all $(i, j) \in [1, N_1] \times [1, N_2]$
        $v(i, j, 0) := v(i, j, 2)$
        $v(i, j, N_3 + 1) := v(i, j, N_3 - 1)$
    end for

Update$(\bar{v}, v, N_1, N_2, N_3, \Delta t, h)$
Input: $\bar{v}, N_1, N_2, N_3, \Delta t, h$
Output: v
    for all $(i, j, k) \in [1, N_1] \times [1, N_2] \times [1, N_3]$
        $v(i, j, k) := \bar{v}(i, j, k)$
        $+ \frac{\Delta t}{h^2} [\bar{v}(i, j, k - 1) + \bar{v}(i, j - 1, k) + \bar{v}(i - 1, j, k)$
        $- 6\bar{v}(i, j, k) + \bar{v}(i + 1, j, k) + \bar{v}(i, j + 1, k) + \bar{v}(i, j, k + 1)]$
    end for

We note that both the functions Enforce_BC and Update are designed to accept different lengths in different dimensions for the $\bar{v}$ and v arrays. This is particularly convenient for the parallelized program to be presented in Section 6.6.3. For the sequential implementation, we use the following pseudo code:

Sequential_Heat$(N, T, \Delta t, v)$
Input: $N, T, \Delta t$
Output: v
    Construct arrays $\bar{v}$ and v
    Construct and fill one-dimensional arrays x, y, and z
    $t := 0$
    for all $(i, j, k) \in [1, N] \times [1, N] \times [1, N]$
        $\bar{v}(i, j, k) := v^0(x(i), y(j), z(k))$
    end for
    Enforce_BC$(\bar{v}, N, N, N)$
    while $t < T$ do
        $t := t + \Delta t$
        Update$(\bar{v}, v, N, N, N, \Delta t, h)$
        Enforce_BC(v, N, N, N)
        $\bar{v} := v$
    end while

6.6.3 Parallelization

Following the domain-based parallelization approach in Section 6.5.1, we divide the global computational points into P rectangular parts. Although writing a general three-dimensional partitioning scheme for any arbitrary value of P is not trivial, it is relatively easy for common values of P, such as 2, 4, 8, etc. Anyway, the partitioning result is that subdomain number p is assigned with $N_{1,p} \times N_{2,p} \times N_{3,p}$ computational points. For the current example, it is sufficient to assume that there exists a function

$$\mathsf{Partitioning}(N, p, P, N_{1,p}, N_{2,p}, N_{3,p}, x_p, y_p, z_p)$$

that takes N, p, and P as input, and produces $N_{1,p}$, $N_{2,p}$, $N_{3,p}$, x_p, y_p, and z_p. The partitioning determines where each subdomain is located (through x_p, y_p, z_p), and implies a mapping from a local triplet of indices (i, j, k) on processor number p to a unique global triplet of indices. This mapping is needed if the final local solutions v_p, which are distributed on the P processors, are desired to be "glued" back into a global solution v.

We also note that the local computational operations on each subdomain are exactly the same as those in the sequential case, so the function Update can be re-used as is. In the following, we list the parallelized program:

Parallel_Heat($P, N, T, \Delta t, v$)
Input: P, N, T, Δt on every processor
Output: v_p on processor number p
    Partitioning($N, p, P, N_{1,p}, N_{2,p}, N_{3,p}, x_p, y_p, z_p$)
    Construct arrays $\bar{v}_p$ and v_p using $N_{1,p}, N_{2,p}, N_{3,p}$
    $t := 0$
    for all $(i, j, k) \in [1, N_{1,p}] \times [1, N_{2,p}] \times [1, N_{3,p}]$
        $\bar{v}_p(i, j, k) := v^0(x_p(i), y_p(j), z_p(k))$
    end for
    Parallel_Enforce_BC($\bar{v}_p, N_{1,p}, N_{2,p}, N_{3,p}$)
    while $t < T$ do
        $t := t + \Delta t$
        Update($\bar{v}_p, v_p, N_{1,p}, N_{2,p}, N_{3,p}, \Delta t, h$)
        Parallel_Enforce_BC($v_p, N_{1,p}, N_{2,p}, N_{3,p}$)
        $\bar{v}_p := v_p$
    end while

Comparing Parallel_Heat against Sequential_Heat, we can see that the main change is the use of a new function Parallel_Enforce_BC, instead of Enforce_BC. It is inside this new function that several one-to-one communication operations are embedded. Here, we assume the general situation of partitioning a 3D FDM lattice grid, i.e., there can be up to six neighbors for each subdomain, one "lower-end" neighbor and one "upper-end" neighbor in each direction. If a supposed neighbor on one side does not exist, it means that the subdomain borders the physical boundary on that side, therefore having to enforce the Neumann condition (6.21). In case of an existing neighbor on that side, one-to-one communication must be carried out to exchange data with the neighbor. Therefore, the function Parallel_Enforce_BC can be programmed as follows:

Parallel_Enforce_BC($v_p, N_{1,p}, N_{2,p}, N_{3,p}$)
Input: $v_p, N_{1,p}, N_{2,p}, N_{3,p}$ on processor number p
Output: v_p on processor number p
    if the lower-end x-neighbor exists, then
        n := rank of the lower-end x-neighbor
        collect the values of $v_p(1, j, k)$ into m_{out}
        Send(m_{out},n)
        Receive(m_{in},n)
        copy m_{in} into the locations of $v_p(0, j, k)$
    else
        for all $(j, k) \in [1, N_{2,p}] \times [1, N_{3,p}]$
            $v_p(0, j, k) := v_p(2, j, k)$
        end for
    end if
    if the upper-end x-neighbor exists, then
        n := rank of the upper-end x-neighbor
        collect the values of $v_p(N_{1,p}, j, k)$ into m_{out}
        Send(m_{out},n)
        Receive(m_{in},n)
        copy m_{in} into the locations of $v_p(N_{1,p} + 1, j, k)$
    else
        for all $(j, k) \in [1, N_{2,p}] \times [1, N_{3,p}]$
            $v_p(N_{1,p} + 1, j, k) := v_p(N_{1,p} - 1, j, k)$
        end for
    end if
    same work for the y-direction
    same work for the z-direction

6.7 Parallel Conjugate Gradient Iterations

We can see from the preceding section that parallelizing an explicit FDM scheme is quite straightforward. But what is the case for parallelizing an implicit FDM scheme? For this purpose, let us modify the numerical strategy from Section 6.6.1 by using the Backward Euler scheme, instead of the Forward Euler scheme. That is, (6.24) is replaced by

$$\frac{u_{i,j,k}^{l+1} - u_{i,j,k}^{l}}{\Delta t} = \tag{6.26}$$
$$\frac{u_{i,j,k-1}^{l+1} + u_{i,j-1,k}^{l+1} + u_{i-1,j,k}^{l+1} - 6u_{i,j,k}^{l+1} + u_{i+1,j,k}^{l+1} + u_{i,j+1,k}^{l+1} + u_{i,j,k+1}^{l+1}}{h^2}$$

for $1 \leq i, j, k \leq N$. Consequently, the work of a sequential solver at each time step now becomes solving a linear system of the following form:

$$Av^{l+1} = v^l, \tag{6.27}$$

which arises from combining the implicit scheme (6.26) and the discretized Neumann boundary conditions (6.25). We note that the values on the ghost boundary points of the global lattice grid are not explicitly included as unknowns in (6.27), which has N^3 equations in total.

The system matrix A in (6.27) consists of one main diagonal and six off diagonals. It is symmetric and positive-definite, making the CG method a good candidate for solving (6.27). In the following, we will first see how the CG iterations can be parallelized in association with the implicit FDM. Then, we will treat the general case of parallelizing the CG iterations associated with finite element discretizations on unstructured grids.

6.7.1 Conjugate Gradient Revisited

Let us first repeat the mathematical formulation of the CG method, which is an iterative solution process. For a linear system of the form:

$$Ax = b,$$

the CG iterations are expressed as follows.

$r = b - Ax$ (matrix-vector product and vector addition)
$p = r$
$\pi^0_{r,r} = (r, r)$ (inner product)
while not converged
  $w = Ap$ (matrix-vector product)
  $\pi_{p,w} = (p, w)$ (inner product)
  $\xi = \pi^0_{r,r}/\pi_{p,w}$
  $x = x + \xi p$ (vector addition)
  $r = r - \xi w$ (vector addition)
  $\pi^1_{r,r} = (r, r)$ (inner product)
  $\beta = \pi^1_{r,r}/\pi^0_{r,r}$
  $p = r + \beta p$ (vector addition)
  $\pi^0_{r,r} = \pi^1_{r,r}$

We note from above that the computational kernels in the CG method are matrix-vector products, inner products between two vectors, and vector additions. These three kernels are also the main building blocks of all the other Krylov subspace methods. This means that if we manage to parallelize these three kernels, we can parallelize almost all the Krylov subspace methods.

6.7.2 Parallel CG and FDM

We can use the same division of data as in Section 6.6.3 when parallelizing an implicit FDM program. That is, the function **Partitioning** is still valid. The only difference is that in an implicit FDM program, the unknowns are organized as a vector v, instead of a three-dimensional array. Similarly, subdomain number p needs to arrange a local vector v_p to contain the values on all its $N_{1,p} \times N_{2,p} \times N_{3,p}$ computational points plus the additional ghost boundary points.

We recall from Section 6.5.1 that each subdomain can independently build its local linear system of form (6.19) through a local discretization. This local discretization traverses only the $N_{1,p} \times N_{2,p} \times N_{3,p}$ computational points, but the ghost boundary points also participate due to the seven-point FDM stencil. Therefore, the local matrix A_p has $N_{1,p} \times N_{2,p} \times N_{3,p}$ rows but more columns. Together, the P local linear systems (6.19) represent the global linear system (6.27). It is important to note that all the matrices and vectors are distributed in this way. The division of work follows the division of data, in that each subdomain only executes local operations on its assigned grid points.

Parallel Vector Addition. Additions such as $r = r - \xi w$ are parallelized by letting each subdomain execute

$$r_p = r_p - \xi w_p$$

on all its local points, including the ghost boundary points. No communication is needed. The reason for including the ghost boundary points is that r_p is related to p_p, which must have correct values on the ghost boundary points for participating in a parallel matrix-vector product, see below.

Parallel Inner Product. We recall that the subdomain computational points arise from a disjoint division of the global computational points. So a global inner product, such as (p, w), can be parallelized by first letting each subdomain execute a local inner product over its computational points, (p_p, w_p), followed by an **AllReduce_Add** function. See Section 6.4.2 for more details.

Parallel Matrix-Vector Product. For parallelizing $w = Ap$, we first execute a local matrix-vector product on each subdomain:

$$w_p = A_p p_p,$$

which only affects the values of w_p on the subdomain computational points. Then, for updating the values of w_p on the ghost boundary points, every subdomain needs to carry out message exchanges with all its neighbors. The message exchange process is similar to that in the function **Parallel_Enforce_BC** from Section 6.6.3.

6.7.3 Parallel CG and FEM

Although the FEM often uses unstructured grids, the parallelization follows the same idea of "divide-and-conquer". The main difference from the FDM case is that communication is more complex, because the internal boundaries are no longer straight lines or planes, and the number of neighbors per subdomain may not be a fixed number.

We recall from Section 6.5.2 that the partitioning of a finite element grid, which is typically unstructured, is based on the idea of dividing the elements between the subdomains, either disjointly or non-disjointly. Correspondingly, this gives rise to non-overlapping or overlapping subgrids. One important observation is that such a partitioning strategy does not produce a disjoint division of the global grid points, which is different from partitioning a finite difference lattice grid. Consequently, there are always some overlapping grid points in each subdomain that have duplicates in the neighboring subdomains.

Since the partitioning typically has an unstructured nature, unlike the straight cutting lines or planes in the case of FDM, the neighbor information is often complex and must be handled with care. Let us denote by $\mathcal{O}_p$ a set that contains the local indices of all the overlapping points in subdomain number p. For non-overlapping subgrids, $\mathcal{O}_p$ contains only the local indices of the internal boundary points. For overlapping subgrids, $\mathcal{O}_p$ contains the local indices of both the internal boundary points and the other overlapping points.

For each overlapping point in $\mathcal{O}_p$, we also denote by $o_{p,k}$ the so-called *degree of duplication*, where k is between 1 and the size of $\mathcal{O}_p$. That is, $o_{p,k}$ is the number of subdomains, including subdomain number p, that share the particular overlapping point. For example, points number 3, 5, and 7 in Figure 6.4 on page 195 have a degree of duplication of two, whereas point number 1 has a degree of duplication of three. Also, it is necessary for each overlapping point on a subdomain to know which neighboring subdomains possess the other duplicates of itself.

Division of Data. As in the case of parallel FDM, all the matrices and vectors in parallel FEM also have a distributed representation. The division of data follows the non-disjoint division of the global grid points, arising from either a non-overlapping or an overlapping grid partitioning. The global system matrix A, which is only logically existing, is represented by the set of P subdomain matrices A_p. On subdomain number p, A_p is built independently by running the finite element discretization and assembly procedure over the subdomain. No communication is needed. We note that the resulting rows in A_p associated with the internal boundary points do not have the same entries as the corresponding rows in A. This is because some contributions from elements belonging to the neighboring subdomains are excluded. (The other rows of A_p have entirely correct entries.) The same situation applies to the corresponding entries in the local right-hand side vector b_p. However, this does not prevent us from obtaining correct results from a distributed matrix-vector product, which needs to be supplemented by communication, as will be explained below. For the other distributed vectors w_p, p_p, and r_p, which are used in a parallelized CG method, it is always required that an entry that is associated with an overlapping point must be correctly replicated on all the possessing subdomains.

We remark that we avoid building the global matrix A first on one processor and then distributing the rows of A to other processors through communication.

Although this approach can produce a disjoint division of data, the excessive memory and communication usage constitute too much overhead.

Parallel Vector Addition. Since any overlapping entry in a subdomain vector, except b_p, is correctly replicated over the possessing subdomains, parallel execution of, e.g.,

$$r_p = r_p - \xi w_p$$

can be done straightforwardly without communication. All the entries in r_p and w_p participate in the local vector addition. We remark that the speedup result may suffer a little due to some duplicated computation overhead. But for a sufficiently large global grid that is partitioned into non-overlapping subgrids, this loss of speedup is normally negligible.

Parallel Inner Product. We recall from Section 6.4.2 that both Parallel_Inner and Parallel_Inner2 assume a disjoint division of data. In our approach to partitioning a finite element grid, however, such a disjoint division is not readily achievable. We note that we can not simply skip the overlapping points when computing a local inner product, because the contribution from the overlapping points will thus be missing on all the subdomains. So the remedy is to either find a disjoint division of the entries in a global vector, based on the existing non-disjoint division, or remove the excessive contribution from the overlapping points. We explain the second approach by the following pseudo program function.

Parallel_Inner3(u_p, v_p, c)
Input: u_p, v_p on processor number p
Output: c on every processor
    $c_p := 0$
    for all j from 1 to length of u_p
        $c_p := c_p + u_p(j) \cdot v_p(j)$
    end for
    for all k from 1 to size of $\mathcal{O}_p$
        $j :=$ entry number k in $\mathcal{O}_p$
        $c_p := c_p - \dfrac{o_{p,k} - 1}{o_{p,k}} \cdot u_p(j) \cdot v_p(j)$
    end for
    AllReduce_Add(c_p, c)

We note that the second for-loop above is namely used to remove the excessive contribution inside c_p from the first-loop. Clearly, some overhead arises both due to duplicated computation in the first for-loop and due to extra computation in the second for-loop.

Parallel Matrix-Vector Product. The subdomain matrix A_p and right-hand side vector b_p arise from a local discretization and assembly process on subdomain number

p. The rows of A_p and entries of b_p, which are associated with the internal boundary points, do not have the same values as those in a logically existing global linear system. This is because each internal boundary point on subdomain number p also needs contributions from elements belonging to the neighboring subdomains. Therefore, to achieve the correct collective result of, e.g., $r = b - Ax$ using the subdomain matrices and vectors, we have to supplement the local computation

$$r_p = b_p - A_p x_p \qquad (6.28)$$

with communication and additional computation. Note we have assumed that the entries in x_p, which are associated with the overlapping points, are already correctly replicated among the neighbors.

Let us first consider the case of non-overlapping subgrids. On subdomain number p, the local index j for an internal boundary point is stored as entry number k in $\mathcal{O}_p$. This internal boundary point belongs to $o_{p,k}$ subdomains including subdomain number p. The value of $r_p(j)$ obtained from the local computation (6.28) needs to be added by $o_{p,k} - 1$ other values possessed by the neighboring subdomains. So communication is needed here. Subdomain number p should prepare, for each of its neighbors, one input array and one output array for exchanging values. The length of one such pair of input and output arrays is the number of internal boundary values shared between two neighboring subdomains. Using such a setup, the parallel computation of $r = b - Ax$ can be done by the following steps:

1. Each subdomain independently carries out the local computation (6.28).

2. For each of its neighbors, subdomain number p fills an output array using some of the values of r_p and sends it out as a message. Then, a corresponding incoming message is received into an input array.

3. Afterwards, each internal boundary value $r_p(j)$ retrieves needed values from the input arrays and adds the values to itself.

Work is needed to create these input/output arrays and prepare information about how each internal boundary value should retrieve needed values from the input arrays. This complicated "book-keeping" task can be termed as *communication preparation*, and needs to be done only once before the CG iterations start.

For overlapping subgrids, there are more overlapping points per subdomain. Still, only the internal boundary values need special treatment. This is because all the other overlapping values obtain correct values after the local computation (6.28). We also recall for overlapping subgrids that each internal boundary point has at least one of its duplicates lying in the interior of a neighboring subdomain. This means that the correct value can be provided entirely by such a neighbor. Therefore, the first two steps involved in a parallel matrix-vector product are the same as above, namely carrying out the local computation (6.28) and exchanging input/output arrays. In the third step, the local values for the internal boundary points are simply discarded, and then replaced with values retrieved from the input arrays.

Of course, the task of communication preparation for overlapping subgrids is slightly different from that for non-overlapping subgrids. Such preparation tasks, together with the work spent on handling the input/out arrays during every parallel matrix-vector product, should preferably be programmed as generic functions that are re-usable for different parallel PDE solvers.

We also mention that the overhead due to duplicated computation, which is present in the above parallel matrix-vector product based on overlapping subgrids, can be eliminated with the help of an enlarged volume of communication. The strategy is to find a disjoint division of the rows in A, on the basis of the existing non-disjoint division that is represented by A_p, $1 \leq p \leq P$. Then, each subdomain only computes an assigned subset of the rows in A_p during the local matrix-vector product. Afterwards, each subdomain has to provide more of its computed values to its neighbors. The number of input and output arrays does not change, but the length of the arrays becomes longer.

6.8 Domain Decomposition as Parallel Preconditioners

Preconditioners are often needed to improve the convergence of Krylov subspace solvers such as CG. That is, an equivalent linear system

$$B^{-1}Ax = B^{-1}b \qquad (6.29)$$

is solved instead of the original system $Ax = b$. During the Krylov iterations, the work of the frequently applied preconditioner B^{-1} is in the following form:

$$\text{Given } r, \text{ compute } z = B^{-1}r. \qquad (6.30)$$

When B is represented explicitly as a matrix, we need to solve a linear system

$$Bz = r, \qquad (6.31)$$

which should be simpler than the original system, to obtain z inside each preconditioning operation. Otherwise, the effect of applying B^{-1} is realized by carrying out a set of linear algebra operations. The main requirements for B^{-1} are good resemblance to A^{-1} plus ease of construction and computation.

Parallel preconditioning is obviously important for a parallel PDE solver to achieve the convergence rate of its sequential counterpart. However, parallelizing a sequential preconditioner strictly following its original mathematical definition is often a difficult task, because many preconditioners are inherently sequential, such as SOR, SSOR, and incomplete LU factorizations. When the matrices B_p and vectors r_p, z_p are already distributed on the subdomains, the most straightforward approach to building a parallel (and modified) preconditioner is to first run the preconditioner locally on all the subdomains in parallel:

$$z_p = B_p^{-1}r_p.$$

Then, we enforce some kind of averaging between neighboring subdomains for z_p to obtain the values associated with the overlapping points. In other words, the basic idea is to use parallel local computations plus communication, similar for the parallel matrix-vector products. The effect of such a parallel preconditioner depends on the choice of B_p^{-1}, and whether there exists some kind of global correction between the subdomains. The DD methods (see Section 4.5) suit nicely the above framework for building parallel preconditioners, and theory and practice have shown that these methods sometimes result in optimal convergence. That is, the required number of iterations for achieving convergence is independent of the number of unknowns.

6.8.1 Additive Schwarz Preconditioner

Throughout the rest of this chapter, we will only consider additive Schwarz iterations, which are one variant of the DD methods. This is because additive Schwarz iterations are parallel by nature, and no special interface solvers are needed.

The mathematical description of an additive Schwarz preconditioner is

$$B^{-1} = \sum_{p=0}^{P} R_p^T \tilde{A}_p^{-1} R_p, \tag{6.32}$$

where $\tilde{A}_p^{-1}$, $1 \leq p \leq P$, means that an approximate subdomain solver is allowed. It is required that the subdomains should have a certain amount of overlap between them. The symbols R_p and R_p^T in (6.32) denote the associated restriction and interpolation matrices, needed by the mathematical formulation. However, since we always work directly with distributed subdomain matrices and vectors, instead of global matrices and vectors, the restriction and interpolation matrices, $1 \leq p \leq P$, are never explicitly needed in the parallel computations. We therefore omit R_p and R_p^T in our subsequent formulas. The special matrix A_0 is associated with a discretization on a very coarse global grid. The use of so-called coarse grid corrections in (6.32) is necessary to achieve convergence independent of the number of subdomains P.

The additive Schwarz iterations are parallel by nature. The reason is that the subdomain solvers, i.e., $\tilde{A}_p^{-1}$, can carry out its local computation completely independent of each other. This fits nicely into the domain-based parallelization strategy presented in Section 6.5. That is, the global solution domain is partitioned into P overlapping subdomains. Each subdomain is assigned to one processor, which is responsible for the local discretization that gives rise to A_p and b_p. During a parallel solution process of $Ax = b$ by preconditioned Krylov subspace iterations, the involved linear algebra operations are parallelized as explained in Section 6.7. A parallel additive Schwarz iteration (6.32) can work as the preconditioner. The following four tasks are involved in each parallel additive Schwarz preconditioning operation:

1. The coarse grid correction

$$A_0 z_0 = r_0 \tag{6.33}$$

is performed. The right-hand side vector r_0 refers to a projection of a global vector r onto the global coarse grid. Since the global vector r is distributed as r_p, $1 \leq p \leq P$, each processor carries out a partial projection from r_p (the subdomain part of a global fine-scale vector) to $r_{p,0}$ (the subdomain part of a global coarse-scale vector). Then, the distributed projection results $r_{p,0}$, $1 \leq p \leq P$, are summed to form r_0 by help of collective communication. We assume that the size of the coarse grid problem is too small for parallelization, so (6.33) is solved on every processor.

2. Each processor solves (approximately) a local linear system:

$$\tilde{A}_p z_p = r_p. \tag{6.34}$$

3. Neighboring subdomains exchange values of z_p for the interior overlapping points, i.e, not including the internal boundary points. The exchange of values is done by one-to-one communication. For every point that is an interior overlapping point in more than one subdomain, an average of the multiple values is used. The values of z_p on the internal boundary points are discarded and replaced with corresponding values provided by the neighbors.

4. The coarse grid correction z_0 is projected back to the subdomain grid, before being added to the above z_p.

Scalability and MG as Subdomain Solver. A desired objective of solving large-scale linear systems $Ax = b$ is to keep the total computational effort linearly proportional to the number of unknowns. In the context of parallel computing, an additional objective is to avoid an increase of the total computational effort as the number of processors grows. To achieve both objectives, the following two requirements must be satisfied:

– The required number of Krylov iterations must remain constant, independent of the number of unknowns and the number of subdomains.

– The computational cost for solving a subdomain problem (6.34) must be a linear function with respect to the number of unknowns in the subdomain.

For certain PDEs, the first requirement can be fulfilled by using the additive Schwarz preconditioner. In this connection, coarse grid corrections (6.33) must be used. For handling the second requirement, MG cycles are a good candidate in general. One MG V-cycle can often work as a sufficiently good subdomain solver. Of course, a hierarchy of subgrids must be built on each subdomain to enable a subdomain MG solver.

Overhead in Additive Schwarz. In fact, even the aforementioned combination of an additive Schwarz preconditioner and a multigrid subdomain solver can not produce the perfect scalability. First, it is difficult for real-life PDE applications to achieve optimal convergence of the Krylov iterations. Second, there exist different types of overhead:

- Overhead due to communication is present in both the first task (coarse grid corrections) and the third task (data exchange between neighbors) of an additive Schwarz iteration.

- Overlap between the neighboring subdomains is required by the mathematical definition of the additive Schwarz preconditioner. However, this results in some overhead due to duplicated computation. Similar type of overhead is also present when the coarse grid problem (6.33) is solved on every processor. Parallelizing the coarse grid solver can reduce the overhead of duplicated computation, but it introduces more communication overhead at the same time.

- The subdomain problems may not have exactly the same size, so some subdomain solvers may have to wait for others to finish. Therefore, overhead due to synchronization may arise.

6.8.2 Parallel DD for the Monodomain Equation

To discretize the Monodomain equation, we can use the technique of operator splitting and an associated θ-rule in the temporal direction, together with finite elements in the spatial direction, see Section 3.2.2. Consequently, the following linear system needs to be solved at each time step:

$$A_{\mathrm{Mo}} v^{l+1} = b_{\mathrm{Mo}}^l,$$
$$\text{where} \qquad A_{\mathrm{Mo}} = I + \theta \Delta t A_{\mathrm{v}}, \qquad (6.35)$$
$$b_{\mathrm{Mo}}^l = -(1 - \theta) A_{\mathrm{v}} v^l.$$

Here, I refers to the mass matrix associated with finite element discretizations, and A_{v} arises from dicretizing the term $-\nabla \cdot (M_i \nabla v)$.

Parallelization relies on an overlapping partitioning of the heart domain H, so that each processor is responsible for a subdomain H_p. The global matrices I and A_{v} are distributed as subdomain matrices I_p and $A_{\mathrm{v},p}$. The additive Schwarz preconditioner can then be defined as

$$B_{\mathrm{Mo}}^{-1} = \sum_{p=0}^{P} B_{\mathrm{Mo},p}^{-1} \quad \text{where } B_{\mathrm{Mo},p}^{-1} \approx (I_p + \theta \Delta t A_{\mathrm{v},p})^{-1}. \qquad (6.36)$$

One or multiple MG V-cycles can be applied as the approximate subdomain solver $B_{\mathrm{Mo},p}^{-1}$.

6.8.3 Parallel DD for the Bidomain Equations

For the Bidomain equations, a similar discretization strategy as above will result in a 2×2 block linear system. That is, the following system needs to be solved at each time step:

$$A_{\mathrm{Bi}} \begin{bmatrix} v^{l+1} \\ u^{l+1} \end{bmatrix} = \begin{bmatrix} b \\ 0 \end{bmatrix}, \quad \text{where } A_{\mathrm{Bi}} = \begin{bmatrix} I + \theta \Delta t A_{\mathrm{v}} & \theta \Delta t A_{\mathrm{v}} \\ \theta \Delta t A_{\mathrm{v}}^T & \theta \Delta t A_{\mathrm{u}} \end{bmatrix}. \qquad (6.37)$$

As before, an overlapping partitioning of H is used. For obtaining a parallel preconditioner for the 2×2 block system (6.37), we suggest a layered structure. First, the approximate inverse of the diagonal block matrix

$$\begin{bmatrix} I + \theta \Delta t A_{\mathrm{v}} & 0 \\ 0 & \theta \Delta t A_{\mathrm{u}} \end{bmatrix} \tag{6.38}$$

works as the overall preconditioning framework. Then, one additive Schwarz iteration works as a parallel approximate solver for the upper diagonal block $I + \theta \Delta t A_{\mathrm{v}}$, and another additive Schwarz iteration handles the lower diagonal block $\theta \Delta t A_{\mathrm{u}}$. More specifically, the parallel preconditioner for (6.37) is defined as

$$B_{\mathrm{Bi}}^{-1} = \begin{bmatrix} \displaystyle\sum_{p=0}^{P} B_{\mathrm{Mo},p}^{-1} & 0 \\ 0 & \displaystyle\sum_{p=0}^{P} B_{\mathrm{u},p}^{-1} \end{bmatrix},$$

$$\text{where} \quad B_{\mathrm{Mo},p}^{-1} \approx (I_p + \theta \Delta t A_{\mathrm{v},p})^{-1}, \tag{6.39}$$

$$B_{\mathrm{u},p}^{-1} \approx \theta \Delta t A_{\mathrm{u},p}^{-1}.$$

One or multiple MG V-cycles can be applied as both the approximate subdomain solvers $B_{\mathrm{Mo},p}^{-1}$ and $B_{\mathrm{u},p}^{-1}$.

6.8.4 Extension to the Torso

We have shown in Section 3.3 that the forward problem, which is an elliptic PDE valid in the torso T, can be discretized together with the Bidomain equations. This results in an "enlarged" 2×2 block system (6.3) that needs to be solved at each time step.

Comparing (6.3) with (6.37), we can see that the main difference is that $\hat{A}_{\mathrm{u}}$ in (6.3) arises from a finite element discretization on the entire body domain $\Omega = H \cup T$, instead of only on H. In addition, the matrices A_{v} and A_{v}^{T} are padded with extra zero columns and rows, respectively, to form $\hat{A}_{\mathrm{v}}$ and $\hat{A}_{\mathrm{v}}^{T}$.

To form an overlapping partitioning of Ω, which is required for the parallel solution of $\hat{A}_{\mathrm{u}}$, we suggest the following partitioning strategy: First, we partition the heart domain H, as before, into the overlapping subdomains H_p. Then, we partition the torso domain T into P overlapping subdomains T_p, $1 \le p \le P$. Afterwards, each pair of H_p and T_p is patched together to work as $\Omega_p = H_p \cup T_p$, see Figure 6.5. The reason for not using a general partitioning of Ω is that we normally can not obtain H_p as a part of Ω_p. In case H_p is not a part of Ω_p, a conflict will arise between the partitions of the equations in $\hat{A}_{\mathrm{v}}$ and $\hat{A}_{\mathrm{v}}^{T}$. This will result in additional communication overhead.

A parallel preconditioner for (6.3) uses the same approach as for the Bidomain equations. That is, we also use (6.38) as the overall preconditioning framework and (6.39) as the detailed definition. The only difference is that the lower diagonal block $\theta \Delta t \hat{A}_{\mathrm{u}}$, which now also includes contributions from T, must use an additive Schwarz iteration based on the overlapping subdomains $\Omega_p = H_p \cup T_p$, $1 \le p \le P$.

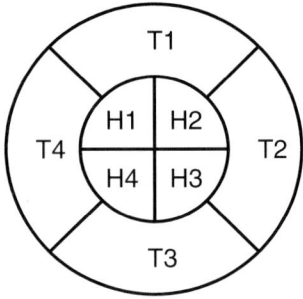

Fig. 6.5. A schematic view of domain partitioning used for the parallel computation.

Table 6.4. The number of CG iterations needed to solve (6.3) during one time step, using the parallel block preconditioner.

Grid levels	# unknowns $(v + u)$	$P = 2$	$P = 4$	$P = 8$	$P = 16$	$P = 32$	$P = 64$
2	302,166	9	9	10	10	12	11
3	1,552,283	9	9	10	10	11	11
4	10,705,353				13	13	14
5	81,151,611				14	15	15

Numerical Experiments. Table 6.4 is concerned with a set of numerical experiments about solving (6.3) by parallel CG iterations, which use the above block preconditioner. The table displays the number of CG iterations, needed during the first time step, for reducing the L_2-norm of the global residual by a factor of 10^4. The used value of θ is 0.5 and Δt has been chosen as 0.125 ms. We observe that the convergence of the parallel CG iterations is clearly independent of P and is quite insensitive to the total number of unknowns. This demonstrates the desired scalability of the parallel block preconditioner.

6.9   Parallelizing Electrocardiac Simulations

In this section, we will describe how to parallelize the electrocardiac simulator from Section 6.1.2 using the parallelization components that are presented in the preceding sections.

6.9.1   The Overall Simulation Process

Roughly, an electrocardiac simulation consists of three stages. The first stage, which is often referred to as *preprocessing*, deals with preparatory work, such as reading user input, choosing the ODE model, determining the values of the parameters, construction of the grids for the heart and torso domain, allocation of the internal data structure, enforcing initial conditions, etc. The second stage, which is often referred to as *computing*, executes a time stepping loop until reaching the desired stopping time for t. The work per time step involves solving the ODE system and the 2×2 block linear system (6.3) that arises from discretizing the PDEs. At the end of some chosen time steps, the computational results of v and u may need to be stored into data files. We note that the 2×2 block matrix in (6.3) normally remains the same for all the time levels, while the right-hand side vector needs to be updated at each time level. We also remark that the ODE system needs to be solved at every heart grid point during each time step. The third stage, which is often referred to as *postprocessing*, summarizes some key computational results and frees the memory allocated for the internal data structure.

Most of the tasks in the above three stages can be parallelized following the domain-based approach. We enforce an overlapping partitioning of H and Ω in connection with the grid construction. Each processor is responsible for one heart subdomain and one body subdomain. There exists almost no physical storage of global data. Instead, all the global data are divided into local units belonging exclusively to the processors. Parallel computing is realized in such a way that each processor mostly executes local operations independently, but occasionally communicates with the other processors. For the few non-parallelizable tasks, we can either replicate them on all the processors or use a master processor, which is responsible for first carrying out a task and then broadcasting the result to all the other processors.

6.9.2   Partitioning the Domains

The basic idea of partitioning H and Ω into overlapping subdomains has been presented in Section 6.8.4. Figure 6.6 shows an example of the partitioning result, where we have used different colors to denote different subdomains.

Recall that we want to use MG V-cycles as subdomain solvers in the additive Schwarz iterations for approximating $(I + \theta \Delta t A_{\mathrm{v}})^{-1}$ and A_{u}^{-1}. This requires two associated hierarchies of subdomain grids, one on H_p and the other on Ω_p. One approach is that we start with a global H grid and a global T grid, both of medium resolution. Then, we partition the two global grids into P overlapping parts, respectively. Afterwards, the subdomain H_p and T_p grids are refined several times on each subdomain independently, giving rise to a hierarchy of subdomain H_p grids and a hierarchy of subdomain T_p grids. Note that a hierarchy of subdomain Ω_p grids arises from a union of the associated hierarchies of subdomain H_p and T_p grids. Based on the subgrids, two sets of communication preparation work have to be carried out, one for the upcoming parallel computations within the heart and the other for the upcoming parallel computations within the body. We remark that the global medium grids for H and Ω are not used in later computations, so they can be discarded immediately after the subdomain grid hierarchies are built.

When the subdomain grid hierarchies are ready and the communication preparation work is done, each processor can concentrate on building its local data structure. This includes allocating subdomain matrices and vectors. During a parallel electro-

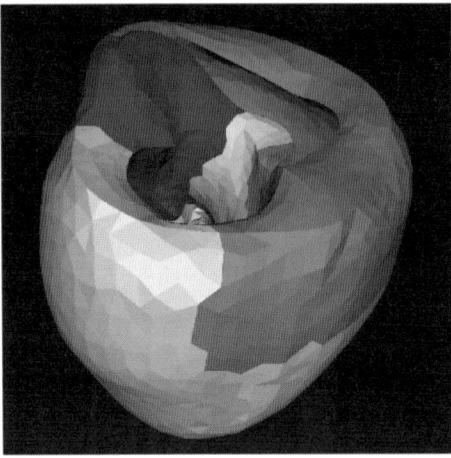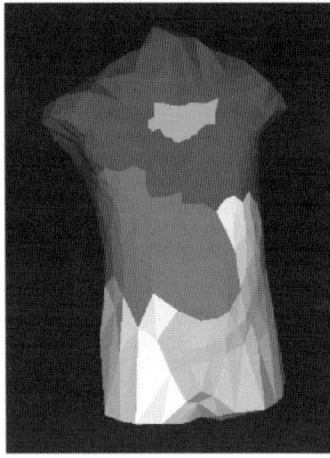

Fig. 6.6. An example of partitioning the heart domain (left) and the body domain (right). (For the color version, see Figure A.10 on page 292).

cardiac simulation, the work on processor number p consists of local computational operations that are restricted to the subdomains H_p and Ω_p.

6.9.3   Straightforward Parallelization Tasks

During each time step, processor number p solves the ODE system associated with every grid point in H_p. Since solving the ODE system on one grid point does not rely on values from the other grid points, the processors can carry out their local ODE computations in parallel. An existing sequential ODE solver can be re-used on each processor. Communication is in principle not needed.

However, if we want to eliminate the overhead due to duplicated ODE computations, recalling that the subdomains H_p are overlapping, we have to find a disjoint division of all the global heart grid points among the processors. Then, each processor only needs to solve the ODE system on an assigned subset of the points in H_p. After these local ODE computations are done, each subdomain has to provide some of the computed S and v values to its neighbors. This exchange of values is achieved by one-to-one communication operations. Therefore, elimination of the duplicated ODE computations comes at the cost of additional communication overhead. Whether it is advantageous to adopt this communication-assisted parallelization of the ODE solver depends on factors such as the amount of overlap, the number of neighbors per subdomain, etc.

Also, the finite element discretization can be parallelized in a straightforward manner, without any need for communication. The subdomain matrices I_p and $A_{v,p}$ arise from discretizations on the finest subdomain heart grid for H_p, whereas $A_{u,p}$ arises from discretizations on the finest subdomain body grid. The existing functionality for sequential finite element discretization should be re-used by all the processors. In addition, we also need to build subdomain matrices associated with the coarser levels of the subdomain grids. These subdomain matrices will be used in the local MG V-cycles, which work as the approximate subdomain solvers in the parallel additive Schwarz preconditioners, see Section 6.8.

Storing the v and u solutions at the end of some chosen time steps is also easy to parallelize. We simply let each processor write its portions of v and u to its own data file, independent of the other processors. Each processor can thus re-use the existing functionality for sequential data output. This avoids a considerable amount of I/O and communication overhead, which will arise if, e.g., the distributed subdomain v solutions are to be stored into a single data file. Later analysis of a set of subdomain data files can be done by a specially designed program.

6.9.4   Solving the Block Linear System in Parallel

At this point, the parallel solution of the 2×2 block linear system (6.3) is the remaining major computational task. As mentioned above, the logically existing global matrices I, A_v, and A_u are distributed as subdomain matrices. In addition, $\hat{A}_v$ in (6.3) is partitioned in the same way as A_v, whereas $\hat{A}_v^T$ in (6.3) is partitioned in the same way as $\hat{A}_u$. All the involved vectors are also partitioned accordingly.

Using such a setup for data distribution, we can follow the idea outlined in Section 6.7.3 to parallelize the Krylov iterations. That is, we use two separate "communicators" during the parallel inner products and matrix-vector products. One communicator is for the v-part associated with the heart subdomains, the other communicator is for the u-part associated with the body subdomains. The preconditioner in form of (6.38), which involves two additive Schwarz iterations, is parallelized as described in Section 6.8, making use of the two separate communicators.

Therefore, during a parallel solution process for (6.3), each processor mostly executes sequential computations within H_p and Ω_p. Inter-processor communication sporadically interleaves these sequential local operations, in form of both one-to-one message changes and collective operations.

6.10   Simulation on a Realistic Geometry

In this section we show some results from a simulation performed on a realistic geometry. The heart grid is based on data from The Bioengineering Research group at The University of Auckland [81]. They have made detailed measurements on a series of dog hearts. In particular the fiber directions have been recorded throughout the myocardium. Figure 6.7 shows a visualization. The torso grid has been constructed from slices of the Visible Human Data, see Figure 6.9 b). The heart has been scaled to a human size.

The depolarization process was initiated at five endocardial locations, corresponding roughly to the activation map published by Dürrer [33]. Some snapshots of the solution are shown in Figures 6.8 and 6.9.

In Figure 6.10 the potentials at chest leads are shown. The QRS complex and the T wave can be seen in most leads. It is clear that the duration of the QRS complex

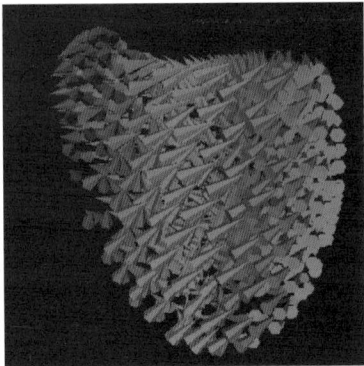 

Fig. 6.7. a) The fiber directions and **b**) the sheet planes. (For the color version, see Figure A.11 on page 292).

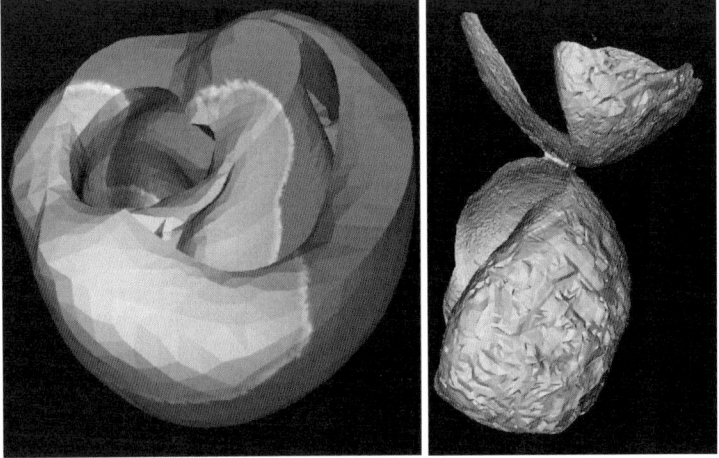

Fig. 6.8. A snapshot of the transmembrane potential during the depolarization phase. The right figure shows an iso-surface of the potential at the time instant when two fronts meet. (For the color version, see Figure A.12 on page 293).

is longer than it should be, yielding a very short ST-segment before the T-wave. The reason is that the conduction velocity in the simulation was too low. This is an effect of the coarse grid used for this simulation.

6.11 Summary

The basic idea behind the parallelization approach explained in this chapter is to let domain partitioning give rise to a division of the global work among P processors. Each processor mostly executes sequential operations that are restricted to its

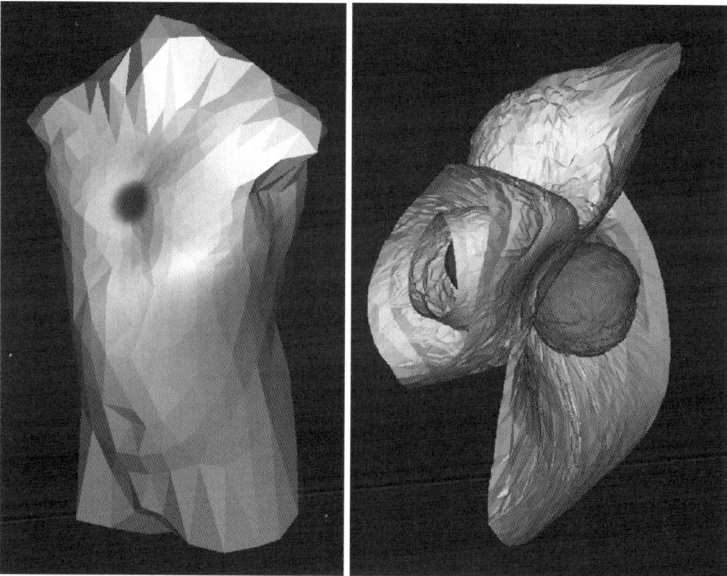

Fig. 6.9. The extra cardiac potential on the torso surface (left) and on some iso-surfaces (right). (For the color version, see Figure A.13 on page 293).

local data structure. Collaboration between the processors happens sporadically in form of inter-processor communication. This generic parallelization approach can be applied to any sequential PDE program, and it allows to a large extent re-use of existing sequential functionality.

To handle the demanding task of ensuring good parallel efficiency when solving the system of PDEs, we have used the combination of three powerful numerical techniques:

1. a block preconditioning strategy that ensures rapid overall convergence of the CG iterations,

2. additive Schwarz iterations (with coarse grid corrections) that efficiently solve the diagonal blocks in parallel, and

3. multigrid V-cycles that approximately solve the subdomain problems with order-optimal complexity.

For electrocardiac simulations, perfect parallel efficiency is difficult to achieve due to different types of overhead, such as the inevitable inter-processor communication and duplicated computation associated with overlapping DD methods. However, satisfactory performance can still be achieved by an electrocardiac simulator that is parallelized by the approach explained in this chapter.

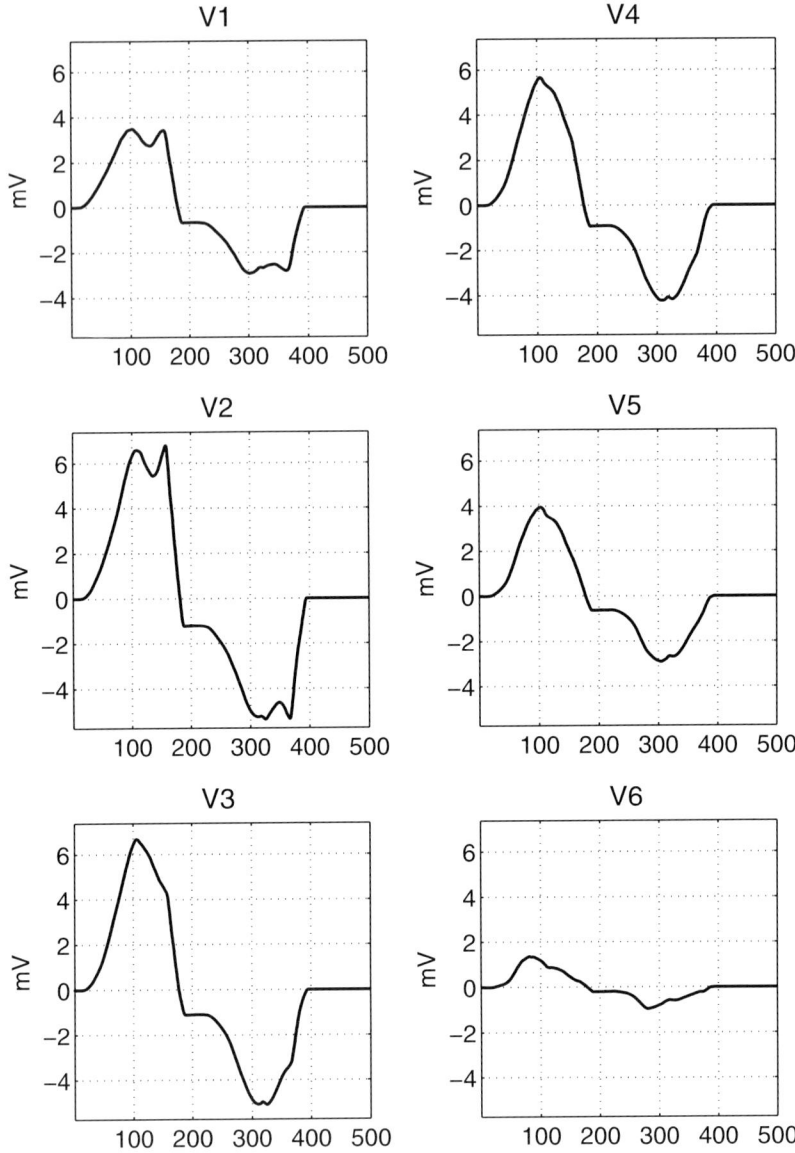

Fig. 6.10. The leads correspond to the six electrodes placed on the chest during a standard 12 lead recording. Time is in milliseconds.

Chapter 7

Inverse Problems

We have seen how mathematical models, numerical methods, software and computers can be used to simulate the electrical activity in the human body. Provided that the physical characteristics of the involved tissues are known, such techniques can, in particular, be applied to compute the electrical potential along the surface of the body generated by the heart. This can be very useful for gaining a better insight into these biological processes, improving the interpretation of ECG recordings, and may serve as a starting point for developing suitable educational tools and training facilities.

From a practical point of view, there are a number of other simulation tasks that are equally important. This chapter is devoted to three of these problems:

a) Is it possible to determine the electrical potential at the epicardial surface from body surface measurements?

b) Can the model (2.65)–(2.72) be simplified? More precisely, is it possible to "replace" the bidomain equations, and thus the model for the electrical activity in the heart, by some sort of simplified source?

c) Can models of the form (2.65)–(2.72) be used to determine the physical characteristics of an infarction in the myocardium?

The potential medical importance of challenges a) and c) is evident, and they will be treated in detail in Sections 7.2 and 7.4, respectively.

The issue introduced in b) is sometimes referred to as the task of determining an "equivalent source". This problem is closely linked to the traditional view of the heart as a source of electrical current positioned in a volume conductor. The aim of such studies is to represent the electrical activity of the heart by a set of suitable current dipoles, and thereby obtain a better qualitative and quantitative understanding of this organ. We will return to this issue in Section 7.3.

Problems a)–c) have in common that they are so-called *inverse problems*. What are *inverse problems*, and how do they differ from *direct problems*? These questions are not easy to answer; see, e.g. Colton, Ewing and Rundell [25], Engl, Hanke and Neubauer [37] and Kirsch [74]. Usually, two problems are referred to as inverse to each other if the formulation of each of them requires full or partial knowledge of the other. Which of the two problems we call direct and which we call inverse is determined by their mathematical properties and the historical development of this subject.

Roughly speaking, the direct problem is usually the one that satisfies the Laplacian paradigm of computing continuously depending effects of a known set of causes. Thus, the direct problem is, in most cases, *well-posed* in the sense of Hadamard[1]:

I)   A solution exists

II)   The solution is unique

III) The solution depends continuously on the data

Note that direct problems are, in some contexts, also referred to as *forward problems*.

By contrast, the solution of an inverse problem determines the cause of an observed effect[2]. As we will see below, attempts to solve an inverse problem will frequently lead to a set of equations that fail to satisfy one, or more, of conditions I)–III). Hence, such problems are often *ill-posed*. In particular, their solutions do not necessarily depend continuously on the data at hand. This means that even very small measurement and/or roundoff errors may become critical for their stable numerical solution. Consequently, they are, in general, much more difficult to solve than their well-posed direct counterparts.

In many cases, the cause we want to identify by solving an inverse problem is described by a set of parameters that appear in one or more equations. For example, the goal could be to determine an unknown coefficient that appears in a differential or integral equation from measurements of the solution of the associated direct problem. Therefore, inverse problems are sometimes referred to as *parameter identification problems*; see Banks and Kunisch [8]. At the present stage, it might be useful to consider some examples.

Example 1. Let us illustrate the concepts of direct and inverse problems for a phenomenon that most readers will have met in earlier studies of partial differential equations. Suppose that we have a uniform rod of unit length that has some sort of initial (time $t = 0$) temperature distribution. As will probably be recalled from basic courses, the temperature evolution is governed by the so-called heat equation. So, by solving this equation, we are able to find the temperature distribution at later times ($t > 0$). This is a typical direct problem. Now suppose, by contrast, that we know the temperature at time $t = 1$, and that we want to know the initial temperature distribution. Is that possible? It will readily be appreciated that this is a really challenging inverse problem. For instance, if the rod has a uniform temperature distribution at $t = 1$ equal to the temperature of its surroundings, it is easy to realize that lots of different initial distributions can give this uniform distribution and thus computing backwards in time has to be very unstable.   ◇

---

[1] Jacques Hadamard (1865–1963), famous French mathematician. He introduced, in the early twentieth century, the concept of a well-posed mathematical problem.

[2] "You see, there is only one constant, one universal, it is the only real truth: Causality. Action. Reaction. Cause and effect." Merovingian in The Matrix Reloaded; see www.zionmainframe.net

Example 2. Assume that the size and location of an infarction is known. The effect of this ischemia on the electrical activity in the body can, as we will see in Section 7.4, be incorporated into the model (2.65)–(2.72) by introducing a suitable set of parameters. Consequently, changes in the ECG recordings, due to the presence of an infarction, can easily be simulated by solving such models. In the present context, this is a direct problem; we compute the effect of a given cause, the ischemia.

Of course, from a medical point of view, it is more interesting to try to compute the physical properties of an infarction based on ECG recordings. That is, to determine the source of an observed effect, the electrical potential at the body surface. Hence, this is an inverse problem.   ◇

Issues a) and b), mentioned above, have been studied thoroughly by several scientists. Our goal is not to give a state-of-the-art presentation of the methodology developed for solving these problems. Instead we will focus on basic principles, the problem formulations and their properties. Thus, this chapter might serve as a starting point for conducting research within this field.

As far as the authors know, challenge c) has not previously been formulated as a parameter identification problem for a system of differential equations. Our goal regarding this problem has been to define a suitable framework for investigating the possibilities for applying mathematics and computers to determine the characteristics of an infarction. Our results so far are summarized in Section 7.4.

The only prerequisites for reading this chapter are some familiarity with calculus, partial differential equations, linear algebra, Fourier analysis and the theory presented in Chapters 1–4. We do not assume that the reader has any experience with inverse problems. Let us therefore start our investigations of inverse problems by analyzing a simple example in detail.

Throughout this chapter it might be useful to keep the following in mind:

– Direct (forward) problem: Compute the effect of a given cause, well-posed
– Inverse problem: Compute the cause of a given effect, ill-posed

7.1   A Simple Example

As mentioned above, no particular knowledge about inverse problems is assumed for reading the present chapter. This section is therefore devoted to the analysis of a simple inverse problem suitable for introducing some of the most important concepts of this field; ill-posedness, output least squares formulations, regularization techniques, etc. Thus, this example has not been included for its physical relevance. Readers already familiar with the theory of inverse problems might, therefore, wish to skip this section and instead turn their attention towards Sections 7.2–7.4. However, the simple model problem studied below will reveal some of the typical features of the *classical inverse problem of electrocardiography* a) in a particularly easy and explicit fashion.

7.1.1 Problem Formulation

Consider the Laplace equation posed on the unit square $\Omega = (0,1) \times (0,1)$

$$\Delta u = 0 \quad \text{for } (x,y) \in \Omega, \tag{7.1}$$

with boundary conditions

$$\nabla u \cdot n = 0 \quad \text{for } (x,y) \in \sigma_1 \cup \sigma_2 \cup \sigma_4, \tag{7.2}$$

$$u = g \quad \text{for } (x,y) \in \sigma_3, \tag{7.3}$$

where n is a vector normal to the boundary of Ω, and

$$\sigma_1 = \{(x,0);\ 0 \le x < 1\},$$
$$\sigma_2 = \{(1,y);\ 0 \le y < 1\},$$
$$\sigma_3 = \{(x,1);\ 0 < x \le 1\},$$
$$\sigma_4 = \{(0,y);\ 0 < y \le 1\},$$

see Figure 7.1. Provided that g is a well-behaved[3] function, equations (7.1)–(7.3) have a unique solution u; see, e.g., Hackbusch [56] or Evans [38]. Note, however, that this solution depends on the function g used to specify the Dirichlet boundary condition (7.3), i.e. $u = u(g)$.

Suppose that Ω represents a passive conductor, with unity conductivity, surrounded by an insulator, e.g. air, rubber or some sort of plastic. If an electrical potential g is applied along the boundary segment σ_3 of $\partial\Omega$, then the resulting electrical potential $u = u(g)$, throughout Ω, is governed by equations (7.1)–(7.3). Throughout this section, it might be helpful to keep this physical phenomenon in mind.

Assume that we, for some reason, are particularly interested in the potential $u(g)|_{\sigma_1}$ along σ_1, and let us introduce a mapping

$$R : S_{\sigma_3} \rightarrow S_{\sigma_1}, \tag{7.4}$$

by defining

$$R(g) = u(g)|_{\sigma_1}. \tag{7.5}$$

Here, S_{σ_3} and S_{σ_1} denote suitable spaces of functions defined along σ_3 and σ_1, respectively. Clearly, $R(g)$ can be computed by the following procedure:

- Solve (7.1)–(7.3), applying g in the Dirichlet boundary condition (7.3), for u
- Record the simulated field $u(g)$ along the boundary segment σ_1. This value is denoted $u(g)|_{\sigma_1}$.

---

[3] Well-behaved means that the function is sufficiently smooth. Precise conditions are given by, e.g., Hackbusch [56].

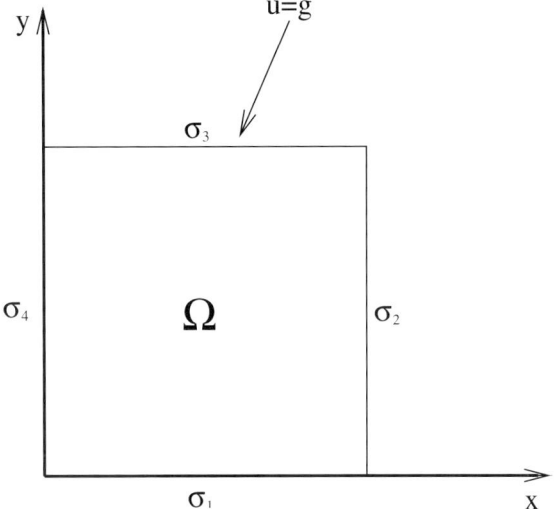

Fig. 7.1. The domain $\Omega = (0, 1) \times (0, 1)$, and its boundary $\partial\Omega = \sigma_1 \cup \sigma_2 \cup \sigma_3 \cup \sigma_4$, of the elliptic model problem (7.1)–(7.3).

In the present context, this is the direct, or forward, problem; we compute the effect $u(g)|_{\sigma_1}$ generated by a given cause g.

From a physical point of view, it seems reasonable that small changes in the boundary date g will only introduce minor changes in the electrical potential throughout Ω, and consequently only small changes in the potential along the boundary segment σ_1. Indeed, it can been shown that the solution $u(g)$ of (7.1)–(7.3) depends continuously on g, provided that the changes in g and $u(g)$ are measured in proper norms. Thus, the direct problem at hand is well-posed. Further information on Laplace's equation and its mathematical properties can be found in, e.g., Hackbusch [56] or Evans [38].

Let us now turn our attention to the inverse problem in the present situation. Assume that we want to recover the function g based on an observed electrical potential $d = u(g)|_{\sigma_1}$ along σ_1. Expressed in mathematical symbols, this problem takes the form: Find a function $g \in S_{\sigma_3}$, defined along σ_3, such that

$$R(g) = d. \qquad (7.6)$$

Note that, if R is invertible, then the solution of (7.6) is given by

$$g = R^{-1}(d).$$

We will now show that the problem (7.6) is ill-posed. More precisely, even though R is invertible, it turns out that the solution g of (7.6) does not depend continuously on small changes in the observation d.

The starting point of our analysis of this problem is to characterize fully the action of both R and R^{-1} in terms of simple trigonometric and hyperbolic functions.

7.1.2   Fourier Analysis

By a straightforward application of the method of separation of variables (see, e.g., Tveito and Winther [139]), it follows that any function of the form

$$N_k(x, y) = \cos(k\pi x)\cosh(k\pi y), \quad k = 0, 1, \ldots$$

satisfies both equations (7.1) and (7.2). Therefore, by the linearity of Laplace's equation and the super position principle, we find that any linear combination[4]

$$u(x, y) = \sum_{k=0}^{\infty} c_k \cos(k\pi x)\cosh(k\pi y) \tag{7.7}$$

of these functions, where $\{c_k\}_{k=0}^{\infty}$ are constants, also fulfills both (7.1) and (7.2).

Along the boundary segment σ_3, a function of the form (7.7) is simply given by the series

$$u(x, 1) = \sum_{k=0}^{\infty} c_k \cos(k\pi x)\cosh(k\pi).$$

Consequently, if the function g can be expressed in terms of a Fourier cosine series, say

$$g(x) = \sum_{k=0}^{\infty} p_k \cos(k\pi x), \tag{7.8}$$

then the solution $u(g)$ of the boundary value problem (7.1)–(7.3) is given by the formula

$$u(g)(x, y) = u(x, y) = \sum_{k=0}^{\infty} \frac{p_k}{\cosh(k\pi)} \cos(k\pi x)\cosh(k\pi y). \tag{7.9}$$

Let S denote the space of functions defined on the unit interval that admits a Fourier cosine series, i.e.

$$S = \{\phi(x) = \sum_{k=0}^{\infty} c_k \cos(k\pi x), \ x \in (0, 1);\ \text{this series converges and allows for}$$

$$\text{term-wise differentiation}\}. \tag{7.10}$$

Recall the definition (7.4)–(7.5) of the solution operator R of the direct problem. By choosing $S_{\sigma_1} = S_{\sigma_3} = S$, it follows from formulae (7.8) and (7.9) that

$$R(g) = R\left(\sum_{k=0}^{\infty} p_k \cos(k\pi x)\right) = u(g)(x, 0) = \sum_{k=0}^{\infty} \frac{p_k}{\cosh(k\pi)} \cos(k\pi x), \tag{7.11}$$

and we have thus derived a very simple expression for the action of $R : S \to S$.

---

[4] We assume that the constants $\{c_k\}_{k=0}^{\infty}$ have been chosen such that the sum (7.7) converges and allows for term-wise differentiation.

Next, suppose that an observation of the electrical potential d along σ_1 is available. Our goal is to use this observation to determine the corresponding potential $g \in S$ along σ_3. That is, we want to compute the source g of the observed effect d. As mentioned above, this can be accomplished by solving the equation

$$R(g) = d \tag{7.12}$$

for g.

Assume that the data d is in S. Thus, d is given in terms of a Fourier series, i.e.

$$d(x) = \sum_{k=0}^{\infty} d_k \cos(k\pi x), \tag{7.13}$$

where $\{d_k\}_{k=0}^{\infty}$ are given such that the series is convergent and can be differentiated term-wise. Is it possible to find constants $p_1, p_2, \ldots$ such that the source term

$$g(x) = \sum_{k=0}^{\infty} p_k \cos(k\pi x)$$

satisfies (7.12)? From formula (7.11), it follows that equation (7.12) can be written in the form: Find $\{p_k\}_{k=0}^{\infty}$ such that

$$\sum_{k=0}^{\infty} \frac{p_k}{\cosh(k\pi)} \cos(k\pi x) = \sum_{k=0}^{\infty} d_k \cos(k\pi x),$$

which clearly holds for

$$p_k = d_k \cosh(k\pi) \quad \text{for } k = 0, 1, \ldots.$$

Hence, we obtain the following simple formula for R^{-1}

$$g(x) = R^{-1}(d) = R^{-1} \left(\sum_{k=0}^{\infty} d_k \cos(k\pi x) \right) = \sum_{k=0}^{\infty} d_k \cosh(k\pi) \cos(k\pi x),$$
$$\tag{7.14}$$

provided that all the involved series converge and allow for term-wise differentiation. In short, given the potential $u = d$ at σ_1, written in the form (7.13), we have found that the source g at σ_3 generating this potential is given by (7.14).

Having found simple expressions for both R and R^{-1}, we are now ready to analyze the properties of these operators in more detail.

7.1.3 Ill-Posedness

Let us first consider the direct problem (7.4)–(7.5). That is, we want to study the properties of $R(g)$ for functions g on the form (7.8). Recall that

$$\cosh(k\pi) = \frac{e^{k\pi} + e^{-k\pi}}{2} \geq 1 \quad \text{for } k = 0, 1, \ldots,$$

and therefore the magnitude of the Fourier coefficients of $R(g)$, cf. (7.11), will be smaller than the magnitude of the corresponding coefficients of g, i.e.

$$\left| \frac{p_k}{\cosh(k\pi)} \right| \le |p_k|, \quad k = 1, 2, \ldots.$$

Furthermore, even for relatively small numbers of k, $\cosh(k\pi)$ is large; for example

$$\cosh(5\pi) \approx 3.32 \cdot 10^6.$$

This means that any oscillatory features present in g are damped efficiently by R. Roughly speaking, $R(g)$ will "always" be very smooth and have low variation.

Example 3. For example, if

$$g(x) = \cos(\pi x),$$
$$g_\delta(x) = \cos(\pi x) + \delta \cos(5\pi x),$$

where $0 < \delta \ll 1$ is a small number, then

$$g(x) - g_\delta(x) = \delta \cos(5\pi x) = O(\delta)$$

and

$$R(g) - R(g_\delta) = \frac{\delta}{\cosh(5\pi)} \cos(5\pi x) = O(10^{-6}\delta).$$

This shows clearly that R has a tendency to diminishing small changes in g. In fact, it is easy to see that any sum of high frequency cosine modes can be added to g without introducing large changes to $R(g)$.

In Figure 7.2 we have plotted the solution $u(g_\delta)$ of (7.1)–(7.3) for $\delta = 0$ (corresponding to $u(g)$), $\delta = 0.5$, $\delta = 1$ and $\delta = 2$. Recall that

$$g_\delta = u(g_\delta)(x, 1) \quad \text{for } x \in (0, 1),$$
$$R(g_\delta) = u(g_\delta)(x, 0) \quad \text{for } x \in (0, 1),$$

and we observe that $R(g_\delta)$ is almost independent of δ. This is clearly in accordance with our theoretical investigations above!   ◇

Next, from (7.11) and simple properties of the cosine function, it follows that

$$\|R(g_2) - R(g_1)\|_{L^2(0,1)} \le \|g_2 - g_1\|_{L^2(0,1)}$$

for any g_1, $g_2 \in S$, see (7.10). This means that R is a continuous mapping. Thus, if g is based on physical measurements, small errors in these measurement will not introduce major changes in the simulation results obtained by solving the forward problem. The direct problem (7.4)–(7.5) is well-posed!

We will now see that these nice features of the direct problem makes it very difficult to solve the inverse problem (7.6). In particular, it turns out that the smoothing[5] property of R, discussed above, makes it very difficult to determine which potential g that was the source of an observed field d along σ_1. This is easily realized by considering a simple example.

---

[5] Strictly speaking, the oscillatory damping effect of R.

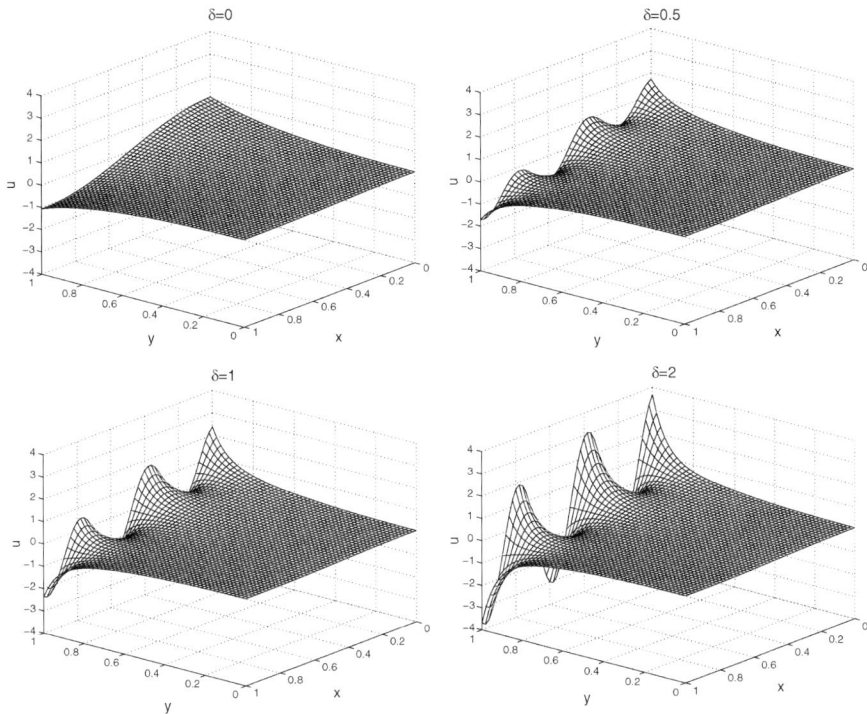

Fig. 7.2. Plots of the solution $u(g_\delta)$ of (7.1)–(7.3) for $\delta = 0$ (corresponding to $u(g)$), $\delta = 0.5$, $\delta = 1$ and $\delta = 2$ in Example 3. Note that, even though $g_\delta = u(g_\delta)(x,1)$ depends heavily on δ, $R(g_\delta) = u(g_\delta)(x,0)$ is almost identical for these four values of δ.

Example 4. Suppose that the data at σ_1 is given by

$$d(x) = \frac{1}{\cosh(\pi)} \cos(\pi x),$$

and that we want to compute the associated source $g = R^{-1}d$. By (7.11)

$$R(\cos(\pi x)) = \frac{1}{\cosh(\pi)} \cos(\pi x),$$

and it follows that the solution of the equation

$$R(g) = d(x)$$

is given by the cosine mode

$$g(x) = \cos(\pi x).$$

Now assume that we introduce a slight perturbation of the data d;

$$d_\delta(x) = d(x) + \delta \cos(5\pi x)$$
$$= \frac{1}{\cosh(\pi)} \cos \pi x + \delta \cos(5\pi x),$$

where δ is a small positive number. According to (7.14), the solution g_δ of the problem

$$R(g_\delta) = d_\delta$$

is a sum of two cosine modes

$$g_\delta(x) = \cos(\pi x) + \delta \cosh(5\pi)\cos(5\pi x)$$
$$\approx g(x) + 3.32 \cdot 10^6 \, \delta \cos(5\pi x).$$

Thus, even if $d_\delta \approx d$, it does not in general follow that g_δ is a good approximation of g. For example, if

$$\|d_\delta - d\|_{L^\infty} = O(10^{-3})$$

then

$$\|g_\delta - g\|_{L^\infty} = O(10^3).$$

This means that a tiny perturbation of the observed data at σ_1 implies a huge difference in the source at σ_3.   ◇

In fact, for any function d in the range of R and any error level $\delta > 0$, it is always possible to find an order δ perturbation d_δ of d that leads to an arbitrary large change in the solution of the inverse problem (7.6). This is easily accomplished by adding high-frequency errors to d. We conclude that equation (7.6) is ill-posed.

Theoretical Considerations. From a more theoretical point of view, this property of the inverse problem can be characterized as follows. Formula (7.11) implies that

$$R(\cos(k\pi x)) = \frac{1}{\cosh(k\pi)}\cos(k\pi x) \quad \text{for } k = 0, 1, \ldots,$$

and consequently, the eigenvalue-eigenfunction pairs of the forward mapping R are easily seen to be

$$\lambda_k = \frac{1}{\cosh(k\pi)}, \quad e_k(x) = \cos(k\pi x) \quad \text{for } k = 0, 1, \ldots. \tag{7.15}$$

Observe that, since $\cosh(k\pi) \to \infty$ as $k \to \infty$, zero is a cluster point[6] for the eigenvalues. This means that the inverse mapping R^{-1} of R is not "well-behaved", i.e. R^{-1} is unbounded. Consequently, even very small observation errors in the data d for the inverse problem will, in most cases, be critical; it is, in general, impossible to bound the error in the solution of (7.6) generated by small changes in d!

In real world simulations, the function d will typically be based on physical measurements of the electrical potential along σ_1. The above discussion shows that even very small measurement errors present in d and/or roundoff errors arising in numerical computations may become critical for solving (7.6).

---

[6] Point of accumulation

We conclude that, unless d has a very simple structure, it is, in general, impossible to determine the exact solution g of (7.6). The nature of this problem is such that even small errors in d have a tendency to lead to large errors in g. The best we can hope for is that it is possible to compute good approximations of g. The next two sections are devoted to this topic.

By formula (7.11) and the property that $\cosh(k\pi) \geq 1$ for any integer k, it follows that R is well-defined for every $g \in S$. On the other hand, equation (7.14) shows that R^{-1} can only be defined on a subset of S, namely in the range

$$
R(S) = \{R(\phi);\ \phi \in S\}
$$
$$
= \left\{ \sum_{k=0}^{\infty} \frac{c_k}{\cosh(k\pi)} \cos(k\pi x); \sum_{k=0}^{\infty} c_k \cos(k\pi x) \in S \right\} \subset S \quad (7.16)
$$

of R. The value of $\cosh(k\pi x)$ increases rapidly as k grows. Hence, a function of the form

$$
d(x) = \sum_{k=0}^{\infty} d_k \cos(k\pi x)
$$

will only belong to $R(S)$ if the Fourier coefficients $\{d_k\}_{k=0}^{\infty}$ decrease very fast as k increases; cf. (7.16).

We conclude that the domain of R^{-1} is "small". Thus, in many applications the right hand side d of equation (7.6) will not belong to the range of R, and consequently, the inverse problem does not have any solution $g \in S$. We will not dwell any further upon this issue for the continuous problem. Instead, we turn our attention to discrete approximations of (7.6).

7.1.4   Discretization and an Output Least Squares Formulation of the Problem

For a fixed positive integer n, we introduce the finite dimensional subspace

$$
S_n = \left\{ \phi(x) = \sum_{k=0}^{n} c_k \cos(k\pi x) \right\} \quad (7.17)
$$

of S, and the associated restriction

$$
R_n = R|_{S_n} \quad (7.18)
$$

of the forward mapping R to S_n. A first approach towards defining an approximation of (7.6) could typically be formulated as follows: Find $g_n \in S_n$ such that

$$
R_n(g_n) = d. \quad (7.19)
$$

However, unless d is in the range

$$
R_n(S_n) = \{R_n(\phi);\ \phi \in S_n\}
$$

of R_n, there does not exist any $g_n \in S_n$ satisfying (7.19). This difficulty can be handled in a number of ways. Let us consider two "natural" possibilities for overcoming this problem:

- Replace d in (7.19) by some sort of approximation $\tilde{d} \in R_n(S_n)$, leading to the equation

$$R_n(g_n) = \tilde{d},$$

  that will have a unique solution $g_n \in S_n$. Typically, $\tilde{d}$ could be determined by computing a suitable projection of d onto the range $R_n(S_n)$ of R_n.

- We can "replace" equation (7.19) by a minimization problem in the form

$$\min_{g_n \in S_n} \|R_n(g_n) - d\|_{L^2(\sigma_1)}^2, \tag{7.20}$$

  and thereby try to minimize the deviation between the observation data d and the simulated electrical potential along σ_1. The formulation (7.20) is commonly referred to as an *Output Least Squares (OLS)* formulation of the problem.

Given proper assumptions, it can be shown that these two approaches are mathematically equivalent. They lead to the same approximate solution of the inverse problem (7.6). We will not dwell upon this issue, but instead turn our attention towards the challenge of solving (7.20).

From definitions (7.17), (7.18), and formula (7.11) we find that

$$R_n(g_n) = R\left(\sum_{k=0}^n p_k \cos(k\pi x)\right) = \sum_{k=0}^n \frac{p_k}{\cosh(k\pi)} \cos(k\pi x) \tag{7.21}$$

for

$$g_n(x) = \sum_{k=0}^n p_k \cos(k\pi x) \in S_n.$$

Thus, it follows that the problem (7.20) can be written in the form

$$\min_{p_0,\dots,p_n \in \mathbf{R}} \left\|\sum_{k=0}^n \frac{p_k}{\cosh(k\pi)} \cos(k\pi x) - d(x)\right\|_{L^2(\sigma_1)}^2 \tag{7.22}$$

or

$$\min_{p_0,\dots,p_n \in \mathbf{R}} J_n(p_0,\dots,p_n),$$

where

$$
\begin{aligned}
J_n &= J_n(g_n) \\
&= \|R_n(g_n) - d\|_{L^2(\sigma_1)}^2 \\
&= J_n(p_0,\dots,p_n) \\
&= \left\|\sum_{k=0}^n \frac{p_k}{\cosh(k\pi)} \cos(k\pi x) - d(x)\right\|_{L^2(\sigma_1)}^2.
\end{aligned} \tag{7.23}
$$

Differentiation of the cost-functional J_n with respect to p_i yields

$$\frac{\partial J_n}{\partial p_i} = 2 \int_{\sigma_1} \left(\sum_{k=0}^{n} \frac{p_k}{\cosh(k\pi)} \cos(k\pi x) - d(x) \right) \frac{1}{\cosh(i\pi)} \cos(i\pi x) \, dx$$

for $i = 0, \ldots, n$. Combining this expression with the first-order necessary condition for a minimum

$$\frac{\partial J_n}{\partial p_i} = 0 \quad \text{for } i = 0, \ldots, n,$$

we obtain the following formula for the solution g_n of (7.20)

$$g_n(x) = \sum_{k=0}^{n} p_k \cos(k\pi x),$$

$$p_k = 2 \cosh(k\pi) \int_{\sigma_1} d(x) \cos(k\pi x) \, dx = d_k \cosh(k\pi) \quad \text{for } k = 1, \ldots, n,$$

$$p_0 = \int_{\sigma_1} d(x) \, dx = d_0,$$

where

$$d_k = 2 \int_{\sigma_1} d(x) \cos(k\pi x) \, dx, \quad k = 1, \ldots, n,$$

$$d_0 = \int_{\sigma_1} d(x) \, dx.$$

We conclude that the discrete output least squares solution of our inverse problem is uniquely determined by the $n + 1$ first Fourier coefficients $\{d_k\}_{k=0}^{n}$ of the observed electrical potential d along σ_1. This result is consistent with the characterization (7.14) of R^{-1}, and should not come as any surprise!

7.1.5 Regularization Techniques

We have seen that the task of identifying the source $g : \sigma_3 \rightarrow \mathbf{R}$ of an observed electrical potential d along σ_1 leads to an ill-posed problem. More precisely, zero is a cluster point for the eigenvalues of the solution operator R of the associated direct problem. Consequently, R^{-1} is unbounded and small errors in the observation data d and/or roundoff errors may become critical. This leads to severe difficulties for designing numerical schemes suitable for computing g such that

$$R(g) = d. \tag{7.24}$$

At the end of Section 7.1.3, we concluded that we have to be content with solving (7.24) approximately. Generally, errors present in d, say of the order δ, will have a tendency to increase and lead to errors in g of order $> \delta$.

This discussion shows that ill-posed problems, in their original form, are not suitable for performing computer simulations. Even though their ill-posedness in many cases reflects properties of the underlying physical process, they must some-how be approximated by well-posed problems to allow stable numerical compu-tations. Methods for approximating ill-posed equations by well-posed problems are referred to as *regularization techniques*. Throughout the last three decades a wide range of regularization methods have been suggested and analyzed; see, e.g. Engl, Hanke, and Neubauer [37], Kirsch [74] or Louis [85]. In what follows, we will briefly discuss two of these techniques, regularization by discretization and Tikhonov regularization, for our model problem (7.24).

Regularization by Discretization. Consider the discrete approximation R_n of R de-fined in equations (7.17) and (7.18). From formula (7.21) we find that the eigenvalue-eigenvector pairs of R_n are

$$\lambda_k^n = \frac{1}{\cosh(k\pi)}, \quad e_k^n(x) = \cos(k\pi x) \quad \text{for } k = 0, 1, \ldots, n.$$

Thus, the eigenvalues are uniformly positive, i.e.

$$\lambda_k^n = \frac{1}{\cosh(k\phi)} \geq \frac{1}{\cosh(n\phi)} > 0 \quad \text{for } k = 0, 1, \ldots, n,$$

and it follows that the inverse

$$R_n^{-1}(d) = R_n^{-1}\left(\sum_{k=0}^{n} d_k \cos(k\pi x)\right) = \sum_{k=0}^{n} d_k \cosh(k\pi) \cos(k\pi x), \qquad (7.25)$$

of R_n is bounded for every fixed positive integer n. Consequently, the approxima-tion of (7.24) defined in terms of the minimization problem

$$\min_{g_n \in S_n} \|R_n(g_n) - d\|_{L^2(\sigma_1)}^2, \qquad (7.26)$$

is well-posed; the solution g_n of (7.26) depends continuously on d, cf. Section 7.1.4. We conclude that the discretization (7.26) defines a regularization[7] of (7.24) for any fixed positive integer n. In this framework, the dimension n of the space S_n serves as a so-called *regularization parameter*. The size of n determines the strength of the regularization; in this method, the degree of regularization increases as n decreases.

Note that, if n is small, then the problem (7.26) is well-behaved; cf. (7.25). On the other hand, such high degrees of regularization may result in a poor approxima-tion of the original problem (7.24). The challenge is to balance the size of the error

---

[7] In this case, the space S_n is spanned by a finite number of eigenfunctions of R. More precisely, S_n is spanned by the constant function 1 and the cosine modes $\cos(\pi x), \ldots, \cos(n\pi x)$. Consequently, the regularization obtained by discretizing (7.24), in terms of S_n, will lead to the same scheme as a technique referred to as *Truncated Singu-lar Value Decomposition (TSVD)*. We will not dwell upon this issue. However, it should be mentioned that, generally, TSVD and discretization techniques lead to different regulariza-tion methods. Further information on the TSVD approach can be found in, e.g., Groetsch [52].

introduced by discretizing (7.24), and choices of n leading to fairly well-behaved approximations of it.

Example 5. Let us reconsider the problem studied in Example 4,

$$d(x) = \frac{1}{\cosh(\pi)} \cos(\pi x), \tag{7.27}$$

$$d_\delta(x) = \frac{1}{\cosh(\pi)} \cos(\pi x) + \delta \cos(5\pi x), \tag{7.28}$$

where d and d_δ represent the "exact" and error-prone observation data defined along σ_1, respectively. Our goal is to recover the source g, defined at σ_3, of the observation d.

If $1 \le n \le 4$, then the solution of

$$R_n(g_{\delta,n}) = d_\delta$$

is identical to the function $g(x) = \cos(\pi x)$ satisfying

$$R(g) = d.$$

Thus, in such cases, this technique provides the correct solution[8]. On the other hand, for $n \ge 5$

$$g_{\delta,n} = g_\delta, \tag{7.29}$$

implying that

$$g_{\delta,n}(x) - g(x) = \delta \cosh(5\pi) \cos(5\pi x) \tag{7.30}$$

and consequently, the regularization does not have any effect.   ◇

This feature is typical for ill-posed problems. The appropriate degree of regularization to apply depends on the size of the errors present in the data. Roughly speaking, the degree of regularization needed increases with the size of the errors. This relation between the regularization parameter and the level of the error is a delicate issue, and has been thoroughly studied by several scientists; see, e.g. Engl, Hanke and Neubauer [37] and references therein. We will not pursue this matter in any detail. However, it is important to notice that, for many ill-posed problems, discretizing the involved equations defines a regularized approximation of the problem at hand.

---

[8] In this example, the error in the date has a very simple form; it is given in terms of a single cosine mode, i.e. $d_\delta(x) - d(x) = \delta \cos(5\pi x)$. Hence, it becomes easy to determine an optimal value for the regularization parameter n. This is, of course, generally not the case. Often, it is extremely difficult to find optimal, or close to optimal, values for n.

Tikhonov Regularization. This section is devoted to a commonly applied and well-known regularization method; namely, Tikhonov regularization. Inspired by the discrete Output Least Squares formulation (7.20), we will now consider a similar approach for the continuous problem (7.24). To this end, let us introduce the function

$$J(g) = \|R(g) - d\|^2_{L^2(\sigma_1)} \quad \text{for } g \in S,$$

cf. definitions (7.4), (7.5), (7.10) and Figure 7.1. With this notation at hand, we can "replace" equation (7.24) by the minimization problem:

$$\min_{g \in S} J(g). \tag{7.31}$$

Of course, merely formulating (7.24) as a minimization problem will not help; the ill-posed of nature of (7.24) is inherited by (7.31). The basic idea of Tikhonov regularization is to approximate (7.31) by a well-behaved problem. This is accomplished by adding a regularization term, preferably small, to the cost-functional J,

$$J_\epsilon(g) = J(g) + \epsilon \|g\|^2_{L^2(\sigma_3)} = \|R(g) - d\|^2_{L^2(\sigma_1)} + \epsilon \|g\|^2_{L^2(\sigma_3)} \quad \text{for } g \in S, \tag{7.32}$$

where $\epsilon > 0$ is a regularization parameter. This leads to the following approximation of (7.31) (and hence of (7.24))

$$\min_{g \in S} J_\epsilon(g). \tag{7.33}$$

Note that, if ϵ is large, then the regularization term in (7.32) becomes dominant, i.e. $J_\epsilon(g) \approx \epsilon \|g\|^2_{L^2(\sigma_3)}$. Furthermore, $g = 0$ solves the problem

$$\min_{g \in S} \epsilon \|g\|^2_{L^2(\sigma_3)}. \tag{7.34}$$

Consequently, minimization of J_ϵ is a compromise between minimizing the norm of the residual, and keeping the size of g small.

For every $\epsilon > 0$, (7.34) is well-posed. We will now show that adding this regularization term to our originally ill-posed problem enforces stability. Recall that every function $g \in S$ can be written in the form

$$g(x) = \sum_{k=0}^{\infty} p_k \cos(k\pi x),$$

for suitable real numbers $p_0, p_1, \ldots$. Thus,

$$J_\epsilon(g) = J_\epsilon(p_0, p_1, \ldots),$$

and the first-order necessary condition

$$\frac{\partial J_\epsilon}{\partial p_i} = 0 \quad \text{for } i = 1, 2, \ldots$$

for a minimum, leads to the formula

$$p_k = \frac{d_k \cosh(k\pi)}{1 + \epsilon \cosh^2(k\pi)}$$

for the Fourier coefficients of the solution g of (7.33). That is,

$$g(x) = \sum_{k=0}^{\infty} \frac{d_k \cosh(k\pi)}{1 + \epsilon \cosh^2(k\pi)} \cos(k\pi x)$$

solves (7.33), where

$$d_0 = \int_{\sigma_1} d(x)\,dx,$$

$$d_k = 2 \int_{\sigma_1} d(x)\cos(k\pi x)\,dx, \quad k = 1, 2, \ldots,$$

are the Fourier coefficients of the observation data d.

Thus, Tikhonov regularization leads to the following approximation R_ϵ^{-1} of the inverse R^{-1} of the transfer mapping[9] R, see (7.4)–(7.5),

$$R_\epsilon^{-1}(d) = R_\epsilon^{-1} \left(\sum_{k=0}^{\infty} d_k \cos(k\pi x) \right)$$

$$= \sum_{k=0}^{\infty} d_k \frac{\cosh(k\pi)}{1 + \epsilon \cosh^2(k\pi)} \cos(k\pi x)$$

$$= \sum_{k=0}^{\infty} d_k \frac{\lambda_k}{\lambda_k^2 + \epsilon} \cos(k\pi x), \qquad (7.35)$$

where

$$\lambda_k = \frac{1}{\cosh(k\pi)} \quad \text{for } k = 0, 1, \ldots,$$

are the eigenvalues of R; cf. (7.15). Comparing this expression, for small values of ϵ, with formula (7.14) for R^{-1}, we observe that

- For the low-frequency components of the data d, the action of R^{-1} and R_ϵ^{-1} are almost identical

- The high-frequency components of d are damped efficiently by R_ϵ^{-1}

This means that the amplification of high-frequency errors, as observed for R^{-1}, is avoided; the regularization term serves as a mollifier. Furthermore, from (7.35) we find that the eigenvalue-eigenvector pairs of R_ϵ^{-1} are

$$\lambda_{\epsilon,k}^{-1} = \frac{\cosh(k\pi)}{1 + \epsilon \cosh^2(k\pi)}, \quad e_{\epsilon,k} = \cos(k\pi x) \quad \text{for } k = 0, 1, \ldots.$$

---

[9] Recall that R can be thought of as a transfer mapping that maps the electrical potential along σ_3 to the corresponding potential along σ_1; cf. Figure 7.1

Clearly,

$$\lambda_{\epsilon,k}^{-1} \le \frac{\epsilon^{-1}(1 + \epsilon \cosh^2(k\pi))}{1 + \epsilon \cosh^2(k\pi)} = \frac{1}{\epsilon}, \quad k = 0, 1, \ldots,$$

$$\lambda_{\epsilon,k}^{-1} > 0, \quad k = 0, 1, \ldots,$$

and it follows that R_ϵ^{-1} is a bounded linear operator[10] for every fixed $\epsilon > 0$.

A bounded linear operator is continuous. Hence, we conclude that the problem (7.33) is well-posed, provided that $\epsilon > 0$. However, as ϵ approaches zero, the ill-posed nature of the limit problem (7.31) will become dominant, and lead to severe difficulties for solving (7.33). The challenge is to find the right balance between the order of the errors present in the data d, and the size of the regularization parameter ϵ.

Example 6. We consider the exact and error-prone data d and d_δ defined in equations (7.27) and (7.28), respectively. Recall that d and d_δ are functions defined along σ_1, and that we want to recover the source g at σ_3 of the electrical potential d; see Figure 7.1.

In this case

$$R^{-1}(d) = \cos(\pi x),$$

$$R_\epsilon^{-1}(d_\delta) = \frac{1}{1 + \epsilon \cosh^2(\pi)} \cos(\pi x) + \frac{\delta \cosh(5\pi)}{1 + \epsilon \cosh^2(5\pi)} \cos(5\pi x),$$

and the Tikhonov scheme thus yields the L^2-error

$$E(\epsilon, \delta) = \|R^{-1}(d) - R_\epsilon^{-1}(d_\delta)\|_{L^2(\sigma_3)}^2$$

$$= \frac{1}{2}\left[\frac{\epsilon \cosh^2(\pi)}{1 + \epsilon \cosh^2(\pi)}\right]^2 + \frac{1}{2}\left[\frac{\delta \cosh(5\pi)}{1 + \epsilon \cosh^2(5\pi)}\right]^2.$$

In Figure 7.3 we have plotted $E(\epsilon, \delta)$, as a function of ϵ, for three different values of δ. Observe that care must be taken when choosing the regularization parameter ϵ: too small a value may result in large errors, leading to the ill-posed nature of the problem becoming dominant; while values of ϵ that are too large will not work properly.

---

[10] As a by-product of this analysis, we find that

$$R_\epsilon(g) = R_\epsilon\left(\sum_{k=0}^{\infty} g_k \cos(k\pi x)\right) = \sum_{k=0}^{\infty} g_k \frac{1 + \epsilon \cosh^2(k\pi)}{\cosh(k\pi)} \cos(k\pi x)$$

provides an approximation of the transfer operator R. Note that the eigenvalues $\{\lambda_{\epsilon,k}\}_{k=0}^{\infty}$ of R_ϵ satisfy the inequality

$$\lambda_{\epsilon,k} \ge \epsilon > 0 \quad \text{for } k = 0, 1, \ldots.$$

Consequently, and contrary to the properties of R, zero is not a cluster point for the eigenvalues of R_ϵ.

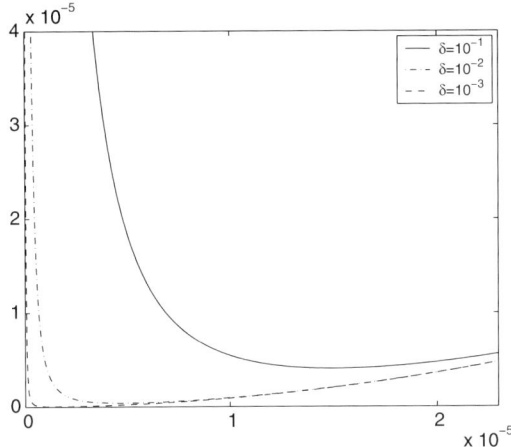

Fig. 7.3. Plots of the error function $E(\epsilon, \delta)$, as a function of ϵ, for three different values of the noise level δ in the observation data.

In the present situation, it can be shown that the optimal value $\hat{\epsilon}$ of ϵ is given approximately by the formula

$$\hat{\epsilon}(\delta) \approx 4.74 \cdot 10^{-5} \sqrt{\delta} \quad \text{for } \delta \in [10^{-6}, 10^{-1}].$$

This optimal value yields the error

$$E(\hat{\epsilon}(\delta), \delta) \approx 4.05 \cdot 10^{-5} \delta,$$

and we conclude that Tikhonov regularization handles this case fairly well.   ◇

Higher-order Tikhonov Regularization. Throughout the last forty years a variety of techniques have been suggested that are closely related to Tikhonov regularization. Two of these are first- and second-order Tikhonov regularization[11], which, in our case, are defined in terms of the cost-functionals

$$J_{1,\epsilon}(g) = \|R(g) - d\|^2_{L^2(\sigma_1)} + \epsilon\|g_x\|^2_{L^2(\sigma_3)} \quad \text{for } g \in S,$$
$$J_{2,\epsilon}(g) = \|R(g) - d\|^2_{L^2(\sigma_1)} + \epsilon\|g_{xx}\|^2_{L^2(\sigma_3)} \quad \text{for } g \in S,$$

respectively.

For both of these methods, the added regularization term will have a strong smoothing effect. For example, minimizing $J_{2,\epsilon}(g)$ with respect to g, leads to the

---

[11] The technique (7.32)–(7.33) is nowadays frequently referred to as zero order Tikhonov regularization.

approximation

$$R_{2,\epsilon}^{-1}(d) = R_{2,\epsilon}^{-1}\left(\sum_{k=0}^{\infty} d_k \cos(k\pi x)\right)$$

$$= \sum_{k=0}^{\infty} d_k \frac{\cosh(k\pi)}{1 + \epsilon(k\pi)^4 \cosh^2(k\pi)} \cos(k\pi x)$$

of R^{-1}. Comparing this expression with formula (7.35) for R_ϵ^{-1}, we observe that $R_{2,\epsilon}^{-1}(d)$ has a stronger smoothing effect on the high frequency components of d than R_ϵ^{-1}.

Example 7. As in Example 6, we use the data given in equations (7.27) and (7.28). Applying the second-order Tikhonov scheme to this problem gives the following error function:

$$E_2(\epsilon, \delta) = \|R^{-1}(d) - R_{2,\epsilon}^{-1}(d_\delta)\|_{L^2(\sigma_3)}^2$$

$$= \frac{1}{2}\left[\frac{\epsilon\pi^4 \cosh^2(\pi)}{1 + \epsilon\pi^4 \cosh^2(\pi)}\right]^2 + \frac{1}{2}\left[\frac{\delta \cosh(5\pi)}{1 + \epsilon(5\pi)^4 \cosh^2(5\pi)}\right]^2.$$

The optimal value of ϵ, as a function of the error level δ, is now given approximately by the formula

$$\hat{\epsilon}(\delta) \approx 1.94 \cdot 10^{-8}\sqrt{\delta}.$$

This value of the regularization parameter yields a least L^2-error of the form

$$E_2(\hat{\epsilon}(\delta), \delta) \approx 6.48 \cdot 10^{-8}\delta,$$

and we can conclude that the second-order scheme provides more accurate results than its zero-order counterpart, cf. Example 6.   ◇

7.2   The Classical Inverse Problem of Electrocardiography

Is it possible to use recordings of the electrical potential along the body surface, mathematical methods, and computer software to compute the epicardial potential distribution? This challenge is commonly referred to as *the inverse problem of electrocardiography*. It has been thoroughly studied by several scientists; see, e.g. Franzone, Taccardi and Viganotti [44], MacLeod and Brooks [93], Rudy and Oster [120], Dössel [31], Greensite and Huiskamp [50], Yamashita and Takahashi [148], Shahidi, Savard and Nadeau [126], Gulrajani [54], Cheng, Bodley and Pullan [24], to name but a few.

Our goal is not to give a state-of-the-art presentation of this field. Instead, we will focus on the relevant mathematical formulations and problems, and illuminate the theory and its challenges by a series of numerical experiments.

7.2.1 Mathematical Formulation

Recall that the electrical potential in the torso is governed by an elliptic partial differential equation in the form

$$\nabla \cdot (M_o \nabla u_o) = 0 \quad \text{in } T,$$

where T, M_o and u_o represent the physical domain occupied by the torso, the conductivity of the involved tissue and the electrical potential, respectively. Moreover, as in Chapter 2, we assume that the body is completely insulated, leading to the boundary condition

$$(M_o \nabla u_o) \cdot n = 0 \quad \text{along } \partial T.$$

Here, ∂T represents the surface of the body, and n the outward directed normal vector of ∂T; see Figure 7.4. Observe that, denoting the epicardial surface by ∂H, the boundary of the torso T is given by the set

$$\partial H \cup \partial T,$$

cf. Figure 7.4. Throughout this section we will, for the sake of ease of notation, write u for u_o.

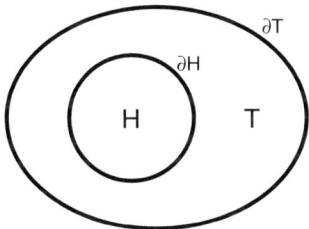

Fig. 7.4. An illustration of the domains H and T.

Assume for a moment that the potential at the epicardial surface ∂H is known and given in terms of a function g. Then we could easily solve the boundary value problem

$$\nabla \cdot (M_o \nabla u) = 0 \quad \text{in } T, \tag{7.36}$$

$$(M_o \nabla u) \cdot n = 0 \quad \text{along } \partial T, \tag{7.37}$$

$$u = g \quad \text{along } \partial H \tag{7.38}$$

for u, and record the electrical potential

$$u|_{\partial T}$$

at the surface of the body. Actually, we could do this for any $g \in H^1(\partial H)$, and thereby define a mapping[12]

$$R : H^1(\partial H) \to L^2(\partial T) \tag{7.39}$$

by

$$R(g) = u(g)|_{\partial T}, \tag{7.40}$$

where $u = u(g)$ is the solution of (7.36)–(7.38) associated with $g \in H^1(\partial H)$.

The function R can be though of as a *transfer mapping* that maps the epicardial potential to the body surface potential. Among applied mathematicians, it is well-known that, for a given $g \in H^1(\partial H)$, (7.36)–(7.38) is a well-posed problem; its solution $u = u(g)$ depends continuously on g and can be computed efficiently by a wide range of numerical methods; see, e.g. Hackbusch [56] or Evans [38].

In the present context, the task of determining the body surface potential $u(g)|_{\partial T}$, for a given epicardial potential g, is referred to as the direct, or forward, problem. From the considerations given above, it follows that this is a well-posed problem, the solution to which present no great difficulty. However, this is not what we want to do; our aim is to try to determine the epicardial potential from measurements of the body surface potential!

In mathematical terms, this problem can be expressed as follows. Let

$$d \in L^2(\partial T)$$

represent the recorded potential along the body surface ∂T at a fixed time t. Can we compute the corresponding epicardial potential

$$g \in H^1(\partial H)$$

at time t, such that
$$u(g)|_{\partial T} = d?$$

Here, $u(g)$ denotes the solution of (7.36)–(7.38) generate by g.

In short, this problem takes the form: Find g such that

$$R(g) = d, \tag{7.41}$$

where R is the operator defined in (7.39)–(7.40). If d is in the range of R, i.e. $d \in R(H^1(\partial H))$, and R is invertible, then

$$g = R^{-1}(d).$$

We will see below, in the section on numerical experiments, that equation (7.41) is ill-posed, and therefore very difficult to solve accurately. In the present situation,

---

[12] Under proper conditions, (7.36)–(7.38) has a solution for every element g in the abstract fractional order Sobolov space $H^{1/2}(\partial H) \subset H^1(\partial H)$. Consequently, from a mathematical point of view, it would have been natural to let $H^{1/2}(\partial H)$ be the domain of the operator R. In the present, more practical situation, this issue is not important. More information about this topic can be found in, e.g., Hackbusch [56], Marti [99] and Adams [1].

(7.41) is referred to as the inverse problem, i.e. *the inverse problem of electrocardiography*.

The similarity between the boundary value problems (7.1)–(7.3) and (7.36)–(7.38) is evident. Consequently, the inverse problem studied in Section 7.1 is closely related to that defined in (7.41); see also the discussion leading to (7.6). However, the conductivity M_o, of the involved tissues, will typically depend on the spatial position, and the domain T is not regular. In general, the method of separation of variables cannot be applied to such problems. Hence, more advanced techniques must be used to analyze (7.36)–(7.38) and the associated inverse problem (7.41).

The Time-Dependent Problem. What is the role of time in this context? Naturally, in many cases we would be interested in how the epicardial potential $g = g(t)$ evolves with time t. Thus, we may want to compute $g(t)$ at several time instances, say for $t = t_0, \ldots, t_M$, based on the corresponding measurements of the electrical potential $d = d(t)$ at the body surface. In this case, a problem of the form (7.41) must be solved for $t = t_\tau$, $\tau = 0, \ldots, M$. This issue, along with a suitable regularization technique, is treated in detail in Section 7.2.4 below. Meanwhile, we will focus on the "stationary" version of equation (7.41), and investigate how this problem can be solved approximately at a fixed time t.

7.2.2 A Linear Problem

Our goal is to show that the mapping R, defined in (7.39)–(7.40), is linear. This will, of course, simplify the mathematical and numerical treatment of the inverse problem (7.41).

Let $u = u(g)$ denote the solution of (7.36)–(7.38) associated with a given epicardial potential $g \in H^1(\partial H)$. For an arbitrary real constant $c \in \mathbf{R}$, consider the function $v = cu$. By straightforward computations, we find that

$$\nabla \cdot (M_o \nabla(v)) = c\nabla \cdot (M_o \nabla(u)) = 0 \quad \text{in } T,$$
$$(M_o \nabla(v)) \cdot n = c(M_o \nabla(u)) \cdot n = 0 \quad \text{along } \partial T,$$

and
$$v = cu = cg \quad \text{along } \partial H.$$

Consequently, $v = cu$ solves the problem

$$\nabla \cdot (M_o \nabla v) = 0 \quad \text{in } T,$$
$$(M_o \nabla v) \cdot n = 0 \quad \text{along } \partial T,$$
$$v = cg \quad \text{along } \partial H,$$

and we conclude that $u(cg) = cu(g)$ for any real constant c.

Next, by the linearity of the restriction operator

$$u \to u|_{\partial T},$$

it follows that

$$R(cg) = cR(g) \quad \text{for any } g \in H^1(\partial H) \text{ and } c \in \mathbf{R}. \tag{7.42}$$

Let g_1, $g_2 \in H^1(\partial H)$ be two arbitrary epicardial potentials, and consider their associated boundary value problems

$$\begin{aligned} \nabla \cdot (M_o \nabla u_1) &= 0 \quad \text{in } T, \\ (M_o \nabla u_1) \cdot n &= 0 \quad \text{along } \partial T, \\ u_1 &= g_1 \quad \text{along } \partial H, \end{aligned}$$

and

$$\begin{aligned} \nabla \cdot (M_o \nabla u_2) &= 0 \quad \text{in } T, \\ (M_o \nabla u_2) \cdot n &= 0 \quad \text{along } \partial T, \\ u_2 &= g_2 \quad \text{along } \partial H. \end{aligned}$$

What sort of problem will the sum $w = u_1 + u_2$ of u_1 and u_2 satisfy? Clearly,

$$\begin{aligned} \nabla \cdot (M_o \nabla w) &= \nabla \cdot (M_o \nabla u_1) + \nabla \cdot (M_o \nabla u_2) = 0 \quad \text{in } T, \\ (M_o \nabla w) \cdot n &= (M_o \nabla u_1) \cdot n + (M_o \nabla u_2) \cdot n = 0 \quad \text{along } \partial T, \end{aligned}$$

and

$$w = u_1 + u_2 = g_1 + g_2 \quad \text{along } \partial H.$$

This means that w is the unique solution of the problem

$$\begin{aligned} \nabla \cdot (M_o \nabla w) &= 0 \quad \text{in } T, \\ (M_o \nabla w) \cdot n &= 0 \quad \text{along } \partial T, \\ w &= g_1 + g_2 \quad \text{along } \partial H, \end{aligned}$$

and it follows that

$$u(g_1 + g_2) = w = u(g_1) + u(g_2) \quad \text{for any } g_1, g_2 \in H^1(\partial H).$$

This fact, along with the linearity of the restriction operator and the property expressed in (7.42), implies that R is a linear mapping, i.e.

$$R(c_1 g_1 + c_2 g_2) = c_1 R(g_1) + c_2 R(g_2) \quad \text{for any } g_1, g_2 \in H^1(\partial H) \text{ and } c_1, c_2 \in \mathbf{R}. \tag{7.43}$$

We conclude that the inverse problem of electrocardiography (7.41), is a linear problem!

7.2.3 Discretization

Our goal is to derive a discrete approximation that is suitable for performing numerical simulations of (7.41). This is accomplished by applying the famous Rayleigh-Ritz-Galerkin methodology to the problem. In the present situation, this may be formulated as follows:

Let

$$g_1, \ldots, g_n \in H^1(\partial H)$$

be a set of linearly independent functions defined on the surface ∂H of the heart, and define the subspace V_n of $H^1(\partial H)$ by

$$V_n = \mathrm{span}\{g_1, \ldots, g_n\} \subset H^1(\partial H).$$

The restriction of the transfer mapping R to V_n is denoted by R_n, i.e.

$$R_n = R|_{V_n} : V_n \to L^2(\partial T). \tag{7.44}$$

Since the linearity of R is inherited by R_n, it is particularly easy to characterize the action of R_n. More precisely, every function $g \in V_n$ can be represented uniquely by n real numbers, say $p_1, \ldots, p_n$,

$$g = \sum_{i=1}^{n} p_i g_i. \tag{7.45}$$

Consequently, if

$$r_i = R_n(g_i) = u(g_i)|_{\partial T} \quad \text{for } i = 1, \ldots, n, \tag{7.46}$$

then the linearity of R_n implies that

$$R_n(g) = \sum_{i=1}^{n} p_i r_i. \tag{7.47}$$

Hence, the range

$$R_n(V_n) = \{R_n(q); \; q \in V_n\} \tag{7.48}$$

of R_n is fully characterized by the functions $r_1, \ldots, r_n$ defined in (7.46). In fact, from these considerations, it follows that

$$R_n(V_n) = \mathrm{span}\left\{r_1, \ldots, r_n\right\}.$$

Moreover, the functions $r_1, \ldots, r_n$ are easily computed by the following procedure: Set $g = g_i$ in (7.38), solve (7.36)–(7.38) for $u = u(g_i)$, and put

$$r_i = u(g_i)|_{\partial T}$$

for $i = 1, \ldots, n$. Thus, the action of the discrete transfer mapping $R_n \approx R$ is completely determined by solving n linear elliptic boundary value problems of the form (7.36)–(7.38).

A first approach towards defining a finite dimensional approximation of the inverse problem (7.41) could typically be as follows: Find $g = \sum_{i=1}^{n} p_i g_i \in V_n$ such that

$$R_n(g) = d. \tag{7.49}$$

However, unless d is in the range $R_n(V_n)$ (see (7.48)), of R_n, there does not exist any $g \in V_n$ solving (7.49). As in Section 7.1.4, this difficulty is handled by "replacing" equation (7.49) by a minimization problem of the form

$$\min_{g \in V_n} \|R_n(g) - d\|^2_{L^2(\partial T)}. \tag{7.50}$$

That is, we will try to minimize the L^2-difference between the observation data d and the simulated electrical potential at the surface ∂T of the body.

Let us now turn our attention towards the problem of solving (7.50) on a computer. So far, we have assumed that the date d is given in terms of a function defined at every point $x \in \partial T$. Recall that d represents physical measurements of the electrical potential at the surface of the body. Most likely, this potential can only be recorded at a finite number of locations, referred to as the leads, along ∂T – say in the regions

$$\Gamma_1, \ldots \Gamma_m \subset \partial T, \tag{7.51}$$

see Figure 7.5. Thus, we assume in what follows that

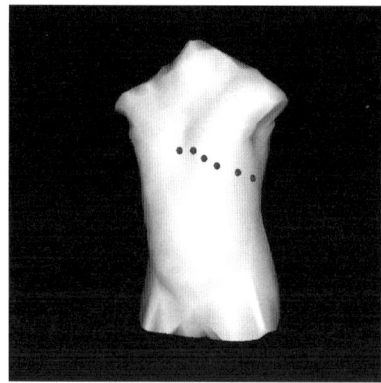

Fig. 7.5. Typical positions of the leads, i.e. the locations used to produce ECG recordings.

$$d : \cup_{j=1}^{m} \Gamma_j \to \mathrm{R} \quad \text{and} \quad d \in L^2(\cup_{j=1}^{m} \Gamma_j).$$

From this perspective, (7.50) should be replaced by the problem

$$\min_{g \in V_n} \|R_n(g) - d\|^2_{L^2(\Gamma)}, \tag{7.52}$$

provided that

$$\Gamma = \cup_{j=1}^{m} \Gamma_j.$$

Every function $g \in V_n$ can be written in the form (7.45), for a suitable set of real numbers $p_1, \ldots, p_n$, and

$$R_n(g) = \sum_{i=1}^{n} p_i r_i,$$

where $r_1, \ldots, r_n$ are the functions given in (7.46). Inspired by this fact, and (7.52), we introduce the cost-functional

$$J = J(p_1, \ldots, p_n) = \left\| \sum_{i=1}^{n} p_i r_i - d \right\|_{L^2(\Gamma)}^2.$$

The problem (7.52) may then be rewritten in the form

$$\min_{p_1, \ldots, p_n \in \mathbf{R}} J(p_1, \ldots, p_n). \tag{7.53}$$

Differentiation of J yields

$$\frac{\partial J}{\partial p_i} = 2 \int_{\Gamma} \left[\sum_{j=1}^{n} p_j r_j - d \right] r_i \, dx \quad \text{for } i = 1, \ldots, n,$$

and then the necessary first-order condition for a minimum point

$$\frac{\partial J}{\partial p_i} = 0 \quad \text{for } i = 1, \ldots, n,$$

gives the $n \times n$ linear system

$$\sum_{j=1}^{n} \int_{\Gamma} r_j r_i \, dx \, p_j = \int_{\Gamma} d r_i \, dx \quad \text{for } i = 1, \ldots, n, \tag{7.54}$$

for the parameters $p_1, \ldots, p_n$. By introducing the notation

$$B = [b_{ij}] \in \mathbf{R}^{n \times n}, \quad b_{ij} = \int_{\Gamma} r_j r_i \, dx, \tag{7.55}$$

$$p = (p_1, \ldots, p_n)^T \in \mathbf{R}^n, \tag{7.56}$$

$$c = \left(\int_{\Gamma} d r_1 \, dx, \ldots, \int_{\Gamma} d r_n \, dx \right)^T \in \mathbf{R}^n, \tag{7.57}$$

(7.54) can be written on matrix-vector form: Find $p \in \mathbf{R}^n$ such that

$$Bp = c. \tag{7.58}$$

Summary. The methodology derived above can be summarized as follows:

a) Pick n linearly independent functions

$$g_1, \ldots, g_n \in H^1(\partial H),$$

defined at the surface ∂H of the heart H

b) For $i = 1, \ldots, n$, set $g = g_i$ in (7.38) and solve (7.36)–(7.38) for $u = u(g_i)$

c) Compute

$$r_i = u(g_i)|_{\partial T}, \quad i = 1, \ldots, n$$

d) Compute the matrix B defined in (7.55)

e) Compute the vector c defined in (7.57)

f) If possible, solve the linear system (7.58)

g) Compute the epicardial potential g by applying formula (7.45)

This algorithm requires the solution of n elliptic boundary value problems of the form (7.36)–(7.38). Having solved these equations, the action of the discrete transfer mapping R_n is fully characterized by the functions $r_1, \ldots, r_n$, cf. step c) and (7.47).

Consequently, if we want to solve (7.58) for several recordings of the potential at the surface of the body, steps a)–d) only have to be executed once. In other words, for each new observation d at the leads $\Gamma_1, \ldots, \Gamma_m$ only steps e)–g) need to be carried out, provided that the basis functions $g_1, \ldots, g_n$ used to describe the epicardial potential are fixed.

7.2.4 The Time-Dependent Problem

As mentioned in Section 7.2.1, the time dependency of the epicardial potential is often of major interest. From a clinical point of view, this information is important for determining the condition of the heart. This section is devoted to incorporating the time dimension into the theory developed above.

Suppose that we have recorded the electrical potential at $M + 1$ chronologically ordered time instances $t_0, \ldots, t_M$, and that the associated data is represented by the functions

$$d^0, \ldots, d^M \in L^2(\Gamma).$$

Here, Γ denotes the union of the leads $\Gamma_1, \ldots, \Gamma_m$, cf. Section 7.2.3 and Figure 7.5.

By the methodology presented in Section 7.2.3, it follows that we might compute the epicardial potential, as a function of time, by solving a minimization problem on the form (7.52) at each time step $t_0, \ldots, t_M$. This leads to a sequence of minimization problems,

$$\min_{g^\tau \in V_n} \| R_n(g^\tau) - d^\tau \|_{L^2(\Gamma)}^2 \quad \text{for } \tau = 0, \ldots, M, \tag{7.59}$$

where R_n is the discrete transfer mapping defined in (7.44). That is, R_n maps the discrete epicardial potential to the discrete body surface potential.

Note that each of the problems in (7.59) can be solved individually. However, as we will see in the numerical experiments section below, each of these problems is ill-posed. This is also what we would expect based on the discussion in Section 7.1.3. Hence, some sort of regularization technique must be applied. As usual for linear problems, there is a wide range of methods available: Tikhonov regularization, iterative regularization schemes, TSVD[13] etc.; see Section 7.1.5. However, for this time dependent problem, there seems to be one particularly attractive approach. For a small parameter $\epsilon \geq 0$, we introduce the following regularized approximation of (7.59)

$$\min_{g^\tau \in V_n} \left[\|R_n(g^\tau) - d^\tau\|^2_{L^2(\Gamma)} + \epsilon \|g^\tau - g^{\tau-1}\|^2_{L^2(\partial H)} \right] \quad \text{for } \tau = 0, \dots, M.$$

(7.60)

That is, expressed in a somewhat rough and ready style, we apply a Tikhonov regularization technique that ensures that the change in the epicardial potential is small from one time step to the next. From a physical point of view, this requirement seems to be reasonable, provided that the time step

$$t_\tau - t_{\tau-1}$$

is small.

Observe that, for $\tau = 0$, an undefined function g^{-1} is involved in the minimization problem (7.60). This function can, of course, be specified in a number of ways. A typical choice could be to apply Tikhonov regularization in its purest form, i.e. defining $g^{-1} = 0$, or the second-order Tikhonov scheme. If the first time step in the simulation process coincides with the start of the depolarization phase, the heart surface at that instance is isoelectric; that is, g^{-1} is constant. For the sake of simplicity, we will throughout this section stipulate $g^{-1} = 0$.

Consider the functional associated with time step t_τ in (7.60). According to equation (7.45), the unknown function $g^\tau \in V_n$ can be expressed as a linear combination of the basis functions of V_n,

$$g^\tau = \sum_{i=1}^n p_i^\tau g_i,$$

(7.61)

where $p_1^\tau, \dots, p_n^\tau$ are real numbers that we want to determine such that g^τ satisfies (7.60). Furthermore, by formula (7.47) we find that, letting $\{r_i\}_{i=1}^n$ be the functions defined in (7.46),

$$R_n(g^\tau) = \sum_{i=1}^n p_i^\tau r_i.$$

Hence, (7.60) may be written in the form

$$\min_{p_1^\tau, \dots, p_n^\tau \in \mathbb{R}} \left[\|\sum_{i=1}^n p_i^\tau r_i - d^\tau\|^2_{L^2(\Gamma)} + \epsilon \|\sum_{i=1}^n p_i^\tau g_i - g^{\tau-1}\|^2_{L^2(\partial H)} \right]$$

(7.62)

---

[13] Truncated Singular Value Decomposition, cf. e.g. Groetsch [52]

for $\tau = 0, \ldots, M$. Note that $g^{\tau-1}$ is already known since

$$g^{\tau-1} = \sum_{i=1}^{n} p_i^{\tau-1} r_i,$$

where $\{p_i^{\tau-1}\}_{i=1}^{n}$ were computed at the previous time-step.

If we define the cost-functionals

$$J_\epsilon^\tau = J_\epsilon^\tau(p_1^\tau, \ldots, p_n^\tau) = \left\| \sum_{i=1}^{n} p_i^\tau r_i - d^\tau \right\|_{L^2(\Gamma)}^2 + \epsilon \left\| \sum_{i=1}^{n} p_i^\tau g_i - g^{\tau-1} \right\|_{L^2(\partial H)}^2$$

for $\tau = 0, \ldots, M$, then (7.62) can be written compactly as

$$\min_{p_1^\tau, \ldots, p_n^\tau \in \mathbf{R}} J_\epsilon^\tau(p_1^\tau, \ldots, p_n^\tau) \quad \text{for } \tau = 0, \ldots, M. \tag{7.63}$$

Let us consider the minimization process (7.63) associated with a fixed time instance t_τ. Suppose that we have already computed $g^{\tau-1}$. By the chain rule of differentiation, it follows that

$$\frac{\partial J_\epsilon^\tau}{\partial p_i^\tau} = 2 \int_\Gamma \left[\sum_{j=1}^{n} p_j^\tau r_j - d^\tau \right] r_i \, dx + 2\epsilon \int_{\partial H} \left[\sum_{j=1}^{n} p_j^\tau g_j - g^{\tau-1} \right] g_i \, dx$$

for $i = 1, \ldots, n$, and the first-order condition for a minimum

$$\frac{\partial J_\epsilon^\tau}{\partial p_i^\tau} = 0, \quad i = 1, \ldots, n,$$

yields the linear system

$$\sum_{j=1}^{n} \left[\int_\Gamma r_j r_i \, dx + \epsilon \int_{\partial H} g_j g_i \, dx \right] p_j^\tau = \int_\Gamma d^\tau r_i \, dx + \epsilon \int_{\partial H} g^{\tau-1} g_i \, dx, \quad i = 1, \ldots, n.$$

By defining

$$B_\epsilon = [b_{\epsilon,ij}] \in \mathbf{R}^{n \times n}, \quad b_{\epsilon,ij} = \int_\Gamma r_j r_i \, dx + \epsilon \int_{\partial H} g_j g_i \, dx, \tag{7.64}$$

$$p^\tau = (p_1^\tau, \ldots, p_n^\tau)^T \in \mathbf{R}^n,$$

$$c_\epsilon^{\tau-1} = \left(\int_\Gamma d^\tau r_1 \, dx + \epsilon \int_{\partial H} g^{\tau-1} g_1 \, dx, \ldots, \int_\Gamma d^\tau r_n \, dx + \epsilon \int_{\partial H} g^{\tau-1} g_n \, dx \right)^T \in \mathbf{R}^n,$$

we may write this system in the matrix-vector form

$$B_\epsilon p^\tau = c_\epsilon^{\tau-1}.$$

To summarize, the minimization problems given in (7.60) can be solved for $g^0, \ldots, g^M$, in a successive manner, by solving the linear systems

$$B_\epsilon p^\tau = c_\epsilon^{\tau-1}, \quad \tau = 0, \ldots, M. \tag{7.65}$$

Recall that the action of the discrete transfer mapping R_n (see (7.44)), is fully characterized by the functions $r_1, \ldots, r_n$ defined in (7.46). This property of R_n is "inherited" by the matrix B_ϵ in the sense that B_ϵ is independent of the time instance t_τ, see (7.64). On the other hand, the right hand side $c_\epsilon^{\tau - 1}$ in (7.65) must be computed for every $\tau = 0, \ldots, M$.

Finally, having computed

$$p^0 = \left(p_1^0, \ldots, p_n^0\right)^T,$$
$$\vdots$$
$$p^M = \left(p_1^M, \ldots, p_n^M\right)^T,$$

we can easily find the corresponding approximate epicardial potentials by applying formula (7.61).

7.2.5 Numerical Experiments

The data for the numerical experiments in this section were taken from the simulations presented in Section 3.4.2. That is, the data were generated by solving the model (2.65)–(2.72), involving the bidomain equations, on a two-dimensional slice of the human body. From these experiments we extracted the potentials both at the heart and the torso surfaces, respectively. In this section, we will try to recover the epicardial potentials using the simulated body surface potentials as observation data.

In the computations below, there are $m = 104$ leads positioned along the surface ∂T of the body; cf. equation (7.51) and Figure 7.5. Moreover, $n = 632$ functions, defined along the surface ∂H of the heart, have been used to discretize (7.41). Further details on this topic can be found in Section 7.2.3.

We consider the simulation results, obtained by solving (2.65)–(2.72), at time[14] $t = 30$ ms. Thus, the observation data consists of the simulated body surface potential at that time instance. Our task is to try to identify the associated extra cellular potential that is present at the epicardial surface.

With no regularization, the transfer matrix B (see equations (7.55)–(7.57)), is so ill-conditioned that the system (7.58) cannot be solved accurately using normal machine precision. In order to be able to compute a solution of this linear system, a very small amount of zero-order Tikhonov regularization ($\epsilon = 10^{-15}$) was applied. The results are shown in Figure 7.6a). Clearly, the approximate solution of the inverse problem (7.41) oscillates wildly in this case. However, the observation data, recorded along the surface ∂T of the body, is reproduced almost exactly; see Figure 7.6b). This illustrates the ill-posed nature of the problem at hand. We are able to find an epicardial potential that reproduces the observed data, but the estimated potentials at the heart surface are not correct.

In the next test, the regularization parameter in the zero-order Tikhonov scheme is increased to $\epsilon = 10^{-3}$. The results obtained in this way are depicted in the second

---

[14] More precisely, we consider the results of these simulation at time 30 ms after the initial stimulation of the tissue

row of Figure 7.6. Clearly, the estimated epicardial potential is now closer to the correct potential, but large oscillations are still present in the solution. We also observe that the match with the body surface potential has begun to deteriorate. This is as expected, since using a "large" regularization parameter will obviously put more weight onto the regularization term in the associated cost-functional. Consequently, the precision of the data-fitting procedure (the output least squares requirement) is reduced.

Recall that the second-order Tikhonov regularization scheme has a stronger smoothing effect on the high-frequency components of the data than the zero-order method; cf. Section 7.1.5. Therefore, since the presence of large oscillations in the estimated epicardial potential is such a major problem, it is natural to try the second-order scheme. In the present situation, this method is implemented by minimizing the following cost-functional:

$$J_{2,\epsilon}(g) = \|R(g) - d\|^2_{L^2(\Gamma)} + \epsilon \|\Delta_{\partial H} g\|^2_{L^2(\partial H)}. \tag{7.66}$$

Here

$$\Delta_{\partial H} g^\tau = \mathrm{curl}_{\partial H}\, \mathrm{curl}_{\partial H} g^\tau$$

denotes the so-called Laplace-Beltrami operator, which can be thought of as a generalization of the Laplacian to functions defined on curved surfaces. We will not dwell any further upon the technical issues of this operator. Further information can be found in, e.g., Dautray and Lions [29].

Figure 7.7 shows the effect of using this technique. With no noise in the observation data, $\epsilon = 10^{-8}$ is a good choice. The solution is smooth and the epicardial potential is computed accurately by the second-order Tikhonov scheme. The second row of this figure shows the results generated by adding 1% noise to the observation data (still recorded along the body surface!). In this case, a larger value of the regularization parameter ϵ is needed in order to get a smooth solution. The depicted solution was generated by choosing $\epsilon = 1$.

A drawback with the regularization strategies used to generate the results shown in Figures 7.6 and 7.7 is that they do not exploit the fact that the epicardial potential is correlated in time. Consequently, the estimated epicardial potential might fluctuate from time step to time step, even though it is smooth in the spatial direction.

Hence, temporal regularization, as described in Section 7.2.4, was tested from $t = 30\,\mathrm{ms}$ to $t = 35\,\mathrm{ms}$, applying 40 time steps in all. The noise level was set as above, i.e. 1% noise was added to the body surface potentials. The results are shown in Figure 7.8 a). The beneficial effect of temporal correlated signals at the epicardial surface was achieved (but this cannot be seen in this "stationary" figure).

A disadvantage of this temporal regularization strategy is that small oscillations in the estimated epicardial potential have a tendency to appear as time evolves. To overcome this difficulty, we tested a "hybrid" scheme, combining temporal and spatial regularization techniques. More precisely, the method (7.60) was modified in order to include a second-order Tikhonov term. This was accomplished by minimizing a cost-functional of the form

$$J^\tau_{\epsilon,\beta}(g^\tau) = \|R_n(g^\tau) - d^\tau\|^2_{L^2(\Gamma)} + \epsilon \|g^\tau - g^{\tau-1}\|^2_{L^2(\partial H)} + \beta \|\Delta_{\partial H} g^\tau\|^2_{L^2(\partial H)}. \tag{7.67}$$

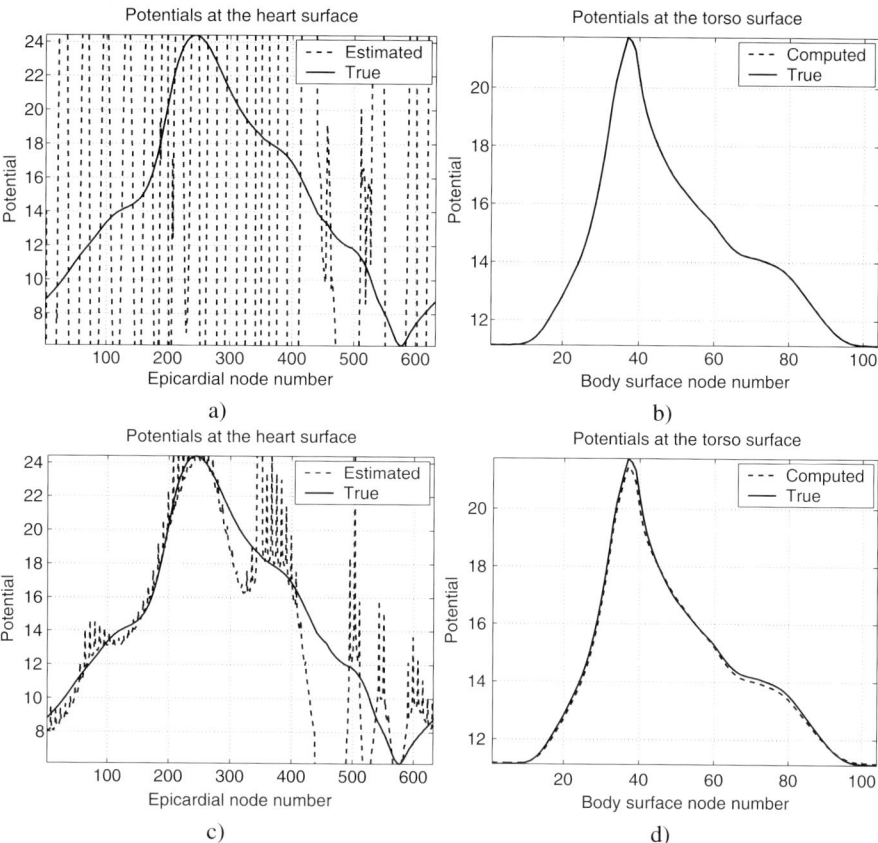

Fig. 7.6. The results obtained by applying the zero-order Tikhonov scheme to the problem studied in Section 7.2.5: **a)** The correct and estimated epicardial potentials with virtually no regularization, $\epsilon = 10^{-15}$. **b)** The body surface potentials generated by the two epicardial potentials shown in Figure a). There is no visible difference between the measured and the computed signal, even though the estimated epicardial values are obviously wrong. **c)** Here $\epsilon = 10^{-3}$, and the estimated epicardial potentials are much closer to the correct signal than in the case considered in Figure a). However, the solution still contains large oscillations. **d)** The electrical potentials along the body surface associated with the epicardial potentials depicted in Figure c). With this amount of regularization, a discrepancy between the measured potential and the body surface potential associated with applying the zero-order Tikhonov scheme is present.

where now both ϵ and β are regularization parameters. The results obtained by applying this technique are shown in Figure 7.8 b). The computed epicardial potential is now smooth as well as correlated in time!

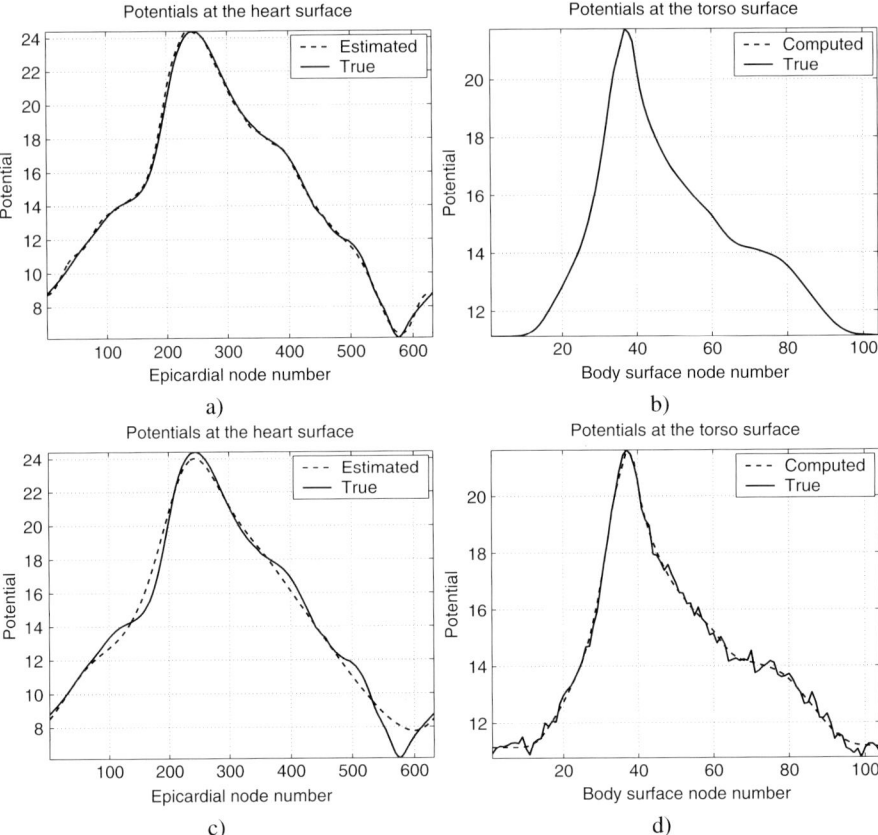

Fig. 7.7. The results obtained by applying the second-order Tikhonov scheme to the problem discussed in Section 7.2.5. In the second row noise has been added to the observation data (still recorded along the body surface ∂T, cf. Figure 7.4). **a)** Estimates obtained by using the regularization parameter $\epsilon = 10^{-8}$. **b)** The body surface potentials associated with the epicardial potentials shown in Figure a). **c)** In this case, 1% noise has been added to the observation data, and $\epsilon = 1$. The figure shows the epicardial potential computed by the second order scheme along with the correct potential. **d)** The noisy observation data and the simulated body surface potential associated with the epicardial potentials depicted in Figure c).

For both the simulation results shown in Figure 7.8, the associated body surface potentials were reproduced accurately. In fact, they were very similar to the potential depicted in Figure 7.7d).

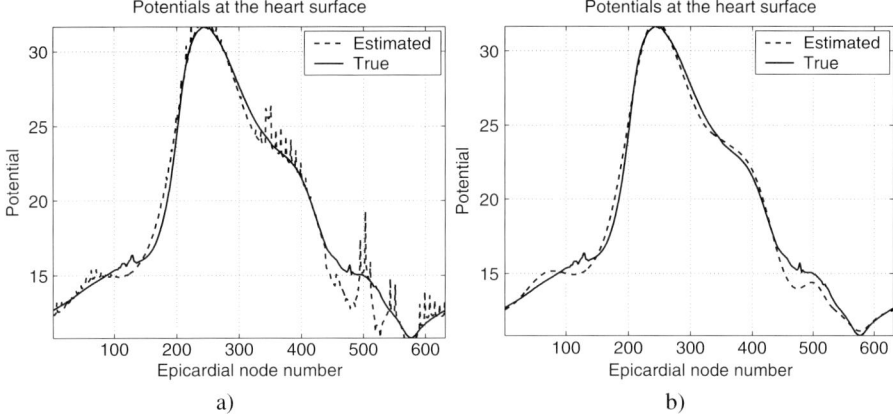

Fig. 7.8. a) Temporal regularization with $\epsilon = 0.01$ in equation (7.60). **b**) "Hybrid" regularization with $\epsilon = 0.01$ and $\beta = 1$ in equation (7.67).

7.3   Identifying a Simplified Source – Computing the Location and Orientation of a Dipole

The electrical signal that can be measured on the surface of the human body is generated by the heart. The heart can thus be viewed as a current source positioned in a volume conductor. In Chapter 2 a simple model of this phenomenon was introduced,:

$$\nabla \cdot (M\nabla u) = f \quad \text{in } \Omega, \tag{7.68}$$

$$(M\nabla u) \cdot n = 0 \quad \text{along } \partial\Omega \,(= \partial T), \tag{7.69}$$

$$\int_{\Omega} u \, dx = 0, \tag{7.70}$$

cf. the discussion preceding equations (2.5)–(2.6). Here u, M, f and Ω denote the electrical potential, the conductivity of the involved tissues, the current source(s) and the region occupied by the human body, respectively. Throughout this section we will focus on this simple elliptic boundary value problem. Note that $\Omega = H \cup T$ (see Figure 7.4), and thus the solution domain of (7.68) consists of both the heart H and its surrounding body T.

In these kinds of models, the function f represents the electrical activity of the heart. As mentioned in Chapter 1, this activity is frequently modelled by a so-called dipole; see Figure 3.4. This means that f is, at least approximately, defined in terms of two opposite point charges[15].

---

[15] In many cases more than two point charges are used to represent the electrical activity of the heart. This leads to models of the form (7.68)–(7.70), involving several dipoles. In the present text, we will only consider the case of a single dipole.

Models of this kind, and the heart vector interpretation of them, go back as far as Einthoven's work in the early 20th century. They have always been a central part of ECG analysis; see, e.g. Keener and Sneyd [72]. The purpose of the present section is not to discuss the physiological or biological reasons for studying models of the form (7.68)–(7.70) in detail. Rather, we will focus on various mathematical and computational properties of these equations.

Assume that the characteristics of M, Ω and f are known. Then we can use the Finite Element Method (FEM), discussed in Chapter 3, to solve (7.68)–(7.70) numerically for u. Consequently, we can simulate ECG recordings along the surface $\partial\Omega$ of the body, and study how various changes in the dipole (heart) influence the electrical signal. In the present situation, this is the direct problem: we compute the effect (ECG) of a dipole (the cause) on the electrical potential.

In this section, our ultimate goal is not to solve this direct problem, but to design methods for computing the characteristics of the current source f from body surface measurements of the electrical potential u. That is, we will try to use observations of $u|_{\partial\Omega}$ to compute the properties of f. From a medical point of view, solving this inverse problem can hopefully improve our qualitative and quantitative understanding of the heart; see, e.g. Gulrajani [54], Pullan, Paterson and Greensite [116], Yamashita and Geselowitz [147], and references therein.

Typically, the position and strength of the point charges in the dipole f will depend on time; they will move around during a heart beat. Consequently, f should ideally be a function of both space $x \in \Omega$ and time $t \in [0, t^*]$, where $[0, t^*]$ represents the time interval of interest. This means that the inverse problem described above must be solved for every $t \in [0, t^*]$. Assuming that we have a method for solving this inverse problem at a fixed time, this can be easily be accomplished by applying similar techniques to those introduced in Section 7.2.4. However, to keep things simple, we will throughout this section focus on the stationary version of this task.

Note that, if u satisfies (7.68)–(7.69), the function $u + c$, where c is an arbitrary constant, also satisfies these two equations. Thus, condition (7.70) has been added to ensure the uniqueness of the solution. The role of the constant c in the present framework will be studied in detail in Section 7.3.5 below.

A Remark. In Chapter 2 we introduced the highly complex model (2.65)–(2.72) for the electrical activity in the human body. Now we want to study the heart by considering the simple system of equations (7.68)–(7.70). Is it possible to justify this approach? Comparing these two models in a precise mathematical sense is, of course, a very challenging scientific problem. Nevertheless, it is actually quite simple to establish a rough link between (2.65)–(2.72) and (7.68)–(7.70). This can be accomplished as follows.

From equation (2.67) in Chapter 2 it follows that

$$\nabla((M_i + M_e)\nabla u_e) = -\nabla(M_i\nabla v) \quad \text{in } H. \qquad (7.71)$$

Moreover, in the tissue surrounding the heart, the propagation of the electrical potential is governed by a model of the form

$$\nabla \cdot (M_o \nabla u_o) = 0 \quad \text{in } T, \tag{7.72}$$

cf. (2.65)–(2.72) and Figure 7.4. Thus, by simply putting

$$M = \begin{cases} M_i + M_e & \text{in } H, \\ M_0 & \text{in } T, \end{cases}$$

$$f = \begin{cases} -\nabla(M_i \nabla v) & \text{in } H, \\ 0 & \text{in } T, \end{cases}$$

$$u = \begin{cases} u_e & \text{in } H, \\ u_0 & \text{in } T, \end{cases} \tag{7.73}$$

we see that u, as defined in (7.73), satisfies an equation on the form (7.68).

A further investigation of the link/relationship between these two models would require a deep and lengthly analysis of the involved PDEs. That topic certainly lies beyond the scope of this text.

7.3.1 Preliminaries

Prior to presenting the mathematics of the inverse problem described above, we will derive a property that the source term f in (7.68)–(7.70) must satisfy; cf. also Hackbusch [56] or Marti [99].

Assume that u solves (7.68)–(7.70), and let ϕ be a smooth function defined on Ω. From (7.68) it follows that

$$\int_\Omega f\phi \, dx = \int_\Omega \nabla \cdot (M\nabla u)\phi \, dx,$$

and by Gauss' divergence theorem, and equation (7.69), we conclude that

$$\int_\Omega f\phi \, dx = -\int_\Omega M\nabla u \cdot \nabla\phi \, dx + \int_{\partial\Omega} \phi M\nabla u \cdot n \, ds = -\int_\Omega M\nabla u \cdot \nabla\phi \, dx. \tag{7.74}$$

Since ϕ was arbitrary, we may choose

$$\phi(x) = 1 \text{ for all } x \in \Omega,$$

and then (7.74) implies that

$$\int_\Omega f \, dx = 0. \tag{7.75}$$

Consequently, if (7.68)–(7.70) has a solution, f must satisfy (7.75). That is, this boundary value problem cannot have a solution unless f fulfills (7.75). Thus, we will throughout this section assume that the involved source term belongs to the following subspace V of the classical $L^2(\Omega)$ space,

$$V = \left\{ f \in L^2(\Omega); \int_\Omega f \, dx = 0 \right\}. \tag{7.76}$$

7.3.2   The Inverse Problem

Our goal is to derive a mathematical formulation of the inverse problem described above. Clearly, the solution u of (7.68)–(7.70) depends on the source term f, i.e.

$$u = u(f).$$

Having computed $u(f)$ we can read off the trace (restriction)

$$u(f)\Big|_{\partial\Omega}$$

of this function along the boundary $\partial\Omega$ of Ω. Thus, we may define the operator

$$Q : V \to L^2(\partial\Omega), \tag{7.77}$$

by

$$Q(f) = u(f)\Big|_{\partial\Omega,} \tag{7.78}$$

where V is the space of functions defined in (7.76). In the present context, the task of computing $Q(f)$, for a given f, is the direct problem. In other words, the direct problem consists of solving (7.68)–(7.70), for a given source f, and thereafter recording the simulated electrical potential u along the boundary $\partial\Omega$ of the torso Ω.

Assume that we want to apply noninvasive methods for characterizing the current source. Only observations of the potential along the surface $\partial\Omega$ of Ω are then available to us as data. That is,

$$u(f)\Big|_{\partial\Omega} \tag{7.79}$$

is the available data and f represents the unknown source. So, as mentioned above, our challenge can be described as follows: Can we use data in the form (7.79) and the model (7.68)–(7.70) to determine f?

In mathematical terms this problem takes the form: Let d be an observation of the potential along $\partial\Omega$, solve the equation

$$Q(f) = d \tag{7.80}$$

for f. Note that, if Q is invertible, then

$$f = Q^{-1}(d).$$

Throughout this section we will assume that f depends on the spatial variable x, and that this function is parameterized by a set of parameters $p_1, \ldots, p_n$, i.e.

$$f = f(x; p_1, \ldots, p_n).$$

Our task is to determine $p_1, \ldots, p_n$ such that (7.80) is (approximately) satisfied. Depending on the sort of parameterization that is used, this will lead to either a linear system of algebraic equations or a nonlinear minimization problem defined in terms of a suitable cost-functional.

7.3.3 Parameterizations Leading to Linear Problems

We will now show that the operator Q, defined in (7.77), is a linear mapping. Thereafter, this property will be used to design a discrete approximation of (7.80), which in turn leads to a linear system of algebraic equations.

Assume that $f \in V$, and let $u = u(f)$ denote the solution of (7.68)–(7.70). For an arbitrary constant $c \in \mathbf{R}$, consider the function $v = cu$. Clearly

$$\nabla \cdot (M\nabla(cu)) = c\nabla \cdot (M\nabla u) = cf, \tag{7.81}$$

$$M\nabla(cu) \cdot n = cM\nabla u \cdot n = 0 \ \text{ along } \partial\Omega, \tag{7.82}$$

and

$$\int_\Omega cu \, dx = c\int_\Omega u \, dx = 0. \tag{7.83}$$

This means that $v = cu$ satisfies the following set of equations

$$\nabla \cdot (M\nabla v) = cf \ \text{ in } \Omega,$$

$$M\nabla v \cdot n = 0 \ \text{ along } \partial\Omega,$$

$$\int_\Omega v \, dx = 0.$$

Thus, since

$$(cu)\Big|_{\partial\Omega} = c \cdot u\Big|_{\partial\Omega},$$

we conclude from (7.81)–(7.83) that

$$Q(cf) = cQ(f) \tag{7.84}$$

for any $c \in \mathbf{R}$ and $f \in V$.

Next, consider the boundary value problems

$$\nabla \cdot (M\nabla u_1) = f_1 \ \text{in } \Omega,$$

$$M\nabla u_1 \cdot n = 0 \ \text{along } \partial\Omega,$$

$$\int_\Omega u_1 \, dx = 0,$$

and

$$\nabla \cdot (M\nabla u_2) = f_2 \ \text{in } \Omega,$$

$$M\nabla u_2 \cdot n = 0 \ \text{along } \partial\Omega,$$

$$\int_\Omega u_2 \, dx = 0,$$

where f_1 and f_2 are functions in V; see (7.76). Clearly,

$$\nabla \cdot (M\nabla(u_1 + u_2)) = \nabla \cdot (M\nabla u_1) + \nabla \cdot (M\nabla u_2) = f_1 + f_2 \ \text{ in } \Omega,$$

$$M\nabla(u_1 + u_2) \cdot n = M\nabla u_1 \cdot n + M\nabla u_2 \cdot n = 0 \ \text{ along } \partial\Omega,$$

$$\int_\Omega u_1 + u_2 \, dx = \int_\Omega u_1 \, dx + \int_\Omega u_2 \, dx = 0,$$

and hence it follows that $v = u_1 + u_2$ solves the problem

$$\nabla \cdot (M\nabla v) = f_1 + f_2 \text{ in } \Omega,$$
$$M\nabla v \cdot n = 0 \text{ along } \partial\Omega,$$
$$\int_\Omega v \, dx = 0.$$

This means that

$$Q(f_1 + f_2) = Q(f_1) + Q(f_2),$$

and by combining this property with (7.84) we conclude that Q is a linear mapping, i.e.

$$Q(a_1 f_1 + a_2 f_2) = a_1 Q(f_1) + a_2 Q(f_2) \tag{7.85}$$

for any real numbers a_1 and a_2 and any functions $f_1, f_2 \in V$.

Discretization. In order to solve the inverse problem (7.80) on a computer, it must be discretized by suitable numerical methods. We will now apply the famous Rayleigh-Ritz-Galerkin idea to this equation. However, prior to doing so, we must discuss a practical issue highly relevant for the framework presented above.

Our goal is to solve, at least approximately, equation (7.80). In this equation, d represents an observation of the electrical potential along the surface $\partial\Omega$ of the torso Ω. Above, we assumed that d was a function defined at every point along the boundary $\partial\Omega$ of the torso. However, from a practical point of view, it is only possible to record the electrical potential at a finite number of locations (referred to as the leads), say, in the regions

$$\Gamma_1, \Gamma_2, \ldots, \Gamma_m \subset \partial\Omega.$$

From now on, we will let d^m denote the observed electrical potential at these locations, i.e.

$$d^m = (d_{\Gamma_1}, d_{\Gamma_2}, \ldots, d_{\Gamma_m})^T \in \mathbf{R}^m,$$

where, for $j = 1, \ldots, m$, and d_{Γ_j} represents the measured potential in the region Γ_j. We assume that each d_{Γ_j} is constant over the area Γ_j.

In Section 7.2.3 a discrete approximation of *the inverse problem of electrocardiography* was derived by minimizing the deviation between the simulated and the observed electrical potentials along the body surface, cf. (7.50). In the present situation, we will apply a somewhat different approach. A discrete approximation of (7.80) will be obtained by comparing the arithmetic average of the simulated potential at the leads with the observation data.

Recall the definition (7.77)–(7.78) of the operator Q. The associated mapping Q^m, with discrete range,

$$Q^m : V \to \mathbf{R}^m, \tag{7.86}$$

is defined by

$$Q^m(f) = (Q(f)_{\Gamma_1}, Q(f)_{\Gamma_2}, \ldots, Q(f)_{\Gamma_m})^T \in \mathbf{R}^m \tag{7.87}$$

for $f \in V$. Here, $Q(f)_{\Gamma_j}$ represents the arithmetic average of $Q(f)$ in Γ_j, i.e.

$$Q(f)_{\Gamma_j} = \frac{1}{|\Gamma_j|} \int_{\Gamma_j} Q(f) \, dx = \frac{1}{|\Gamma_j|} \int_{\Gamma_j} u(f) \, dx \quad \text{for } j = 1, \ldots, m.$$

By straightforward considerations, it follows that the linearity of Q is inherited by Q^m.

How can $Q^m(f)$, for a given f, be computed? This can easily be accomplished by first solving the boundary value problem (7.68)–(7.70) for u, and thereafter recording the function values of u in the regions $\Gamma_1, \ldots, \Gamma_m$, positioned along the boundary $\partial\Omega$ of the torso Ω.

With this notation to hand, we can approximate equation (7.80) by the problem of finding $f \in V$ such that

$$Q^m(f) = d^m. \tag{7.88}$$

The Rayleigh-Ritz-Galerkin methodology can now be applied to this latter problem as follows. Let

$$g_1, g_2, \ldots, g_n \in V$$

be n linearly independent functions in V, and define

$$V_n = \text{span}\{g_1, g_2, \ldots, g_n\}. \tag{7.89}$$

If Q_n^m denotes the restriction of Q^m to V_n, i.e.

$$Q_n^m = Q^m\Big|_{V_n} : V_n \to \mathbf{R}^m, \tag{7.90}$$

then the fully discrete approximation of (7.80) is given by this problem: Find $f_n \in V_n$ such that

$$Q_n^m(f_n) = d^m. \tag{7.91}$$

Since Q_n^m is a linear operator on a discrete space, we can realize (7.91) in terms of a matrix-vector linear system.

Matrix-Vector Representation. For every function $f_n \in V_n$ there exists a unique set of real numbers $p_1, p_2, \ldots, p_n$ such that

$$f_n = p_1 g_1 + p_2 g_2 + \cdots + p_n g_n. \tag{7.92}$$

Let $p = (p_1, p_2, \ldots, p_n)^T$ and define the linear mapping

$$B_n^m : \mathbf{R}^n \to \mathbf{R}^m$$

by

$$B_n^m(p) = Q_n^m(p_1 g_1 + p_2 g_2 + \cdots + p_n g_n),$$

where Q_n^m is the operator introduced in (7.90).

Clearly, since Q_n^m is linear, it follows that B_n^m is a linear mapping between Euclidean spaces, and it can thus be represented by a matrix-vector product:

$$B_n^m(p) = Bp,$$

where

$$B = [b_{kl}], \quad k = 1, \ldots, m \text{ and } l = 1, \ldots, n,$$

is a matrix and

$$p = (p_1, p_2, \ldots, p_n)^T$$

is a vector of length n. More precisely, let

$$e_1, e_2, \ldots, e_n$$

be the standard Euclidean basis for $\mathbf{R}^n$. Then the columns of B are given by the formula

$$\begin{pmatrix} b_{1l} \\ b_{2l} \\ \vdots \\ b_{ml} \end{pmatrix} = B_n^m(e_l) \text{ for } l = 1, \ldots, n,$$

see any introductory text on linear algebra.

Thus, we conclude that each column of the matrix B can be computed by solving (7.68)–(7.70) with a source term f that equals the l'th basis function of V_n; see (7.89). That is, by solving (7.68)–(7.70) with

$$f = g_l$$

for $u = u(g_l)$, and by putting

$$b_{kl} = \frac{1}{|\Gamma_k|} \int_{\Gamma_k} u(g_l) \, dx \quad \text{for } k = 1, \ldots, m.$$

This means that n elliptic boundary value problems have to be solved in order to determine the matrix entries of B.

Having introduced this formalism, we may replace equation (7.91) by the linear system

$$Bp = d^m, \tag{7.93}$$

where $B \in \mathbf{R}^{m,n}, p \in \mathbf{R}^n, d^m \in \mathbf{R}^m$.

Recall that m represents the number of locations along the boundary of the torso at which we record the electric potential, and that n is the number of parameters used to describe the unknown source f. Of course, in most applications $m \neq n$, and the system (7.93) may be over- or underdetermined.

For the sake of simplicity, let us now assume that

$$m \geq n.$$

In this case, it is natural to consider the normal equations associated with (7.93),

$$B^T Bp = B^T d^m, \tag{7.94}$$

see, e.g., Golub [49]. This is a system of n equations with n unknowns $p_1, p_2, \ldots, p_n$, and $B^T B$ is a symmetric matrix with non-negative eigenvalues.

In the numerical experiments section below, we will see that the system (7.94) is extremely ill-conditioned, thus reflecting the ill-posed nature of the inverse problem (7.80) at hand. Consequently, some sort of regularization scheme must be applied. Let us consider the zero-order Tikhonov method for this problem.

It is well-known (see, e.g., Golub [49]), that the solution of (7.94) also solves the minimization problem

$$\min_{p \in \mathbf{R}^n} \| Bp - d^m \|^2,$$

where $\| \cdot \|$ represents the Euclidean norm on $\mathbf{R}^m$. Therefore, Tikhonov regularization can be applied by considering a cost-functional of the form

$$J(p) = \| Bp - d^m \|^2 + \epsilon \|p\|^2 \quad \text{for } p \in \mathbf{R}^n, \tag{7.95}$$

where ϵ is a small positive parameter. Minimizing J leads to the following regularized approximation of (7.94),

$$(B^T B + \epsilon I)p = B^T d^m, \tag{7.96}$$

see, e.g., Hämmelin and Hoffmann [60].

7.3.4   A Numerical Example

To illustrate the techniques described above we study the following problem. The domain is a square $\Omega = [-50, 50]^2$, and the conductivity is set to $M = 1$. The observations are recorded at $m = 124$ equidistant locations around the boundary $\partial \Omega$ of Ω. The basis V_n (see equation (7.89)), for the dipole consists of functions of the form

$$g(x; x_i) = \exp^{-\|x - x_i\|^2 / \gamma^2} \quad \text{for } i = 1, \ldots, n,$$

where the centres $\{x_i\}_{i=1}^n$ are given points. The basis functions are positioned in a uniform lattice defined over Ω. If the spacing of the lattice is h, the width is set to $\gamma = h/2\sqrt{\log(2)}$. This ensures that neighbouring functions have suitable overlapping supports, specifically

$$g((h/2, 0); (0, 0)) = g((h/2, 0); (h, 0)) = \frac{1}{2}.$$

Figure 7.9 shows a one-dimensional cross section at $y = 0$, with $h = 1$, of the basis functions centreed at $(1, 0)$ and $(2, 0)$.

First, we investigate the case where the dipole can be expressed exactly as a linear combination of the basis functions in V_n. We use a spacing $h = 20$ and $n = 36$ basis functions. The basis functions centred at $x_i = (-10, -10)$ and $x_j =$

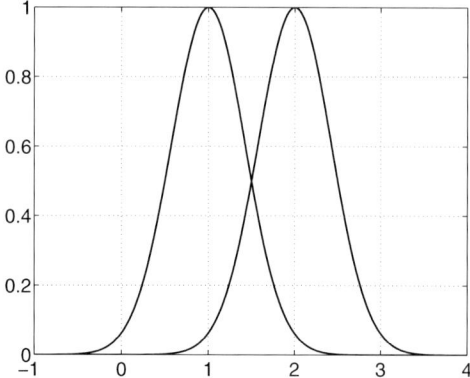

Fig. 7.9. A one-dimensional view of two neighbouring basis functions.

$(10, 10)$ are given the weights $p_i = 1$ and $p_j = -1$, respectively. The rest of the entries of $p = (p_1, \ldots, p_n)^T$ are set to zero; see equation (7.92). Figure 7.11a) shows this dipole f, and the corresponding solution of the direct problem (7.68)–(7.70) is shown in Figure 7.10.

In this case, the approximate solution of the inverse problem (7.80), defined by the normal equations (7.94), reproduces p almost exactly. The error in the nonzero components of $p = (p_1, \ldots, p_n)^T$ is less than 0.1%. This is as expected, since the data is noise free.

Is it also possible to recover the position of the dipole in the case of noisy data? We denote the noisy signal by $\hat{d}^m$, i.e.

$$\hat{d}^m = d^m + e_\rho, \tag{7.97}$$

where the elements of e_ρ are drawn from a normal distribution with zero mean and standard deviation

$$\sigma = \rho\sqrt{\pi/2} \cdot |\bar{d}|,$$

where

$$|\bar{d}| = \frac{1}{m} \sum_{i=1}^{m} |d_i|.$$

This ensures that the expected magnitude of the noise, $E(|e_\rho|)$, is a given fraction ρ of the magnitude of the signal, i.e.

$$E(|e_\rho|) = \rho|\bar{d}|.$$

Figure 7.10b) shows the body surface potential, i.e. the observation data d^m, along with the added 1% noise (corresponding to $\rho = 0.01$). The estimated dipole obtained by solving the normal equations (7.94) is depicted in Figure 7.11b). It is clear that, in this case, this method does not work well when noise is present. The ill-posed nature of the problem becomes dominant.

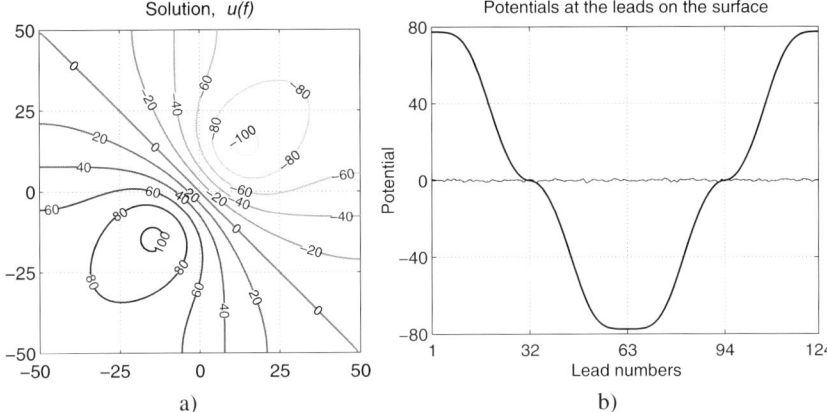

Fig. 7.10. a) The potential distribution throughout $\Omega = [-50, 50]^2$ generated by the source term shown in Figure 7.11a). **b)** The body surface potential generated by the source term shown in Figure 7.11a) (thick line). Noise is added to this signal as described in equation (7.97). The thin line represents the noise (1%). Leads 1, 32, 63, and 94 are located at the corners of the domain Ω.

Next, we apply the Tikhonov scheme derived in equations (7.95)–(7.96) to this problem. Recall that the level of regularization is controlled by the parameter $\epsilon > 0$. In general, it is not easy to determine a good value for ϵ; see, e.g. Skipa, Sachse and Dössel [127], Franzone, Guerri, Taccardi and Viganotti [42], Franzone, Guerri, Tentonia, Viganotti and Baruffi [43], Miller [103] and Hansen and O'Leary [61]. However, in the present case, we are fortunate in that the exact solution is known to us! We can, therefore, evaluate the performance of this regularization technique with respect to various choices of ϵ. To this end, we consider the relative error

$$r(\epsilon) = \frac{\int_\Omega |f(x; p(\epsilon; \hat{d}^m)) - f(x; p^*)| \, dx}{\int_\Omega |f(x; p^*)| \, dx}$$

of the estimated dipole. Here, p^* represents the parameters of the true/correct dipole and $p = p(\epsilon; \hat{d}^m)$ is the estimate that depends upon the noisy data $\hat{d}^m$ and the regularization parameter ϵ.

Figure 7.12a) shows this relative error of the estimated dipole as a function of ϵ. For large values of ϵ, the solution becomes too smooth, and the relative error is large. An optimal value seems to be given by $\epsilon \approx 2.4 \cdot 10^{-4}$. Even for this choice of ϵ, $r(\epsilon)$ is not small. In fact, 1% noise in the observation data leads to an error of at least $\approx 38\%$ in the computed position of the dipole; the problem is ill-posed! Figure 7.11c) shows the estimated dipole for the optimal value of ϵ.

Finally, for various values of the noise level ρ, we computed the relative error generated by the optimal choice of ϵ. The results are shown in Figure 7.12b). Observe that, even with an optimal regularization parameter, the error becomes large

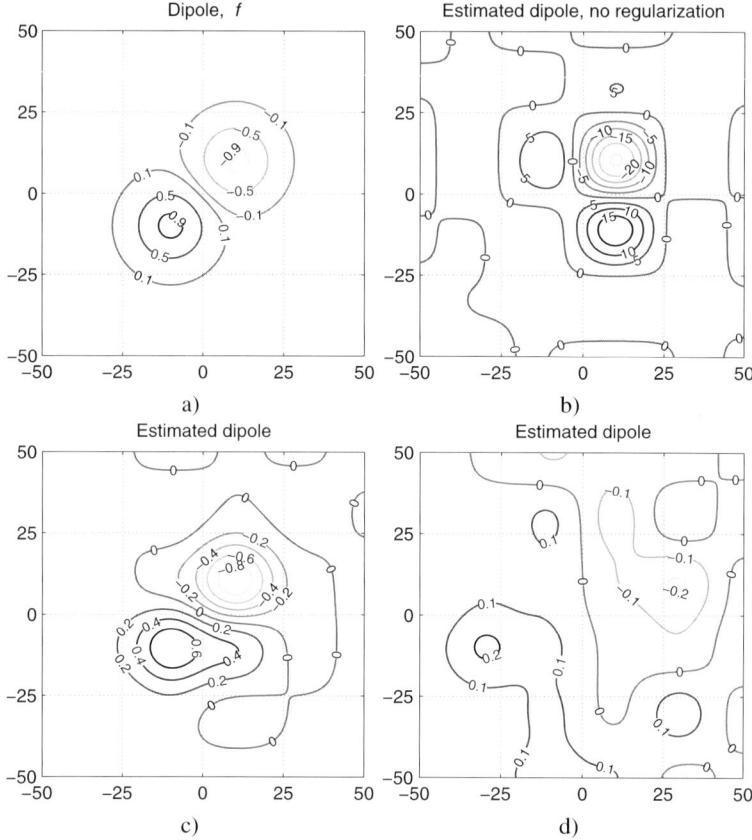

Fig. 7.11. a) The chosen dipole f. **b)** The computed dipole with 0.1% noise added to the observation data, no regularization. **c)** As in b), but Tikhonov regularization with $\epsilon = 2.4 \cdot 10^{-4}$ has been used. **d)** The computed dipole with 1% noise added to the observation data. Optimal Tikhonov regularization, $\epsilon = 6.3$, has been applied.

for relatively moderate levels of noise. Figure 7.11d) shows the estimated dipole in the case of 1% noise.

We may conclude that parameterizing a dipole in terms of a set of basis functions covering Ω may not provide very accurate results for the inverse problem at hand. The performance of the numerical schemes derived in this fashion will tend to be dominated by the ill-posed nature of equation (7.80). Even with only 1% noise, we were not able to accurately recover the position of the source term f. Furthermore, the solutions we computed did not have the typical characteristic form of a dipole. In the next section, a parameterization that enforces a dipole structure is proposed. The drawback of this approach is that it leads to a system of nonlinear equations for (approximately) solving (7.80).

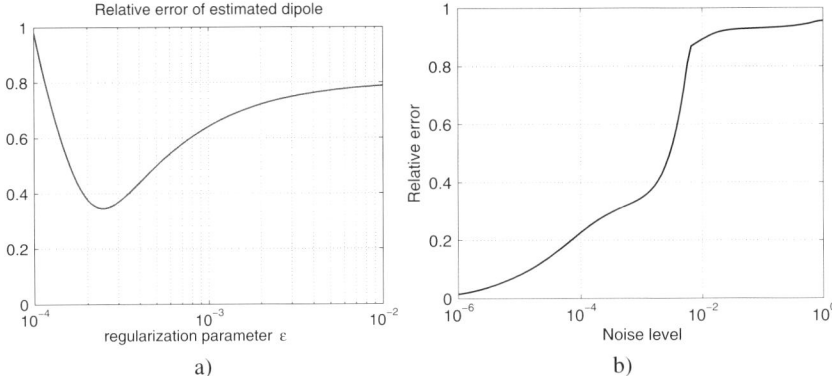

Fig. 7.12. a) The relative error of the computed dipole as a function of the regularization parameter ϵ. **b)** The relative error of the estimated dipole as a function of the noise level ρ. For each value of ρ, the error generated by the optimal value of ϵ is plotted. The "sudden shift" that appears close to $\rho = 0.01$ in the graph is caused by a jump in the optimal value for ϵ at that point.

7.3.5 Parameterizations Leading to Nonlinear Problems

We have seen that the solution u of (7.68)–(7.70) depends linearly on the source term f. Consequently, if f is approximated by a finite linear sum of the form (7.92), then we may formulate the inverse problem (7.80) as a linear system of the form (7.94). However, in order to compute the matrix entries of B, we have to solve n elliptic boundary value problems of the form (7.68)–(7.70), where n represents the number of basis functions that are used to discretize f. Consequently, if a large number of basis functions are needed to represent f accurately, this methodology might become quite CPU-expensive. In addition, as n grows, the condition number of $B^T B$ becomes very large and the ill-posed nature of the problem becomes dominant. Thus, it seems to be important to keep the number of parameters $p_1, p_2, \ldots, p_n$ at a minimum.

Let us now consider a somewhat different approach to this problem. Keep in mind that we are searching for a dipole, and not some sort of general source term. Somehow, it should be possible to exploit this *a priori* information available to us. We will now use this insight to reduce the number of parameters n that are used to characterize the dipole.

For example, assuming that we consider a two-dimensional problem, it seems to be reasonable that a dipole can be accurately represented by

$$f(x,y) = e^{-[(x-p_1)^2+(y-p_2)^2]/\gamma^2} - e^{-[(x-p_3)^2+(y-p_4)^2]/\gamma^2}, \qquad (7.98)$$

where (p_1, p_2) and (p_3, p_4) represent the centres of the two poles in the dipole, and γ their radius. However, if parameterizations of this kind are applied, f cannot be expressed as an elementary linear sum of basis functions. That is, f cannot in general be approximated by a function f_n of the form (7.92). Hence, the mathematical framework presented above cannot be used. The purpose of this section is to derive methods suitable for implementing this kind of strategy.

As above, the dipole f is parameterized by the parameters $p_1, p_2, \ldots, p_n$, i.e.

$$f = f(x; p_1, \ldots, p_n).$$

We will also assume that this function is sufficiently regular that, for $i = 1, \ldots, n$,

$$\frac{\partial}{\partial p_i} \int_\Omega f\phi \, dx = \int_{\Omega_{p_i}} f_{p_i}\phi \, dx$$

for every test function ϕ. Here, f_{p_i} denotes

$$\frac{\partial f}{\partial p_i}, \quad i = 1, \ldots, n.$$

Recall the definition (7.77)–(7.78) of the mapping Q. A reasonable criterion for determining the parameters $p_1, p_2, \ldots, p_n$, seems to be to require that the error in the simulated electrical potential should be as small as possible at the leads $\Gamma_1, \ldots, \Gamma_m$. With this in mind, we introduce the cost-functional

$$J = J(p_1, \ldots, p_n) = \sum_{j=1}^m \int_{\Gamma_j} [Q(f(p_1, \ldots, p_n))(x) - d_j]^2 \, dx$$

$$= \sum_{j=1}^m \int_{\Gamma_j} [u(x; p_1, \ldots, p_n) - d_j]^2 \, dx, \quad (7.99)$$

and the associated minimization problem

$$\min_{(p_1, \ldots, p_n) \in \mathbf{R}^n} J(p_1, \ldots, p_n). \quad (7.100)$$

Hence, our goal is to solve (7.100), using a moderate number n of parameters, and thereby recover the position of the dipole f.

The Role of the Undetermined Constant in the Solution of the Neumann Problem. Assume that u solves the homogeneous Neumann problem

$$\nabla \cdot (M\nabla u) = f \quad \text{in } \Omega, \quad (7.101)$$
$$(M\nabla u) \cdot n = 0 \quad \text{along } \partial\Omega. \quad (7.102)$$

Then any function of the form $u + c$, where c is an arbitrary constant, also satisfies these two equations. The condition

$$\int_\Omega u \, dx = 0 \quad (7.103)$$

was added above to this boundary value problem in order to obtain a complete set of equations, i.e. to enforce the uniqueness of its solution. From a mathematical point of view, (7.103) is convenient. If this condition is used along with a parameterization of the form (7.92), then the operator Q, defined in (7.77)–(7.78), becomes linear. Consequently, the methodology developed in Section 7.3.3 can be applied.

However, from a biological point of view, there is no reason to require that (7.103) should hold. This means that we are "free" to use the constant c to minimize the deviation between the observation data and the simulated potential along $\partial\Omega$.

Let u be an arbitrary integrable function defined in $\overline{\Omega} = \Omega \cup \partial\Omega$, and consider the function

$$\hat{u} = u + c,$$

where $c \in \mathbf{R}$. Inspired by (7.99), we introduce a function on the form

$$g(c) = \sum_{j=1}^{m} \int_{\Gamma_j} [\hat{u} - d_j]^2 \, dx = \sum_{j=1}^{m} \int_{\Gamma_j} [u + c - d_j]^2 \, dx \quad \text{for } c \in \mathbf{R}.$$

From the criterion

$$g'(c) = 0$$

for a minimum, we obtain the following formula:

$$c = \frac{1}{m|\Gamma_1|} \left[\sum_{j=1}^{m} \int_{\Gamma_j} d_j \, dx - \sum_{j=1}^{m} \int_{\Gamma_j} u \, dx \right], \qquad (7.104)$$

provided that the leads are of uniform size, i.e.

$$|\Gamma_1| = \ldots = |\Gamma_m|.$$

Next, the sum of the integrals of the function

$$\hat{u} = u + \frac{1}{m|\Gamma_1|} \left[\sum_{j=1}^{m} \int_{\Gamma_j} d_j \, dx - \sum_{j=1}^{m} \int_{\Gamma_j} u \, dx \right]$$

over the leads is

$$\sum_{j=1}^{m} \int_{\Gamma_j} \hat{u} \, dx = \sum_{j=1}^{m} \int_{\Gamma_j} d_j \, dx.$$

Consequently, we will throughout this section apply a condition of the form

$$\sum_{j=1}^{m} \int_{\Gamma_j} u \, dx = \sum_{j=1}^{m} \int_{\Gamma_j} d_j \, dx \qquad (7.105)$$

instead of (7.103). That is, we consider the boundary value problem that consists of equations (7.101)–(7.102) and (7.105).

A condition of the form (7.105) cannot be applied in the technique described in Section 7.3.3. More precisely, if (7.105) is used, the linearity of Q is lost. It is not sufficient to simply use a parameterization of the form (7.92), cf. the mathematical derivations of the linearity of the operator Q. On the other hand, an approximation of a dipole, already computed by the scheme (7.94), can easily be improved by adding the constant (7.104). Thus, our analysis is invoked as a post-process. We will not dwell further upon this issue, but instead focus on the nonlinear approach introduced above.

Differentiation of the Cost-Functional. Many minimization algorithms for solving problems of the form (7.100) require the partial derivatives

$$\frac{\partial J}{\partial p_1}, \ldots, \frac{\partial J}{\partial p_n}$$

of the cost-functional J. This is, for instance, the case for the Method of Steepest Descent and for Newton's method; see, e.g. Luenberger [86]. Of course, these partial derivatives can be computed in a straightforward manner by finite differences. That is, if $\{\Delta p_i\}_{i=1}^n$ are small numbers, then we may define the approximations

$$\frac{\partial J}{\partial p_i} \approx \frac{J(p_1, \ldots, p_i + \Delta p_i, \ldots, p_n) - J(p_1, \ldots, p_i, \ldots, p_n)}{\Delta p_i}$$

for $i = 1, \ldots, n$. However, this approach requires the solution of $n + 1$ boundary value problems of the form (7.101), (7.102), (7.105); see (7.99). More precisely, the following values

$$J(p_1, \ldots, p_n), \ J(p_1 + \Delta p_1, \ldots, p_n), \ J(p_1, \ldots, p_n + \Delta p_n),$$

of the cost-functional are needed to approximately determine the gradient ∇J of J at the point $(p_1, \ldots, p_n)$. For fairly large values of n, this may lead to an unacceptable workload.

Can we perform this task in a more efficient way? Yes, it turns out that by introducing a single auxiliary problem, all of the partial derivatives of J can be computed by solving one extra elliptic boundary value problem. This approach, frequently referred to as the adjoint method, can be derived as follows.

From (7.99) we find that

$$\frac{\partial J}{\partial p_i} = 2 \sum_{j=1}^{m} \int_{\Gamma_j} [u(x) - d_j] \, u_{p_i}(x) \, dx, \tag{7.106}$$

where u_{p_i} denotes the partial derivative of u with respect to p_i, i.e.

$$u_{p_i} = \frac{\partial u}{\partial p_i}.$$

Let $H^1(\Omega)$ denote the classical Sobolev space of square-integrable functions with square-integrable weak derivatives defined on Ω, i.e.

$$H^1(\Omega) = \left\{ \phi \in L^2(\Omega); \int_{\Omega} |\nabla \phi|^2 \, dx < \infty \right\},$$

where $\nabla\phi$ is defined in terms of the distributional derivatives of ϕ. Then, the weak form of the problem (7.101), (7.102), (7.105) can be written in the following form. Find u such that

$$\int_\Omega \nabla\phi \cdot (M\nabla u)\, dx = -\int_\Omega f\phi\, dx \quad \text{for all } \phi \in H^1(\Omega), \qquad (7.107)$$

$$\sum_{j=1}^m \int_{\Gamma_j} u\, dx = \sum_{j=1}^m \int_{\Gamma_j} d_j\, dx. \qquad (7.108)$$

Next, if we differentiate with respect to p_i in (7.107)–(7.108), we find that u_{p_i} must satisfy the equations

$$\int_\Omega \nabla\phi \cdot (M\nabla u_{p_i})\, dx = -\int_\Omega f_{p_i}\phi\, dx \quad \text{for all } \phi \in H^1(\Omega), \qquad (7.109)$$

$$\sum_{j=1}^m \int_{\Gamma_j} u_{p_i}\, dx = 0. \qquad (7.110)$$

Note that (7.109)–(7.110) hold for $i = 1, \ldots, n$, and thus the equations satisfied by the partial derivatives $u_{p_1}, \ldots, u_{p_n}$ of u with respect to the parameters $p_1, \ldots, p_n$ are very similar; only the right hand side of (7.109) depends on which of the derivatives are considered! We will now see how this observation can be exploited to derive an efficient scheme for computing all of the partial derivatives of our cost-functional J.

Consider the auxiliary problem: Find w such that

$$\int_\Omega \nabla\phi \cdot (M\nabla w)\, dx = 2\sum_{j=1}^m \int_{\Gamma_j} [u(x) - d_j]\,\phi(x)\, dx \text{ for all } \phi \in H^1(\Omega), \; (7.111)$$

$$\int_\Omega w\, dx = 0. \qquad (7.112)$$

If w solves this problem, then by putting $\phi = u_{p_i}$ in (7.111) it follows that

$$\int_\Omega \nabla u_{p_i} \cdot (M\nabla w)\, dx = 2\sum_{j=1}^m \int_{\Gamma_j} [u(x) - d_j]\, u_{p_i}(x)\, dx, \qquad (7.113)$$

and from (7.106) we find that

$$\frac{\partial J}{\partial p_i} = 2\sum_{j=1}^m \int_{\Gamma_j} [u(x) - d_j]\, u_{p_i}(x)\, dx = \int_\Omega \nabla u_{p_i} \cdot (M\nabla w)\, dx. \qquad (7.114)$$

Now, recall that M is a symmetric tensor. Therefore

$$\frac{\partial J}{\partial p_i} = \int_\Omega \nabla w \cdot (M\nabla u_{p_i})\, dx,$$

and by choosing $\phi = w$ in equation (7.109) we conclude that

$$\frac{\partial J}{\partial p_i} = -\int_\Omega f_{p_i} w \, dx \quad \text{for } i = 1, \ldots, n. \tag{7.115}$$

This means that all of the partial derivatives

$$\frac{\partial J}{\partial p_1}, \ldots, \frac{\partial J}{\partial p_n}$$

of the cost-functional J, at a point $(p_1, \ldots, p_n)$, can be computed by the following procedure:

1. Solve equations (7.107)–(7.108) for u

2. Solve equations (7.111)–(7.112) for w

3. Use formula (7.115) to compute the partial derivatives of J

7.3.6 A Numerical Experiment

The numerical experiment described here is similar to the one studied in Section 7.3.4. A source term f for (7.68) of the form (7.98) is selected by picking values for the parameters $p_1, \ldots, p_4$. Thereafter, the corresponding forward problem (7.101), (7.102), (7.105) is solved and the solution of these equations at the boundary $\partial\Omega$ provides the "artificial" observation data d for the inverse problem (7.80). As explained above, this inverse problem is not linear with the parameterization given in (7.98). Solving (7.100) thus leads to a nonlinear optimization problem.

Several types of minimization algorithm exist. Here, we will compare the steepest decent algorithm with Nelder-Mead's simplex method. The first method uses gradient information; the latter does not. Since we have derived an efficient method for computing the gradient of the cost-functional J, it will be interesting to compare these two methods.

A reference solution was generated by choosing $p = (-1, 0, 1, 0)$, i.e. a dipole with centres at $(-1, 0)$ and $(1, 0)$. The minimization algorithms can be sensitive to the initial guess, or the starting point, of the search. They might get stuck in a local minimum instead of converging to the global solution. To investigate the sensitivity of the starting points, ten different initial guesses were drawn randomly from a normal distribution with zero mean and a standard deviation of 10.

The Simplex method converged to the correct solution in 50% of the cases. The average number of function evaluations for the successful searches was 340. The steepest decent method converged for all 10 initial guesses, and about 50 function evaluations were needed to obtain convergence.

The experiments were repeated twice, with 1% and 10% noise added to the "artificial" data d. The results were very similar to those obtained in the noise-free cases. The main difference was that the steepest decent algorithm failed to identify the correct parameters in one out of the ten cases.

Overall, the steepest decent algorithm performed very well. The reason for that is that the cost-functional J is well-behaved; Figure 7.13 shows two cross sections of this function.

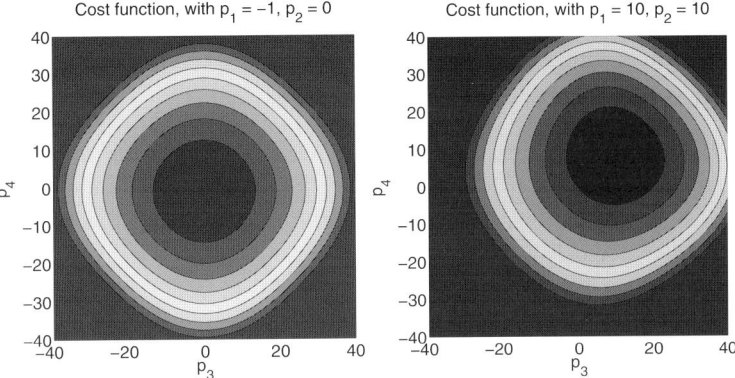

Fig. 7.13. The cost-functional J depends upon four parameters: p_1, p_2, p_3, p_4. In order to make a 2D visualization of J, p_1 and p_2 are kept fixed. The plot on the left shows a cross section with the correct values for p_1 and p_2, i.e. $p_1 = -1$ and $p_2 = 0$. The plot on the right shows a cross section with $p_1 = p_2 = 10$, i.e. very different from the correct values. A steepest decent algorithm will converge quickly to the correct values for p_3 and p_4 in the first case, but not in the latter. This shows that the parameters cannot be computed independently. Both plots suggest that the four-dimensional cost-functional J is convex and ideally suited for a steepest descent search. (For the color version, see Figure A.14 on page 294).

7.4 Computing the Size and Location of a Myocardial Infarction

In the Western world, ischemic heart disease is one of the most widespread illnesses, killing approximately one million people every year [3,143,106]. In many cases, traditional ECG based methods will fail to detect the ischemia [80]. Moreover, ECG recordings provide only a rather crude picture of the position and size of infarctions [12]. Thus, there is a need for improving this technology.

Based on the theory presented in the previous chapters, one might feel tempted to ask the following question:

> *Is it possible to use mathematical models and computer simulations to*
>
> *determine the size and location of a myocardial infarction?*   (7.116)

This section is devoted to this extremely challenging and important problem. More precisely, we will present a conceptual framework, based on the models presented in Chapter 2, for investigating the possibilities of computing the physical characteristics of an ischemia.

To the knowledge of the authors, the challenge (7.116) has not previously been formulated as an inverse problem for the bidomain equations. Thus, our goal is to derive a parameter identification problem for (2.65)–(2.72), and hence propose a methodology for analyzing (7.116). This will lead to a highly nonlinear and ill-posed optimization problem. The potential of this framework will be illuminated through

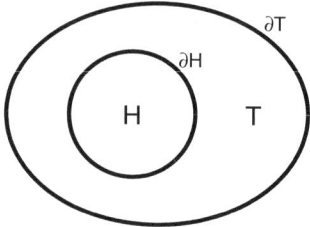

Fig. 7.14. The domains H and T.

a series of numerical experiments. Moreover, we will derive an efficient technique for computing the partial derivatives, needed by several optimization schemes, of the involved cost-functional.

We would like to emphasize that we do not provide any complete solution to the challenge (7.116). This section merely contains some preliminary investigations and results concerning future possibilities for using computers to detect and characterize myocardial infarctions. Further information about this issue can be found in [89,91].

7.4.1   Modelling Infarctions, the Direct (Forward) Problem

Consider the model (2.65)–(2.72) for the electrical activity in the heart H and in the surrounding body T, cf. Figure 7.14. For the sake of simplicity, we will throughout this section use a somewhat simplified model, and assume that the ionic current across the cell membrane I_{ion} depends only on the transmembrane potential v, i.e.

$$I_{\mathrm{ion}} = I_{\mathrm{ion}}(v).$$

In principle, it is not very difficult to do similar considerations in the presence of complex cell-models, but we prefer the simplified model in order to reduce the complexity. This leads to the mathematical model

$$\frac{\partial v}{\partial t} + I_{\mathrm{ion}}(v) = \nabla \cdot (M_i \nabla v) + \nabla \cdot (M_i \nabla u_e) \quad \text{in } H, \qquad (7.117)$$

$$\nabla \cdot (M_i \nabla v) + \nabla \cdot ((M_i + M_e)\nabla u_e) = 0 \quad \text{in } H, \qquad (7.118)$$

$$\nabla \cdot (M_o \nabla u_o) = 0 \quad \text{in } T, \qquad (7.119)$$

$$(7.120)$$

and suitable interface and boundary conditions specified along ∂H and ∂T, respectively.

These equations, as explained in Chapter 2, may serve as a model for the propagation of the electrical potential in an individual with a healthy heart. In order to simulate the effects of an ischemia on the ECG signal, (7.117)–(7.119) must be changed. To this end, let us introduce the n parameters

$$p_1, p_2, \dots, p_n$$

to represent the geometrical characteristics of an infarction. These variables will, in what follows, be referred to as the *infarction parameters*.

According to biological observations, the cells in an infarcted region in the heart are not excitable. This means that the tissue in such a zone behaves like a passive conductor. Consequently, an ischemia can be incorporated into the model (7.117)–(7.119) by removing the ion transport in the infarcted area. In mathematical terms, we will use a function

$$g = g(x, t; p_1, p_2, \ldots, p_n)$$

to model this effect. More precisely, the term I_{ion}, in equation (7.117), is simply replaced by $gI_{\text{ion}}(v)$, which leads to the model

$$\frac{\partial v}{\partial t} + gI_{\text{ion}}(v) = \nabla \cdot (M_i \nabla v) + \nabla \cdot (M_i \nabla u_e) \quad \text{in } H, \qquad (7.121)$$

$$\nabla \cdot (M_i \nabla v) + \nabla \cdot ((M_i + M_e) \nabla u_e) = 0 \quad \text{in } H, \qquad (7.122)$$

$$\qquad (7.123)$$

$$(M_i \nabla v + M_i \nabla u_e) \cdot n = 0 \quad \text{on } \partial H, \qquad (7.124)$$

$$u_e = u_o \quad \text{on } \partial H, \qquad (7.125)$$

$$M_e \nabla u_e \cdot n = M_o \nabla u_o \cdot n \quad \text{on } \partial H, \qquad (7.126)$$

$$\qquad (7.127)$$

$$\nabla \cdot M_o \nabla u_o = 0 \quad \text{in } T, \qquad (7.128)$$

$$M_o \nabla u_o \cdot n = 0 \quad \text{on } \partial T. \qquad (7.129)$$

Typically, the function g will be very small in infarcted regions, and elsewhere approximately equal to one. Throughout this section we will assume that g is a given smooth function that depends on both the spatial position x and the infarction parameters $p_1, \ldots, p_n$.

Example 8. Let the infarcted area of the heart be denoted by M, and assume that it can characterized by the function $F(x; p_1, p_2, \ldots, p_n)$ in the following way:

$$M = \{x : F(x) < 0\}.$$

For convenience, we further assume that the minimal value of F is -1. In view of the discussion above, the function g should be equal to one in the healthy part of the heart tissue and zero in the centre of the infarcted area. A possible choice for g is then:

$$g(x) = \frac{1 + \tanh(F(x)/\tau)}{2}$$

Here, $\tau > 0$ is a small parameter that controls the steepness of the transition between health and infarcted tissue.

In the numerical experiments below, we will let M be a circle with radius r and centred at the point (a, b), i.e.

$$F(x) = \left(\frac{x_1 - a}{r}\right)^2 + \left(\frac{x_2 - b}{r}\right)^2 - 1.$$

◇

The direct problem, in the present context, can now be described as follows. For a given set of infarction parameters $p_1, p_2, \ldots, p_n$, compute the effect of the infarction on the ECG signal by solving (7.121)–(7.129). Note that the solution of this direct problem depends on the parameters used to describe the properties of the infarction, i.e.,

$$v = v(x, t; p_1, p_2, \ldots, p_n),$$
$$u_e = u_e(x, t; p_1, p_2, \ldots, p_n),$$
$$u_o = u_o(x, t; p_1, p_2, \ldots, p_n).$$

In this setting, we compute the effect of a given cause, namely the infarction. Thus, we refer to this procedure as the direct, or forward, problem.

For the sake of ease of notation, let us introduce the mapping

$$R : [0, t^*] \times \mathbf{R}^n \to L^2(\partial T),$$

by defining

$$R(t, p_1, p_2, \ldots, p_n) = u_o(x, t; p_1, p_2, \ldots, p_n)|_{\partial T},$$

where $[0, t^*]$ represents the time interval under consideration. That is, R is the solution operator of the forward problem discussed above. Generally, R is a nonlinear function of $n + 1$ real variables. For a given set of infarction parameters $p_1, p_2, \ldots, p_n$, $R(t, p_1, p_2, \ldots, p_n)$ can be determined by a two-step procedure:

- Solve (7.121)–(7.129) in the time interval $[0, t^*]$
- Record, for every $t \in [0, t^*]$, the simulated electrical potential u_o along the boundary ∂T of the body

7.4.2 Modelling Infarctions, the Inverse Problem

Given a set of ECG recordings of a patient, can we determine whether or not this individual suffers from a myocardial infarction? If so, can we somehow recover the geometrical characteristics of the infarction? In the present context, this challenge is what we refer to as the inverse problem (or the parameter identification problem); we want to find the cause of a potentially observable effect.

In mathematical terms, this means that we want to use the data from the ECG recordings to compute the parameters $p_1, \ldots, p_n$ describing the properties of the infarction. The purpose of this section is to define a suitable, and rather general, mathematical framework for doing so. Ideally, this problem could be solved by inverting the operator R defined above. However, computing R^{-1} is, if not impossible, an extremely difficult task. It seems like we must be content with minimizing, with respect to the infarction parameters, the deviation between the simulated potential at the leads and the observation data.

To this end, we need to introduce some notation. Let

$$d_1(t), d_2(t), \ldots, d_m(t)$$

be functions of time, $t \in [0, t^*]$, representing the data obtained from the ECG measurements. Next, let

$$\Gamma_1, \Gamma_2, \ldots, \Gamma_k \subset \partial T$$

denote the leads, used to produce the ECG recordings, positioned along the surface ∂T of the body. In this section, the mean values of a function $\phi(x, t)$, defined along a lead Γ_i, will be denoted by $\phi(\Gamma_i, t)$, i.e.

$$\phi(\Gamma_i, t) = \frac{1}{|\Gamma_i|} \int_{\Gamma_i} \phi(x, t) \, dx \quad \text{for } i = 1, \ldots, k. \tag{7.130}$$

The data recorded by the ECG measurements will typically be of the form[16]

$$d_i(t) \approx \tilde{u}_o(\Gamma_{i+k/2}, t) - \tilde{u}_o(\Gamma_i, t) \quad \text{for } i = 1, \ldots, m, \tag{7.131}$$

assuming that there is an even number k of leads and that $m = k/2$. Here, $\tilde{u}_o(x, t)$, for $x \in \partial T$, represents the actual (physical) electrical potential measured at the surface of the patient's body.

In order to recover the characteristics of a potential infarction from the observation data $\{d_i(t)\}_{i=1}^m$ we introduce an objective function

$$J = J(p_1, \ldots, p_n) : \mathbf{R}^n \to \mathbf{R}^n,$$

by defining

$$J(p_1, \ldots, p_n) = \int_0^{t^*} \sum_{i=1}^m \left[d_i(t) - \left\{ u(\Gamma_{i+\frac{k}{2}}, t) - u(\Gamma_i, t) \right\} \right]^2 dt. \tag{7.132}$$

For a given set of infarction parameters $p_1, \ldots, p_n$, $J(p_1, \ldots, p_n)$ measures the error, or deviation, between the simulation results, obtained by solving the model (7.121)–(7.129), and the data $\{d_i(t)\}_{i=1}^m$ of the ECG recordings. Consequently, we can try[17] to compute the characteristics of the myocardial infarction by solving a minimization problem of the form

$$\min_{(p_1, \ldots, p_n) \in P} J(p_1, \ldots, p_n). \tag{7.133}$$

Here, P represents an admissible set of infarction parameters.

---

[16] ECG recording devices typically measure the difference between the potential at two leads (or the difference between the potential at one specific lead and the average of the potential at the remaining leads). In Sections 7.2 and 7.3 we assumed, for the sake of simplicity, that the observation data was of a simpler form; more precisely, that the electrical potential itself at the leads were measured. This is, however, not usually the case. In the present situation, we want to show that we are able to handle data of the form (7.131). The theory presented in Sections 7.2 and 7.3 can also, in a straightforward manner, be modified to incorporate such cases.

[17] The purpose of the present section is to investigate the possibilities for characterizing an ischemia by this approach!

7.4.3 Differentiation of the Objective Function

In order to solve the problem (7.133), we must apply some sort of optimization procedure. As mentioned in Section 7.3.5, such minimization algorithms often require the partial derivatives of the involved objective function with respect to all of its variables. That is, we need to compute

$$\frac{\partial J}{\partial p_1}, \frac{\partial J}{\partial p_2}, \ldots, \frac{\partial J}{\partial p_n}. \tag{7.134}$$

This can, as already discussed in Section 7.3.5, be accomplished by a straightforward finite difference approach, i.e. by applying the approximations

$$\frac{\partial J}{\partial p_i} \approx \frac{J(p_1, \ldots, p_i + \Delta p_i, \ldots, p_n) - J(p_1, \ldots, p_i, \ldots, p_n)}{\Delta p_i} \tag{7.135}$$

for $i = 1, \ldots, n$. Note that, to compute the fraction (7.135) we must solve our model problem (7.121)–(7.129), not only in the case of the parameter set $p_1, p_2, \ldots, p_n$, but also for the perturbed "infarction" parameters $p_1, \ldots, p_i + \Delta p_i, \ldots, p_n$. Thus, $n + 1$ coupled problems, of the form (7.121)–(7.129), must be solved in order to compute all of the partial derivatives of J at a single point. This leads to an extremely CPU-demanding procedure. For a fairly large number n of infarction parameters, this cannot, within reasonable time limits, be done by even the fastest computers available today.

The Adjoint Problem Approach

Our aim is to develop an efficient technique suitable for differentiating J with respect to the infarction parameters. It turns out that all of the n partial derivatives (7.134) of J can be computed by solving a single auxiliary problem. Magic? No, it follows by a mathematical trick involving the so-called adjoint problem of (7.121)–(7.129). In fact, the derivation presented below, is similar to the argument presented in Section 7.3.5.

In the following, we will write $I(v)$ instead of $I_{ion}(v)$, and we define the function

$$u = \begin{cases} u_e & \text{in } H, \\ u_0 & \text{in } T. \end{cases}$$

With this notation at hand, we may write the model (7.121)–(7.129) on its variational form, also integrating in time,

$$\int_0^{t^*} \int_H v_t \phi \, dx \, dt + \int_0^{t^*} \int_H M_i \nabla v \cdot \nabla \phi \, dx \, dt + \int_0^{t^*} \int_H M_i \nabla u \cdot \nabla \phi \, dx \, dt$$

$$+ \int_0^{t^*} \int_H g I(v) \phi \, dx \, dt = 0 \quad \text{for all } \phi \in V(H), \tag{7.136}$$

$$\int_0^{t^*} \int_H M_i \nabla v \cdot \nabla \psi \, dx \, dt + \int_0^{t^*} \int_H (M_i + M_e) \nabla u \cdot \nabla \psi \, dx \, dt$$

$$+ \int_0^{t^*} \int_T M_o \nabla u \cdot \nabla \psi \, dx \, dt = 0 \quad \text{for all } \psi \in V(H \cup T), \tag{7.137}$$

where $V(H)$ and $V(H\cup T)$ represent suitable spaces of functions defined on H and $H \cup T$, respectively. The derivation of weak forms of systems of partial differential equations was discussed in detail in Chapter 3. Hence, we will not dwell any further upon this issue in the present context.

As mentioned above, since the function $g = g(x,t;p_1,\ldots,p_n)$ depends on the infarction parameters, the solution (u,v) of the system (7.136)–(7.137) will also depend on $p_1,\ldots,p_n$. Let $j \in \{1,\ldots,n\}$ be arbitrary and define, for the sake of simple notation, $p = p_j$. Our goal is to compute the partial derivative of the cost-functional J with respect to the variable p. From (7.132) it follows that

$$\frac{\partial J}{\partial p} = -2\int_0^{t^*} \sum_{i=1}^m \left[d_i(t) - \left\{ u(\Gamma_{i+\frac{k}{2}},t) - u(\Gamma_i,t) \right\} \right] \left[u_p(\Gamma_{i+\frac{k}{2}},t) - u_p(\Gamma_i,t) \right] dt,$$
(7.138)

where

$$u_p(\Gamma_i,t) = \frac{1}{|\Gamma_i|} \int_{\Gamma_i} u_p(x,t)\,dx \quad \text{for } i = 1,\ldots,k,$$
(7.139)

cf. (7.130). In (7.139), $u_p(x,t)$ denotes the partial derivatives of $u(x,t)$ with respect to the infarction parameter p, i.e.

$$u_p(x,t) = \frac{\partial u}{\partial p}(x,t).$$

We differentiate (7.136)-(7.137) with respect to p and find that

$$\int_0^{t^*}\int_H v_{pt}\phi\,dx\,dt + \int_0^{t^*}\int_H M_i\nabla v_p \cdot \nabla\phi\,dx\,dt$$

$$+ \int_0^{t^*}\int_H M_i\nabla u_p \cdot \nabla\phi\,dx\,dt + \int_0^{t^*}\int_H g_p I(v)\phi\,dx\,dt$$

$$+ \int_0^{t^*}\int_H g I'(v)v_p\phi\,dx\,dt = 0 \quad \text{for all } \phi \in V(H),$$
(7.140)

$$\int_0^{t^*}\int_H M_i\nabla v_p \cdot \nabla\psi\,dx\,dt + \int_0^{t^*}\int_H (M_i + M_e)\nabla u_p \cdot \nabla\psi\,dx\,dt$$

$$+ \int_0^{t^*}\int_T M_o\nabla u_p \cdot \nabla\psi\,dx\,dt = 0 \quad \text{for all } \psi \in V(H\cup T)$$
(7.141)

If we define the operator

$$A : [V(H) \times V(H\cup T)]^2 \to \mathbf{R}$$

by

$$A\left[(r,\tau),(\phi,\psi)\right] = \int_0^{t^*}\int_H r_t\phi\,dx\,dt + \int_0^{t^*}\int_H M_i\nabla r\cdot\nabla\phi\,dx\,dt$$
$$+ \int_0^{t^*}\int_H M_i\nabla\tau\cdot\nabla\phi\,dx\,dt + \int_0^{t^*}\int_H gI'(v)r\phi\,dx\,dt$$
$$+ \int_0^{t^*}\int_H M_i\nabla r\cdot\nabla\psi\,dx\,dt + \int_0^{t^*}\int_H (M_i+M_e)\nabla\tau\cdot\nabla\psi\,dx\,dt$$
$$+ \int_0^{t^*}\int_T M_o\nabla\tau\cdot\nabla\psi\,dx\,dt,$$

then it follows from (7.140)-(7.141) that (v_p, u_p) satisfies

$$A\left[(v_p, u_p),(\phi,\psi)\right] = -\int_0^{t^*}\int_H g_pI(v)\phi\,dx\,dt \qquad (7.142)$$

for all $(\phi,\psi)\in V(H)\times V(H\cup T)$.

Next, consider the auxiliary problem: find (w, q) such that

$$A\left[(\phi,\psi),(w,q)\right] = \qquad\qquad\qquad\qquad\qquad (7.143)$$
$$-2\int_0^{t^*}\sum_{i=1}^m \left[d_i(t) - \left\{u(\Gamma_{i+\frac{k}{2}},t) - u(\Gamma_i,t)\right\}\right]\left[\psi(\Gamma_{i+\frac{k}{2}},t) - \psi(\Gamma_i,t)\right]dt$$

for all $(\phi,\psi)\in V(H)\times V(H\cup T)$. The right hand side of this equation should be compared with the expression (7.138) for the partial derivative of J with respect to p. Clearly, by putting $(\phi,\psi)=(v_p, u_p)$ in (7.143), we conclude from (7.138) that

$$\frac{\partial J}{\partial p} = A\left[(v_p, u_p),(w,q)\right],$$

provided that (w, q) solves (7.143). Moreover, choosing $(\phi,\psi)=(w,q)$ in (7.142) yields that

$$\frac{\partial J}{\partial p} = A\left[(v_p, u_p),(w,q)\right] = -\int_0^{t^*}\int_H g_pI(v)w\,dx\,dt.$$

Recall that the index $j\in\{1,\ldots,n\}$, and hence $p=p_j$, was arbitrary. Therefore, we conclude that

$$\frac{\partial J}{\partial p_j} = -\int_0^{t^*}\int_H g_{p_j}I(v)w\,dx\,dt \quad\text{for } j=1,\ldots,n, \qquad (7.144)$$

where

$$g_{p_j} = \frac{\partial g}{\partial p_j}.$$

Thus, it follows that it is sufficient to solve equations (7.136)-(7.137) and the auxiliary problem (7.143) in order to compute all of the partial derivatives

$$\frac{\partial J}{\partial p_1}, \ldots, \frac{\partial J}{\partial p_n}$$

of J by the expression (7.144). Consequently, the workload for computing all of these derivatives is almost independent of the number n of infarction parameters that is used in the model (7.121)–(7.129). Only the number of integrals in (7.144) that must be calculated will grow as n increases.

The Adjoint Problem. From these considerations, it follows that we might want to solve problems of the form (7.143). Yet what sort of problem is this auxiliary equation? The next step is to investigate this issue in detail, and derive a form of this problem suitable for applying numerical discretization procedures.

By choosing $\psi = 0$ in (7.143) we find that

$$A\left[(\phi, 0), (w, q)\right] = 0 \quad \text{for all } \phi \in V(H),$$

or

$$\int_0^{t^*} \int_H \phi_t w \, dx \, dt + \int_0^{t^*} \int_H M_i \nabla \phi \cdot \nabla w \, dx \, dt + \int_0^{t^*} \int_H gI'(v)\phi w \, dx \, dt$$

$$+ \int_0^{t^*} \int_H M_i \nabla \phi \cdot \nabla q \, dx \, dt = 0 \quad \text{for all } \phi \in V(H). \quad (7.145)$$

From Gauss' divergence theorem, and by choosing appropriate test functions ϕ, we conclude from this expression that w and q must satisfy the equation

$$-w_t + gI'(v)w = \nabla \cdot (M_i \nabla w) + \nabla \cdot (M_i \nabla q) \quad \text{in } H. \quad (7.146)$$

Next, choosing $\phi = 0$ in (7.143) yields that

$$A\left[(0, \psi), (w, q)\right] =$$

$$-2 \int_0^{t^*} \sum_{i=1}^{m} \left[d_i(t) - \left\{ u(\Gamma_{i+\frac{k}{2}}, t) - u(\Gamma_i, t) \right\} \right] \left[\psi(\Gamma_{i+\frac{k}{2}}, t) - \psi(\Gamma_i, t) \right] dt$$

for all $\psi \in V(H \cup T)$, or

$$\int_0^{t^*} \int_H M_i \nabla \psi \cdot \nabla w \, dx \, dt + \int_0^{t^*} \int_H (M_i + M_e)\nabla \psi \cdot \nabla q \, dx \, dt$$

$$+ \int_0^{t^*} \int_T M_o \nabla \psi \cdot \nabla q \, dx \, dt \qquad (7.147)$$

$$= -2 \int_0^{t^*} \sum_{i=1}^{m} \left[d_i(t) - \left\{ u(\Gamma_{i+\frac{k}{2}}, t) - u(\Gamma_i, t) \right\} \right] \left[\psi(\Gamma_{i+\frac{k}{2}}, t) - \psi(\Gamma_i, t) \right] dt.$$

All the leads $\Gamma_1, \ldots, \Gamma_k$ are located along the boundary ∂T of the body. Hence, by choosing test functions ψ with support in the interior of H and T, respectively, we conclude from (7.147) that (w, q) must satisfy the equations

$$\nabla \cdot (M_i \nabla w) + \nabla \cdot ((M_i + M_e)\nabla q) = 0 \quad \text{in } H, \tag{7.148}$$

$$\nabla \cdot (M_o \nabla q) = 0 \quad \text{in } T. \tag{7.149}$$

So far, we have seen that a smooth solution (w, q) of (7.143) must satisfy the partial differential equations (7.146), (7.148), and (7.149). The next step, in order to obtain a complete set of equations, is to derive suitable temporal and boundary conditions for w and q.

From equation (7.145) and Gauss' divergence theorem we find that

$$\int_H \phi(x, t^*)w(x, t^*)\, dx - \int_H \phi(x, 0)w(x, 0)\, dx$$

$$- \int_0^{t^*} \int_H \phi w_t\, dx\, dt + \int_0^{t^*} \int_H M_i \nabla \phi \cdot \nabla w\, dx\, dt$$

$$+ \int_0^{t^*} \int_H gI'(v)\phi w\, dx\, dt + \int_0^{t^*} \int_H M_i \nabla \phi \cdot \nabla q\, dx\, dt = 0. \tag{7.150}$$

Let $\tilde{t} \approx t^*$, and pick a test function $\tilde{\phi}$ such that

$$\tilde{\phi}(x, t) = 0 \quad \text{for } t \in [0, \tilde{t}].$$

Inserting $\phi = \tilde{\phi}$ into (7.150) implies that

$$\int_H \tilde{\phi}(x, t^*)w(x, t^*)\, dx \approx 0,$$

and thus we have derived the following final condition for w

$$w(x, t^*) = 0 \quad \text{for } x \in H. \tag{7.151}$$

The interface and boundary conditions along ∂H and ∂T, respectively, can be derived in a similar manner. More precisely, by choosing suitable test functions, with support only in the vicinity of ∂H and ∂T, we find that the auxiliary problem (7.143) can be written in the "classical" form: find (w, q) such that

$$-w_t + gI'(v)w = \nabla \cdot (M_i \nabla w) + \nabla \cdot (M_i \nabla q) \text{ in } H, \tag{7.152}$$

$$\nabla \cdot (M_i \nabla w) + \nabla \cdot ((M_i + M_e)\nabla q) = 0 \quad \text{in } H,$$

$$w(x, t^*) = 0 \quad \text{in } H,$$

$$(M_i \nabla w + M_i \nabla q) \cdot n = 0 \quad \text{on } \partial H,$$

$$(M_e \nabla q) \cdot n = (M_o \nabla q) \cdot n \quad \text{on } \partial H,$$

$$\nabla \cdot (M_o \nabla q) = 0 \quad \text{in } T,$$

and

$$
\int_{\partial T} \psi(M_o \nabla q) \cdot n \, dx =
$$

$$
-2 \sum_{i=1}^{m} \Big[d_i(t) - [u(\Gamma_{i+\frac{k}{2}}, t) - u(\Gamma_i, t)] \Big] \Big[[\psi(\Gamma_{i+\frac{k}{2}}, t) - \psi(\Gamma_i, t)] \Big] \tag{7.153}
$$

for all smooth test functions $\psi \in C^{\infty}(\partial T)$ and $t \in [0, t^*]$. The boundary condition (7.153) involves the mean values of both the potential u and the test function ψ along the leads $\Gamma_1, \ldots, \Gamma_k$. Therefore, this equation cannot be written in a "classical", point-wise form.

The similarity between our original model (7.121)–(7.129), for the forward problem, and (7.152)–(7.153) is striking. In mathematical terms, we refer to (7.152)–(7.153) as the adjoint problem of (7.152)–(7.153). Note that condition (7.151) is specified at time $t = t^*$, i.e. (7.151) is a final condition for w. This is in "agreement" with the minus sign that appears in front of the term involving the time derivative of w in (7.152).

We have seen that all of the partial derivatives of the cost-functional J can be computed efficiently by solving the adjoint problem. More precisely, the procedure for doing this is as follows:

– For a given set $p_1, \ldots, p_n$ of infarction parameters, solve (7.121)–(7.129) for v and

$$
u = \begin{cases} u_e \text{ in } H, \\ u_0 \text{ in } T \end{cases}
$$

– Solve the adjoint problem (7.152)–(7.153) for w and q

– Compute the partial derivatives of J by applying formula (7.144)

The CPU time needed by this algorithm will thus be almost independent of the number n of infarction parameters used to describe the ischemia! Observe that, as soon as v and u have been determined, the adjoint problem (7.152)–(7.153) is, in contrast to (7.121)–(7.129), linear.

7.4.4 Numerical Experiments

Numerical experiments were carried out on an idealized, circular geometry. The heart domain H was modelled by a disk with radius 5 cm and the torso T had a radius of 7 cm. A small radius of the torso was chosen in order to reduce the computational load. Initially the tissue was at rest, but a stimulation current was applied inside a circle. It was centred 3 cm off the origin of the heart-disk, and had a radius of 1 cm. Figure 7.15 shows a schematic. The extracardiac potential was recorded at 100 locations on the torso surface. The infarcted area was characterized by a circle with a given position (a, b) and radius r. Each simulation was carried out for 100 ms, which in most cases was enough time for all of the tissue to be depolarized. The ionic current was modelled by applying the formula

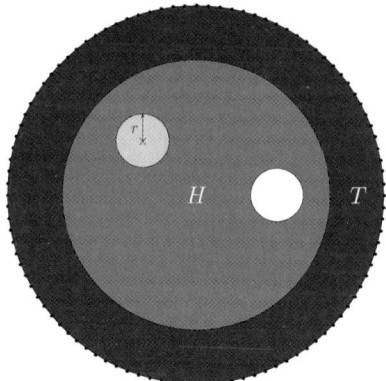

Fig. 7.15. The figure shows the idealized geometry used in the experiments in Section 7.4.4. The white circle represents the area in which stimulus current was applied. It was fixed throughout all the simulations. The yellow area represents an area with no active propagation. It had the same shape in all the simulations, but varied in size and location. The dots on the outer boundary represent leads, i.e. locations where the extra cellular potential was recorded. (For the color version, see Figure A.15 on page 294).

$$I_{\text{ion}}(v) = -0.04^2(v + 85)(v + 65)(v - 40),$$

which means that there was no repolarisation phase in the simulations.

Figure 7.16 shows the solution of our modelled problem (7.121)–(7.129) at three time instances. In this case, an infarction was located at the point $(-2, 2)$ and had a radius of 1.5 cm. The propagation of the electrical potential was (see Figure 7.16), clearly influenced by the presence of the infarction!

In Figure 7.17 the extra cardiac potential at two locations at the body surface ∂T is plotted. Note the relationship between this figure and the second row in Figure 7.16. For example, the potential at $(0,7)$ changed from being positive at time $t = 15$ ms to negative at time $t = 25$ ms, while the potential at $(7,0)$ stayed negative throughout most of the simulation. Simulation results generated without an infarcted area are plotted for comparison. Note that large deviations in the surface potentials did not occur until the wavefront reached the infarcted area, about 15 ms into the simulation.

Let us now turn our attention to the cost-functional J defined in (7.132). Synthetic ECG recordings were generated by the following procedure:

- A circular infarction with radius 1 cm and centre $(3, 3)$ was incorporated into the model (7.121)–(7.129), cf. also Example 8
- We computed the solution of the resulting forward problem
- The electrical potential at the surface ∂T of the body was recorded and "ECG" measurements were produced by applying formula (7.131)

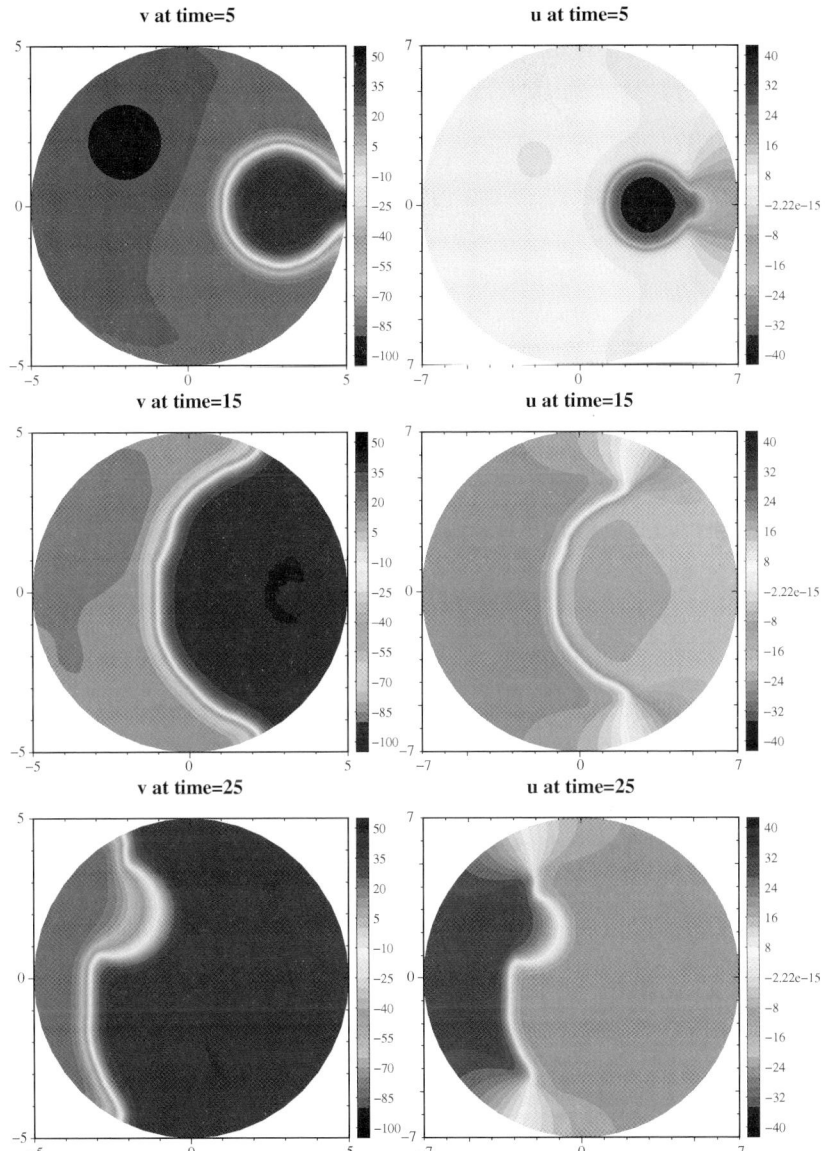

Fig. 7.16. Simulation results generated with an infarction located at $(-2,2)$. The left column shows the transmembrane potential in the heart, and the right column illustrates the extra cellular and extra cardiac potential in $\Omega = H \cup T$. At time $t = 25$ ms, the presence of the infarction is clearly visible in the propagation pattern. (For the color version, see Figure A.16 on page 295).

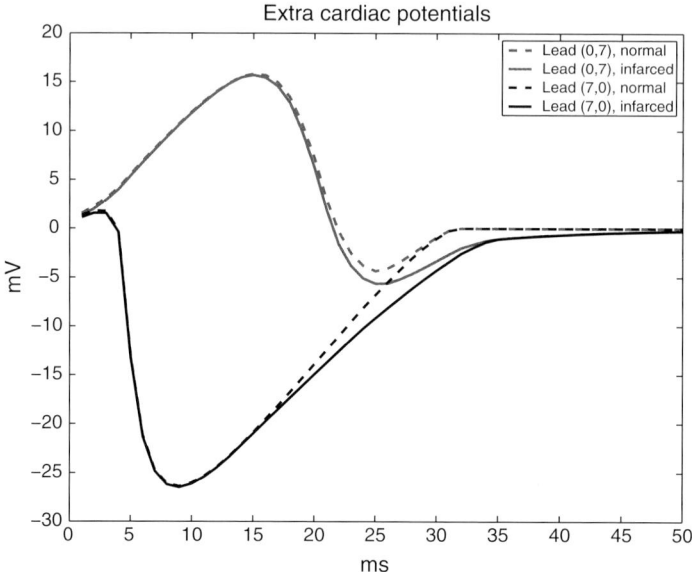

Fig. 7.17. "ECG data" from the simulation results shown in Figure 7.16. The data was recorded at the leads positioned at $(0, 7)$ and $(7, 0)$. Measurements were performed for both a healthy (solid lines) and an infarcted (dashed lines) heart. (For the color version, see Figure A.17 on page 296).

In this procedure for generating "observation data", $k = 100$ uniformly distributed leads positioned along ∂T were used; see Figure 7.15. Consequently, we obtained $m = k/2 = 50$ artificial ECG recordings of the form (7.131).

Let p_1 and p_2 be two infarction parameters that represent the centre (p_1, p_2) of an infarction with radius 1 cm. This leads to the following version of the objective function J defined in (7.132):

$$J(p_1, p_2) = \int_0^{t^*} \sum_{i=1}^{50} [d_i(t) - \{u(\Gamma_{i+25}, t) - u(\Gamma_i, t)\}]^2 \, dt. \qquad (7.154)$$

Here,

$$u = u(x, t; p_1, p_2)$$

denotes the simulated electrical potential obtained by solving (7.136)–(7.137), and $\Gamma_1, \ldots, \Gamma_{100}$ the involved leads. Hence, in order to compute $J(p_1, p_2)$, for a given centre (p_1, p_2), the model (7.121)–(7.129) must be solved. Note that, since the data $\{d_i\}_{i=1}^{50}$ was produced by the procedure described above,

$$J(3, 3) = 0.$$

Figure 7.18 shows three plots of $J(p_1, p_2)$ for various levels of noise in the observation data. Clearly, our objective function seems to be well-behaved:

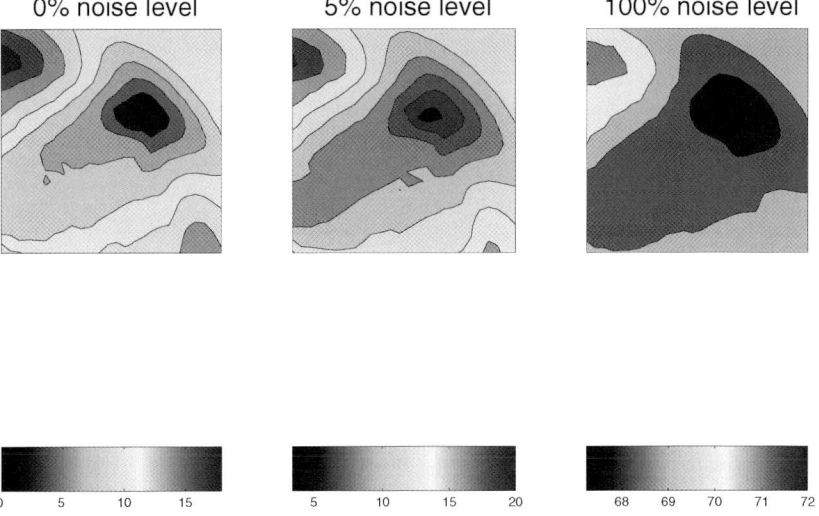

Fig. 7.18. Plots of the cost-functional $J = J(p_1, p_2)$ as a function of the centre (p_1, p_2) of the infarction for various levels of noise in the observation data. The size of the parameters p_1 and p_2 are represented by the 'x' and 'y' axes, respectively. (For the color version, see Figure A.18 on page 296).

- In the cases of no and 5% noise in the data, J is fairly convex and has a unique minimum. Moreover, the minimum is located at the right position $\approx (3, 3)$
- Even with a noise level of 100%, our cost-functional seems to capture the basic properties of the problem at hand. However, in such cases, it seems to be extremely difficult to actually compute the minimum of J.

Considering these results, one might ask the question; *is the problem (7.133) well-posed?* As explained in Section 7.1.5, discretizing an ill-posed system of equations often leads to a regularized approximation of the underlying problem. Based on physical considerations, it seems to be reasonable to conclude that the continuous counterpart, using infinitely many infarction parameters, of (7.133) is ill-posed. Thus, the number n of infarction parameters used in (7.133) defines a regularization parameter. In the present example $n = 2$, and consequently we obtain a well-behaved minimization problem (7.154).

We have seen that if the number n of infarction parameters is low, then the objective function J is well-behaved. Thus, it seems to be important to investigate the number of parameters needed to accurately characterize an infarction. In addition, efficient minimization algorithms, using the adjoint problem approach described in Section 7.4.3, for solving (7.133) should be implemented and tested. Such methods will typically require the solution of (7.121)–(7.129) and (7.152)–(7.153) for many different choices of the infarction parameters $p_1, \ldots, p_n$. Both of these problems are extremely CPU-demanding. Hence, the methods for solving the direct problem modelling the electrical activity in the human body must be further improved. We hope to return to these issues in our future work.

Appendix A

Color Figures

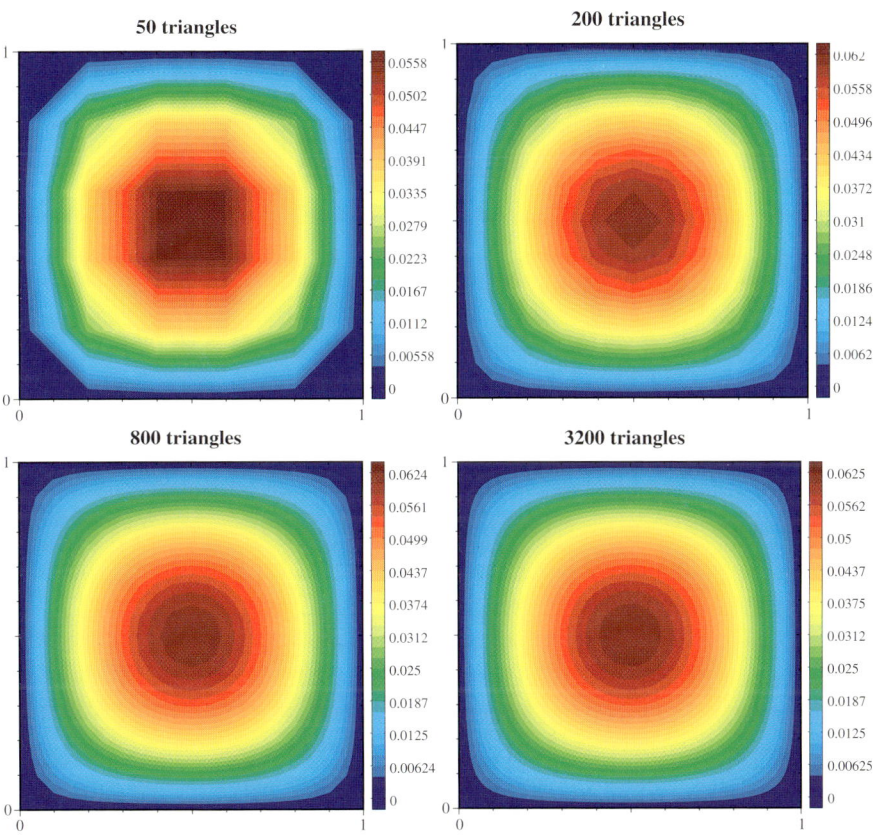

Fig. A.1. Plots of the solution u of (3.1)–(3.2) for four different levels of grid refinement. (This is a color version of Figure 3.3 on page 64).

dipole

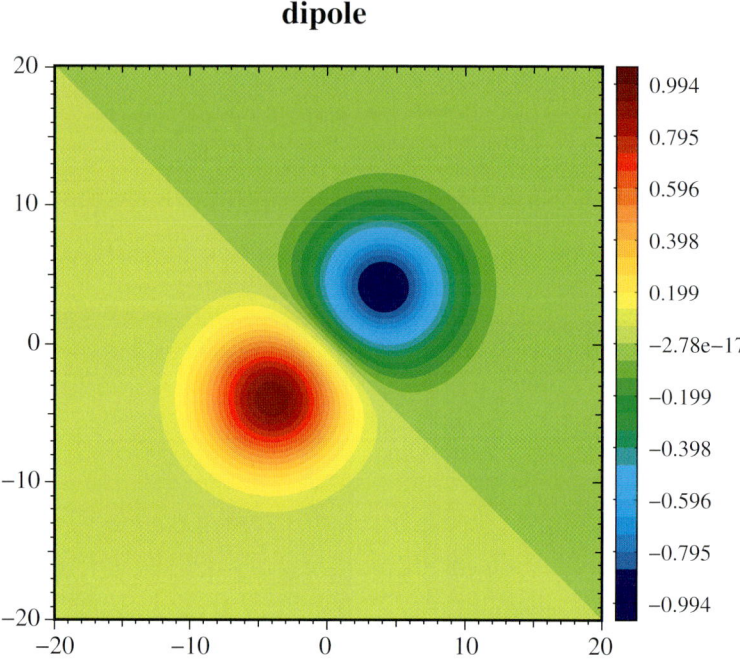

Fig. A.2. An example of the function $f(\mathbf{x})$ in (3.15).(This is a color version of Figure 3.4 on page 65).

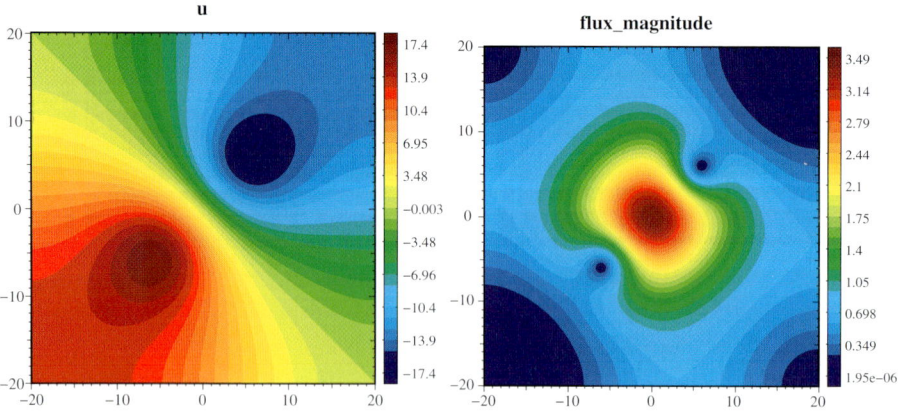

Fig. A.3. The solution resulting from the source term function $f(\mathbf{x})$ given above. The left panel shows the potential, while the right panel shows the magnitude of the current. (This is a color version of Figure 3.5 on page 67).

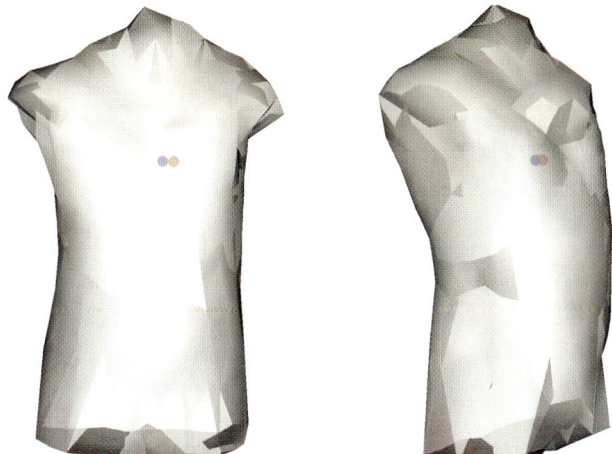

Fig. A.4. The location of the poles are shown as red and blue spheres. (This is a color version of Figure 3.6 on page 68).

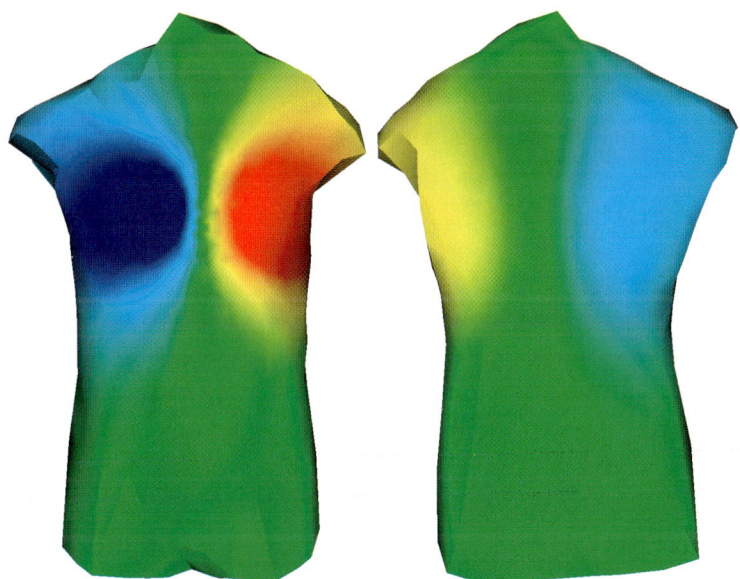

Fig. A.5. The electrical potential set up by the dipole. The colour scale is from −4 mV to 4 mV. Note that the field is stronger on the front (left). (This is a color version of Figure 3.7 on page 68).

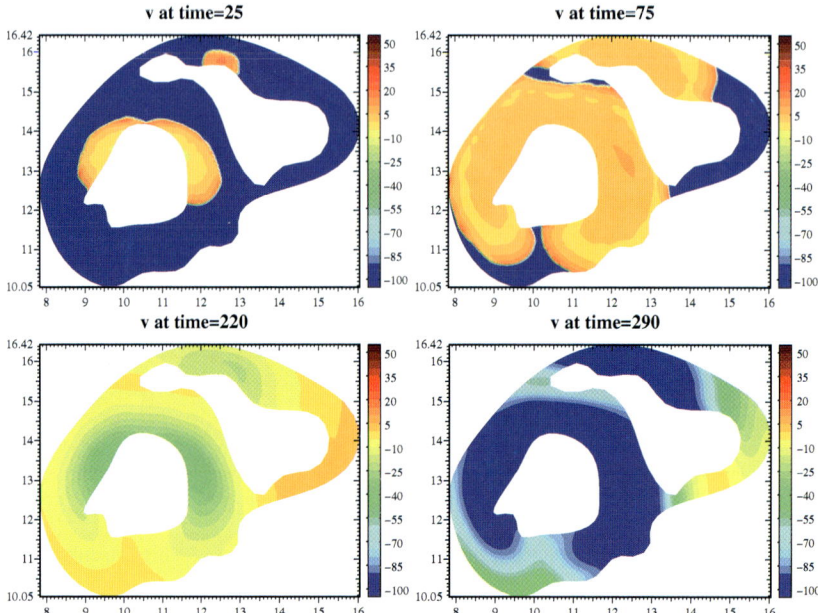

Fig. A.6. The transmembrane potential (mV) at four stages during normal propagation. (This is a color version of Figure 3.11 on page 94).

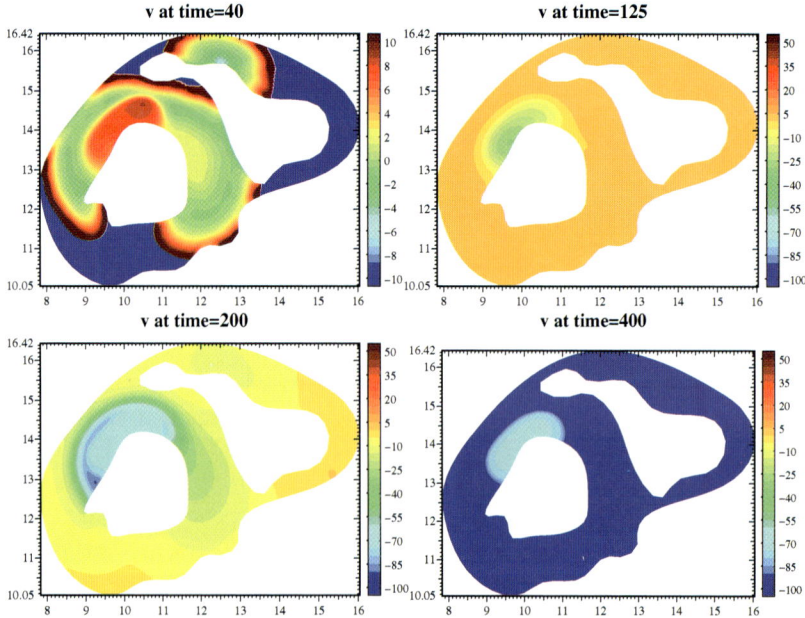

Fig. A.7. Four stages during ischemic propagation. (This is a color version of Figure 3.14 on page 96).

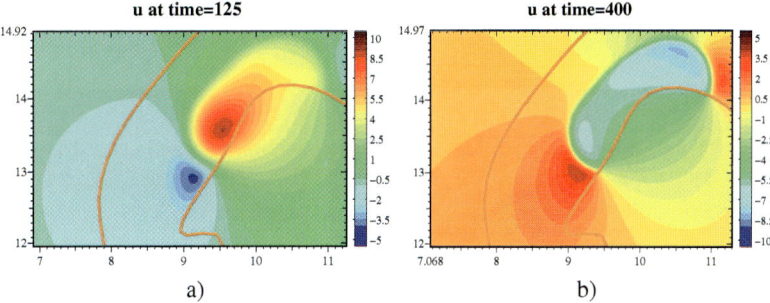

u at time=125   **u at time=400**

a)   b)

Fig. A.8. The extracellular potential around the ischemic area during a) the ST segment b) the TP segment. The heart boundary is indicated by the solid line. (This is a color version of Figure 3.15 on page 97).

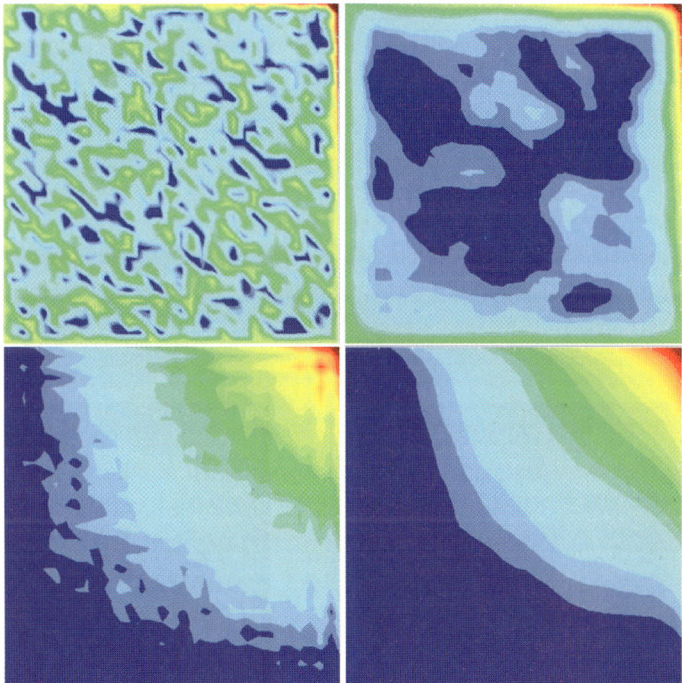

Fig. A.9. The upper left picture shows the initial vector. It is a random vector that should contain "all possible" errors. In the upper right picture the solution after one symmetric Gauss–Seidel sweep is displayed. It is clear that the random high–frequency behaviour in the initial solution has been effectively removed. The picture down to the left shows the solution after the coarse grid correction. The smooth components of the solution have improved dramatically. In the last picture the solution after the post smoothing is displayed. The solution is now very close to the actual solution. (This is a color version of Figure 4.5 on page 129).

Fig. A.10. An example of partitioning the heart domain (left) and the body domain (right). (This is a color version of Figure 6.6 on page 214).

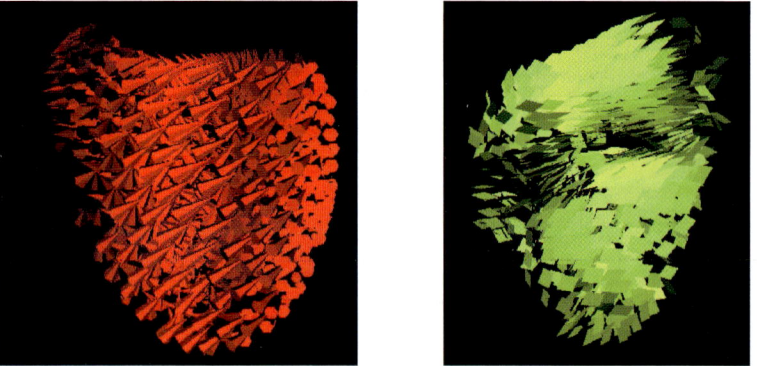

Fig. A.11. a) The fiber directions and b) the sheet planes. (This is a color version of Figure 6.7 on page 216).

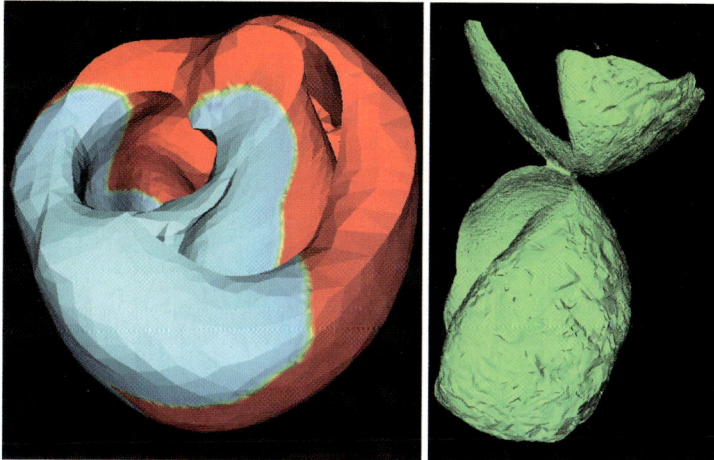

Fig. A.12. A snapshot of the transmembrane potential during the depolarization phase. The right figure shows an iso-surface of the potential at the time instant when two fronts meet. (This is a color version of Figure 6.8 on page 216).

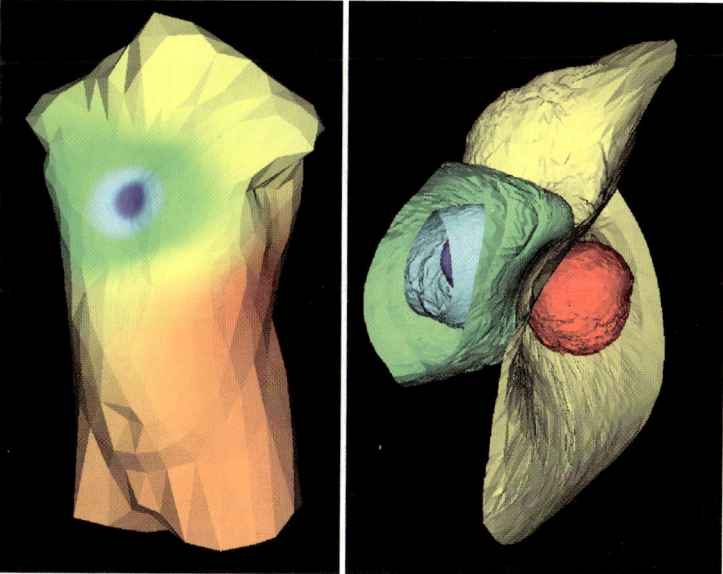

Fig. A.13. The extra cardiac potential on the torso surface (left) and on some iso-surfaces (right). (This is a color version of Figure 6.9 on page 217).

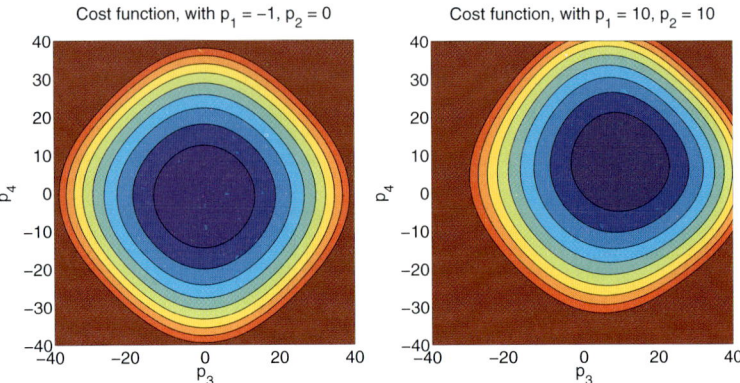

Fig. A.14. The cost-functional J depends upon four parameters: p_1, p_2, p_3, p_4. In order to make a 2D visualization of J, p_1 and p_2 are kept fixed. The plot on the left shows a cross section with the correct values for p_1 and p_2, i.e. $p_1 = -1$ and $p_2 = 0$. The plot on the right shows a cross section with $p_1 = p_2 = 10$, i.e. very different from the correct values. A steepest decent algorithm will converge quickly to the correct values for p_3 and p_4 in the first case, but not in the latter. This shows that the parameters cannot be computed independently. Both plots suggest that the four-dimensional cost-functional J is convex and ideally suited for a steepest descent search. (This is a color version of Figure 7.13 on page 271).

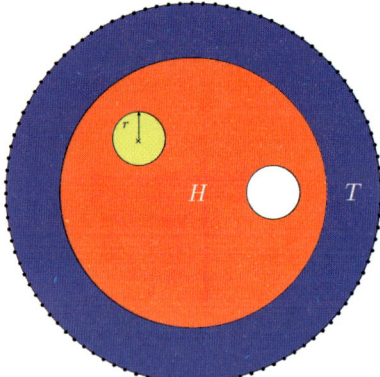

Fig. A.15. The figure shows the idealized geometry used in the experiments in Section 7.4.4. The white circle represents the area in which stimulus current was applied. It was fixed throughout all the simulations. The yellow area represents an area with no active propagation. It had the same shape in all the simulations, but varied in size and location. The dots on the outer boundary represent leads, i.e. locations where the extra cellular potential was recorded. (This is a color version of Figure 7.15 on page 282).